Laboratory Experiments in Microbiology

EIGHTH EDITION

Ted R. Johnson
St. Olaf College

Christine L. Case
Skyline College

San Francisco Boston New York
Cape Town Hong Kong London Madrid Mexico City
Montreal Munich Paris Singapore Sydney Tokyo Toronto

EXECUTIVE EDITOR: Leslie Berriman
ASSISTANT EDITOR: Blythe Robbins
MANAGING EDITOR: Wendy Earl
PRODUCTION EDITOR: David Novak
COPY EDITOR: Michelle Gossage
PROOFREADER: Martha Ghent
INDEXER: Sylvia Coates
COMPOSITOR: The Left Coast Group, Inc.
COVER DESIGNER: Yvo Riezebos
COVER IMAGE: Steve Taylor/Stone/Getty Images, Inc.
SENIOR MANUFACTURING BUYER: Stacey Weinberger
EXECUTIVE MARKETING MANAGER: Lauren Harp

ISBN 0-8053-8292-5
3 4 5 6 7 8 9 10—QWD—10 09 08
www.aw-bc.com

Contents

Preface

Laboratory Experiments in Microbiology, Eighth Edition, is designed to supplement any nonmajors microbiology textbook. Co-authored by Christine L. Case, this manual is an ideal companion to *Microbiology: An Introduction*, Ninth Edition, by Tortora, Funke, and Case. Our goal is to provide a manual of basic microbiological techniques with applications for undergraduate students in diverse areas, including the biological sciences, the allied health sciences, agriculture, environmental science, nutrition, pharmacy, and various preprofessional programs. This manual contains 57 thoroughly class-tested exercises covering every area of microbiology. Most exercises require about 1½ hours of laboratory time. By selecting an appropriate combination of exercises, the instructor can provide learning experiences to meet the needs of a particular course.

Our Approach

Our goal in writing this manual has been twofold—to teach microbiological techniques and to show students the importance of microbes in our daily lives and their central roles in nature. *All* of the laboratory techniques, thinking skills, and safety procedures recommended by the American Society for Microbiology are included in this manual.

Laboratory safety is our primary concern. Many students are preparing for work in a clinical environment and need to learn to handle biologically contaminated materials. Students preparing for work in biotechnology and other laboratories must master aseptic techniques to avoid contamination of their samples. Students must not only learn but also practice these safety techniques so that safety becomes part of their professional behavior. We have followed the Centers for Disease Control and Prevention guidelines for the safe handling of microbes and human body fluids. Students are instructed to work only with their own body fluids. We have included a safety contract that students can hand in to their instructors to indicate that they understand safety requirements. See the section entitled Specific Hazards on page xii and the following sections of the Introduction for specific safety suggestions. At key points in the exercises, safety boxes appear. These are marked with either a biohazard logo ☣ or a general safety logo ⚠ indicating appropriate safety techniques.

Almost every exercise includes an actual experiment requiring students to analyze data. We hope in this way to promote *analytical reasoning* and to make laboratory sessions interesting for students as well as to provide a variety of opportunities for *reinforcement* of the technical skills they have learned. Each exercise has a series of Critical Thinking questions to enhance students' investigative skills. To further demonstrate some practical uses of microbiology, we have frequently included material with direct application to procedures performed in clinical and commercial laboratories.

Scope and Sequence

This manual is divided into 13 parts. The introduction to each part explains the unifying theme for that part. Exercises in the first three parts provide sequential development of fundamental techniques. The remaining exercises are as independent as possible to allow the instructor to select the most desirable sequence. A Techniques Required section preceding each experiment lists prerequisites from earlier exercises.

Part One, Microscopy, emphasizes observation through the microscope. Practice in use and care of the microscope is followed by observations of microbes, to familiarize students with their sizes. Phase-contrast and brightfield microscopy are used to examine living material.

Part Two, Staining Methods, begins with an explanation of how to handle bacterial cultures and teaches the use of the most common stains, including preparation of stained samples and their examination. A highly reliable capsule stain is included. Exercise 8, Morphologic Unknown, introduces the concept of unknowns and can be used to test knowledge of staining methods.

Part Three, Cultivation of Bacteria, stresses aseptic technique and covers the isolation of bacterial strains and the maintenance of bacterial cultures. An exercise dealing with special media prepares the student for the next part.

Part Four, Microbial Metabolism, includes five exercises on bacterial metabolism that provide the tools for Exercises 18, 33, and 51, consisting of unknown identifications. These exercises are especially useful for learning principles of metabolism. Exercise 18, Rapid Identification Methods, demonstrates numerical identification.

Part Five, Microbial Growth, deals with the effects of environmental conditions, such as temperature and the presence of oxygen, on growth. Biofilms are grown and observed in Exercise 21.

Part Six, Control of Microbial Growth, provides practical applications of concepts of microbial growth. Methods of controlling unwanted microbes in food or in a clinical environment are examined in this part.

Part Seven, Microbial Genetics, is of particular interest because of recent notable advances in this field. Exercise 27 is a unique experiment in which students look at expression in two operons using green fluorescent protein. Exercise 28 demonstrates the isolation of bacterial mutants. Exercise 29 allows students to isolate DNA and transform cells without special equipment. DNA fingerprinting is introduced in Exercise 30. Transformation by an antibiotic resistance plasmid is performed in Exercise 31. Restriction enzyme digestion and agarose gel electrophoresis are used to analyze the plasmid. In Exercise 32, suspected chemical carcinogens are tested by the Ames test.

Part Eight, The Microbial World, examines the diversity of microorganisms. Information and techniques learned in previous exercises are used to identify a bacterial unknown in Exercise 33. Cyanobacteria are included in Exercise 36, Phototrophs: Algae and Cyanobacteria. The morphology and ecological niches of free-living eukaryotic organisms studied in microbiology are examined in Exercises 34 through 37.

Part Nine, Viruses, provides an opportunity to isolate, cultivate, and quantify bacteriophages. Students determine the host range of a plant virus in Exercise 39.

Part Ten, Interaction of Microbe and Host, introduces basic concepts of epidemiology. Methods of tracking and identifying causes of infectious diseases are practiced in the exercises in this part.

Part Eleven, Immunology, covers the host's response to infectious disease with exercises on innate immunity and adaptive immunity (agglutination tests and the ELISA technique).

Part Twelve, Microorganisms and Disease, emphasizes procedures employed in the clinical laboratory. In Exercise 51, students identify an unknown from a simulated clinical sample.

Part Thirteen, Microbiology and the Environment, includes standard methods for the examination of food. The MUG test is used to examine water for bacteriological quality. Exercise 55, Microbes Used in the Production of Foods, offers an opportunity to study food microbiology. Exercise 56, Microbes in Soil: The Nitrogen and Sulfur Cycles, provides an example of biogeochemical cycles and soil microbiology. Exercise 57, Microbes in Soil: Bioremediation, is a unique exercise illustrating the use of bacteria for cleaning up environmental pollution.

The appendices at the end of the manual provide a convenient reference to techniques required in several exercises. Appendix H includes six identification keys to bacteria.

Organization of Each Exercise

The exercises in this manual may involve mastering a skill or procedure or understanding a particular concept. Most of the exercises are investigative by design, and the student is asked to analyze the experimental results and draw conclusions. The exercises are organized as follows:

- Each exercise begins with a section called *Objectives,* which lists skills or concepts to be mastered in that exercise. The objectives can be used to test mastery of the new material after completing the exercise.
- The *Background* provides definitions and explanations for each exercise. The student is expected to refer to a textbook for more detailed explanations of the concepts introduced in the laboratory exercises.
- *Materials* lists include supplies, media, and equipment needed for the exercise.
- *Cultures* lists identify the living organisms required for the exercise.
- *Techniques Required* gives a list of techniques needed to complete each exercise.
- *Procedure* then provides step-by-step instructions, stated as simply as possible and frequently supplemented with diagrams. Questions are occasionally asked in the Procedure section to remind the student of the rationale for a step.
- The *Laboratory Report* is designed to help students learn to collect and present data in a systematic fashion. Students are asked to write the *Purpose* of the exercise so they can relate the laboratory activity to their learning. Where appropriate, exercises ask students to formulate their *Hypotheses*. Tables are provided to record *Data* or *Results*. Occasionally, students are asked to write their *Expected Results* using information provided in the Background and their own experience. Students may be asked to write their *Conclusions* following the Data or Results. Usually, *Questions* in the Laboratory Report ask for interpretation of results. The Questions are designed to lead the student from a collection of data or observations to a conclusion. In most instances, the results for each student team will be unique; they can be compared with the information given in the Background and other references but will not be identical to those references. The range of questions in each exercise requires students to think about their results, recall facts, and then use this information to answer the questions. *Critical Thinking* questions are designed to help students use their new knowledge and practice analytic skills.

Illustrations

This book is generously illustrated with diagrams of procedures and 16 pages of photographs in full color. This section has been expanded and updated in this edition to include larger and more-detailed photos. There are two photo quizzes in the color section in which students identify an unknown.

For additional photographs of lab results, students can visit the laboratory supplement on the Microbiology Place web site (http://www.microbiologyplace.com). Photographs of media and techniques help students interpret their results without giving answers and actual lab results.

Preparation Guide

The comprehensive *Preparation Guide* (ISBN 0-8053-8293-3) provides the instructor with all the information needed to set up and teach a laboratory course with this manual. It includes the following:

- General instructions for setting up the lab.
- Information on obtaining and preparing cultures, media, and reagents.
- A master table showing the techniques required for each exercise.
- Cross-references for each exercise to specific pages in *Microbiology: An Introduction*, Ninth Edition.
- For each exercise: helpful suggestions, detailed lists of materials needed, and answers to all the questions in the student manual.

To make *Laboratory Experiments in Microbiology*, Eighth Edition, easy to use, we have carefully designed the experiments to use inexpensive, readily available, nonhazardous materials. Moreover, the exercises have been thoroughly tested in our classes in Minnesota and California by students with a wide variety of talents and interests. Our students have enjoyed their microbiology laboratory experiences; we hope yours will, too!

Acknowledgments

We are most grateful to the following individuals for their time, talent, and interest in our work. Each person carefully read and edited critical parts of the manuscript.

Chuck Hoover of the University of California, San Francisco, for making us aware of the new techniques used in dental microbiology for Exercise 48.

Nick Kapp of Skyline College for providing a novel transformation experiment for Exercise 29.

Anne Jayne of the University of San Francisco for offering suggestions and a capsule stain.

Patricia Carter of Skyline College for her invaluable assistance in preparing materials and proofreading.

We are indebted to St. Olaf College and Skyline College for providing the facilities and resources in which innovative laboratory exercises can be developed.

We would like to commend the staff at Benjamin Cummings for their support. In particular, we thank Leslie Berriman, our Executive Editor; Blythe Robbins, for her editorial skill and meticulous care; and David Novak, for expertly guiding this manual through the production process.

Last, but not least, our gratitude goes to Michelle Johnson, who gave her professional insights and was a sustaining presence, and Don Biederman, who provided timely encouragement and support.

Reviewers

Joel Adams-Stryker, Evergreen Valley College
Bernard Arulanandam, University of Texas at San Antonio
Elaina Bleifield, North Hennepin Community College
Sheila Brady-Root, Nazareth College
Beverly Brown, Nazareth College
John Chikwem, Lincoln University
Iris Cook, Westchester Community College
Michael Davis, Central Connecticut State University
Joe Francis, The Master's College
Dave Gilmore, Arkansas State University
Ann Hefner-Gravink, Solano Community College
Keith Hench, Kirkwood Community College
Judy Kaufman, Monroe Community College
Anne LaGrange Loving, Passaic County Community College
John Lammert, Gustavus Adolphus College
Shyamal Majumdar, Lafayette College
Paul Melchior, North Hennepin Community College
Bonnie Okonek, University of San Francisco
Amy Sprenkle, Salem State College
Tony Yates, Seminole State College

A Special Note to Students

This book is for you. The study of microbiology is dynamic because of the diversity of microbes and the variability inherent in every living organism. Outside of the laboratory—on a forest walk or tasting a fine cheese—we experience the activities of microbes. We want to share our excitement for studying these small organisms. Enjoy!

TED R. JOHNSON
CHRISTINE L. CASE

Introduction

Life would not long remain possible in the absence of microbes.

LOUIS PASTEUR

Welcome to microbiology! Microorganisms are all around us, and as Pasteur pointed out over a century ago, they play vital roles in the ecology of life on Earth. In addition, some microorganisms provide important commercial benefits through their use in the production of chemicals (including antibiotics) and certain foods. Microorganisms are also major tools in basic research in the biological sciences. Finally, as we all know, some microorganisms cause disease—in humans, other animals, and plants.

In this course, you will have firsthand experience with a variety of microorganisms. You will learn the techniques required to identify, study, and work with them. Before getting started, you will find it helpful to read through the suggestions on the next few pages.

Suggestions to Help You Begin

1. Science has a vocabulary of its own. New terms will be introduced in **boldface** throughout this manual. To develop a working vocabulary, make a list of these new terms and their definitions.
2. Because microbes are not visible without a microscope, common names have not been given to them. The word *microbe*, now in common use, was introduced in 1878 by Charles Sedillot. The microbes used in the exercises in this manual are referred to by their *scientific names*. The names will be unfamiliar at first, but do not let that deter you. Practice saying them aloud. Most scientific names are taken from Latin and Greek roots. If you become familiar with these roots, the names will be easier to remember.
3. Microbiology usually provides the first opportunity undergraduate students have to experiment with *living organisms*. Microbes are relatively easy to grow and lend themselves to experimentation. Because there is variability in any population of living organisms, not all the experiments will "work" as the lab manual says. The following exercise will illustrate what we mean:

   Write a description of *Homo sapiens* for a visitor from another planet: ______________________

   ____
   ____
   ____
   ____
   ____
   ____

   After you have finished, look around you. Do all your classmates fit the description exactly? Probably not. Moreover, the more detailed you make your description, the less conformity you will observe. During lab, you will make a detailed description of an organism and probably find that this description does not match your reference exactly.
4. Microorganisms must be cultured or grown to complete most of the exercises in this manual. Cultures will be set up during one laboratory period and will be examined for growth in the next laboratory period. Accurate record keeping is therefore essential. Mark the steps in each exercise with a bright color or a bookmark so you can return to complete your Laboratory Report on that exercise. *Accurate records* and *good organization* of laboratory work will enhance your enjoyment and facilitate your learning.
5. *Observing* and *recording* your results carefully are the most important parts of each exercise. Ask yourself the following questions for each experiment:

   What did the results indicate?

   Are they what I expected? If not, what happened?
6. If you do not master a technique, try it again. In most instances, you will need to use the technique again later in the course.
7. Be sure you can answer the questions that are asked in the Procedure for each exercise. These questions are included to reinforce important points that will ensure a successful experiment.
8. Finally, carefully study the general procedures and safety precautions that follow.

General Procedures in Microbiology

In many ways, working in a microbiology laboratory is like working in the kitchen. As some famous chefs have said:

> *Our years of teaching cookery have impressed upon us the fact that all too often a debutant cook will start in enthusiastically on a new dish without ever reading the recipe first. Suddenly an ingredient, or a process, or a time sequence will turn up, and there is astonishment, frustration, and even disaster. We therefore urge you, however much you have cooked, always to read the recipe first, even if the dish is familiar to you. . . . We have not given estimates for the time of preparation, as some people take half an hour to slice three pounds of mushrooms, while others take five minutes.**

1. Read the laboratory exercises *before* coming to class.
2. *Plan* your work so that all experiments will be completed during the assigned laboratory period. A good laboratory student, like a good cook, is one who can do more than one procedure at a time—that is, one who is efficient.
3. Use only the *required* amounts of materials, so that everyone can do the experiment.
4. *Label* all of your experiments with your name, date, and lab section.
5. Even though you will do most exercises with another student, you must become familiar with all parts of each exercise.
6. Keep *accurate* notes and records of your procedures and results so that you can refer to them for future work and tests. Many experiments are set up during one laboratory period and observed for results in the next laboratory period. Your notes are essential to ensure that you perform all the necessary steps and observations.
7. *Demonstrations* will be included in some of the exercises. Study the demonstrations and learn the content.
8. Let your instructor know if you are color-blind; many techniques require discrimination of colors.
9. Keep your cultures current; discard old experiments.
10. *Clean up* your work area when you are finished. Leave the laboratory clean and organized for the next student. Remember:

    Stain and reagent bottles should be returned to their original locations.

    Slides should be washed and put back into the box clean.

    All markings on glassware (e.g., Petri plates and test tubes) should be removed before putting glassware into the marked autoclave trays.

    Glass Petri plates should be placed agar-side down in marked autoclave containers.

    Swabs and pipettes should be placed in the appropriate disinfectant jars or biohazard containers.

    Disposable plasticware should be placed in marked autoclave containers.

    Used paper towels should be discarded.

Biosafety

The most important element for managing microorganisms is strict adherence to standard microbiological practices and techniques, which you will learn during this course. There are four biosafety levels (BSLs) for working with live microorganisms; each BSL consists of combinations of laboratory practices and techniques, safety equipment, and laboratory facilities. See Table 1 on page xiii. Each combination is specifically appropriate for the operations performed, the documented or suspected routes of transmission of the microorganisms, and the laboratory function or activity.

Biosafety Level 1 represents a basic level of containment that relies on standard microbiological practices with no specific facilities other than a sink for hand washing. When standard laboratory practices are not sufficient to control the hazard associated with a particular microorganism, additional measures may be used.

Biosafety Level 2 includes hand washing, and an autoclave must be available. Precautions must be taken for handling and disposing of contaminated needles or sharp instruments. BSL 2 is appropriate when working with human body fluids. A lab coat should be worn.

Biosafety Level 3 is used in laboratories where work is done with pathogens that can be transmitted by the respiratory route. BSL 3 requires special facilities with self-closing, double doors and sealed windows.

Biosafety Level 4 is applicable for work with pathogens that may be transmitted via aerosols and for which there is no vaccine or safety. The BSL 4 facility is generally a separate building with specialized ventilation and waste management systems to prevent release of live pathogens to the environment.

Which biosafety level is your lab? ______________

Specific Hazards in the Laboratory

Procedures marked with this safety icon should be performed carefully to minimize risk of exposure to chemicals or fire.

Alcohol

Keep containers of alcohol away from open flames.

Glassware Not Contaminated with Microbial Cultures

1. If you break a glass object, sweep up the pieces with a broom and dustpan. Do not pick up pieces of broken glass with your bare hands.
2. Place broken glass in one of the containers marked for this purpose. The one exception to this rule concerns broken mercury thermometers; consult your instructor if you break a mercury thermometer.

---

*J. Child, L. Bertholle, and S. Beck. *Mastering the Art of French Cooking*, Vol. 1. New York: Knopf, 1961.

Table 1

Biosafety Levels

Biosafety Level	Practices	Safety Equipment (Primary Barriers)	Facilities (Secondary Barriers)
1	Standard microbiological practices	None required	Open benchtop sink
2	BSL1 plus • Limited access • Biohazard warning signs • "Sharps" precautions • Safety manual of waste-decontamination policies	Lab coat; gloves, as needed	BSL1 plus autoclave
3	BSL 2 plus • Controlled access • Decontamination of clothing before laundering	BSL 2 plus protective lab clothing; enter and leave lab through clothing changing and shower rooms	BSL 2 plus self-closing, double-door access
4	BSL 3 plus • Separate building	BSL 3 plus full-body air-supplied, positive pressure personnel suit	BSL 3 plus separate building and decontamination facility

Electrical Equipment

1. The basic rule to follow is this: Electricity and water don't mix. Do not allow water or any water-based solution to come into contact with electrical cords or electrical conductors. Make sure your hands are dry when you handle electrical connectors.
2. If your electrical equipment crackles, snaps, or begins to give off smoke, do not attempt to disconnect it. Call your instructor immediately.

Fire

1. If *gas burns* from a leak in the burner or tubing, turn off the gas.
2. If you have a *smoldering sleeve*, run water on the fabric.
3. If you have a *very small fire*, the best way to put it out is to smother it with a towel or book (not your hand). Smother the fire quickly.
4. If a *larger fire* occurs, such as in a wastebasket or sink, use one of the fire extinguishers in the lab to put it out. Your instructor will demonstrate the use of the fire extinguishers.
5. In case of a *large fire* involving the lab itself, evacuate the room and building according to the following procedure:
   a. Turn off all gas burners, and unplug electrical equipment.
   b. Leave the room and proceed to ____________ ____________________________.
   c. It is imperative that you assemble in front of the building so that your instructor can take roll to determine whether anyone is still inside. Do not wander off.

Accidents and First Aid

1. Report all accidents immediately. Your instructor will administer first aid as required.
2. For spills in or near the eyes, use the eyewash immediately.
3. For large spills on your body, use the safety shower.
4. For heat burns, chill the affected part with ice as soon as possible. Call your instructor.
5. Place a bandage on any cut or abrasion.

Power Outage

If the electricity goes off, be sure to turn off your gas jet. When the power is restored, the gas will come back on.

Earthquake

Turn off your gas jet and get under your lab desk during an earthquake. Your instructor will give any necessary evacuation instructions.

Orientation Walkabout

Locate the following items in the lab:

Broom and dustpan
Eyewash
Fire blanket
Fire extinguisher
First-aid cabinet
Fume hood
Instructor's desk
Reference books
Safety shower
To Be Autoclaved area
Biohazard containers

Special Practices

1. Keep laboratory doors closed when experiments are in progress.
2. The instructor controls access to the laboratory and allows access only to people whose presence is required for program or support purposes.
3. Place contaminated materials that are to be decontaminated at a site away from the laboratory into a durable, leakproof container that is closed before being removed from the laboratory.
4. An insect and rodent control program is in effect.
5. A needle should not be bent, replaced in the sheath, or removed from the syringe following use. Place the needle and syringe promptly in a puncture-resistant container and decontaminate, preferably by autoclaving, before discarding them.
6. Inform your instructor if you are pregnant, are taking immunosuppressive drugs, or have any other medical condition (e.g., diabetes, immune deficiency) that might necessitate special precautions in the laboratory.
7. Potential pathogens used in the exercises in this manual are classified in Class 1 by the U.S. Public Health Service. These bacteria present a minimal hazard and require ordinary aseptic handling conditions (Biosafety Level 1). No special competence or containment is required. These organisms are the following:

   Enterobacter species
   Mycobacterium species
   Proteus species
   Pseudomonas aeruginosa
   Salmonella enterica Typhimurium
   Serratia marcescens
   Staphylococcus species
   Streptococcus species

Laboratory Facilities

1. Interior surfaces of walls, floors, and ceilings are water resistant so that they can be easily cleaned.
2. Benchtops are impervious to water and resistant to acids, alkalis, organic solvents, and moderate heat.
3. Windows in the laboratory are closed and sealed.
4. An autoclave for decontaminating laboratory wastes is available, preferably within the laboratory.

Contact with Blood and Other Body Fluids

The following procedures should be used by all health care workers, including students, whose activities involve contact with patients or with blood or other body fluids. While these procedures were developed by the CDC* to minimize the risk of transmitting HIV in a health care environment, adherence to these guidelines will minimize transmission of *all* infections.

1. Wear gloves for touching blood and body fluids, mucous membranes, or nonintact skin and for handling items or surfaces soiled with blood or body fluids. Change gloves after contact with each patient.
2. Wash hands and other skin surfaces immediately and thoroughly if they are contaminated with blood or other body fluids.
3. Wear masks and protective eyewear or face shields during procedures that are likely to generate droplets of blood or other body fluids.
4. Wear gowns or aprons during procedures that are likely to generate splashes of blood or other body fluids.
5. Wash hands and other skin surfaces immediately after gloves are removed.
6. Mouthpieces, resuscitation bags, or other ventilation devices should be available for use in areas in which the need for resuscitation is predictable. Emergency mouth-to-mouth resuscitation should be minimized.
7. Health care workers who have exudative lesions or weeping dermatitis should refrain from all direct patient care and from handling patient care equipment.
8. Pregnant health care workers are not known to be at greater risk of contracting HIV infection than health care workers who are not pregnant; however, if a health care worker develops HIV infection during pregnancy, the infant is at risk of infection. Because of this risk, pregnant health care workers should be especially familiar with and strictly adhere to precautions to minimize the risk of HIV transmission.
9. In a laboratory exercise where human blood is used, students should wear gloves or work only with their own blood and should dispose of all slides and blood-contaminated materials immediately after use. Any cuts or scrapes on the skin should be covered with a sterile bandage.

*CDC. "Recommendations for Prevention of HIV Transmission in Health-Care Settings." *MMWR* 38(S2), 1989.

General Safety in the Laboratory

During your microbiology course, you will learn how to safely handle fluids containing microorganisms. Through practice you will be able to perform experiments so that bacteria, fungi, and viruses remain in the desired containers, uncontaminated by microbes in the environment. These techniques, called **aseptic techniques,** will be a vital part of your work if you are going into health care or biotechnology.

1. Do not eat, drink, smoke, store food, or apply cosmetics in the laboratory.
2. Wear shoes at all times in the laboratory.
3. Tie back long hair.
4. Disinfect work surfaces at the beginning and end of every lab period and after every spill. The disinfectant used in this laboratory is ________________.
5. Wash your hands before and after every laboratory period. Because bar soaps may become contaminated, use liquid or powdered soaps.
6. Use mechanical pipetting devices; do not use mouth pipetting.
7. Place a disinfectant-soaked paper towel on the desk while pipetting.
8. Wash your hands immediately and thoroughly if they become contaminated with microorganisms.
9. Cover spilled microbial cultures with paper towels, and saturate the towels with disinfectant. Leave covered for 20 minutes, and then clean up the spill and dispose of the towels.
10. Do not touch broken glassware with your hands; use a broom and dustpan. Place broken glassware contaminated with microbial cultures or body fluids in the To Be Autoclaved container. (See p. xii for what to do with broken glassware that is not contaminated.)
11. Place glassware and slides contaminated with blood, urine, and other body fluids in disinfectant.
12. Work only with your own body fluids and wastes in exercises requiring saliva, urine, blood, or feces, to prevent transmission of disease. The Centers for Disease Control and Prevention (CDC) state that "epidemiologic evidence has implicated only blood, semen, vaginal secretions, and breast milk in transmission of HIV." *Biosafety in Microbiological and Biomedical Laboratories*, www.cdc.gov.
13. Don't perform unauthorized experiments.
14. Don't use equipment without instruction.
15. Don't engage in horseplay in the laboratory.
16. If you got this far in the instructions, you'll probably do well in lab. Enjoy lab and make a new friend.

Procedures marked with this biohazard icon should be performed carefully to minimize the risk of transmitting disease.

I have read the above laboratory safety rules and agree to abide by them when in the laboratory.

Name: ______________________ Date: ____________

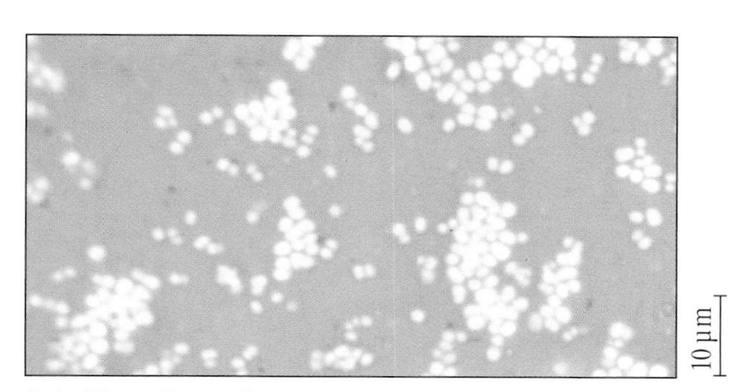

I.1 Negative stain.
Staphylococcus aureus (Exercise 4).

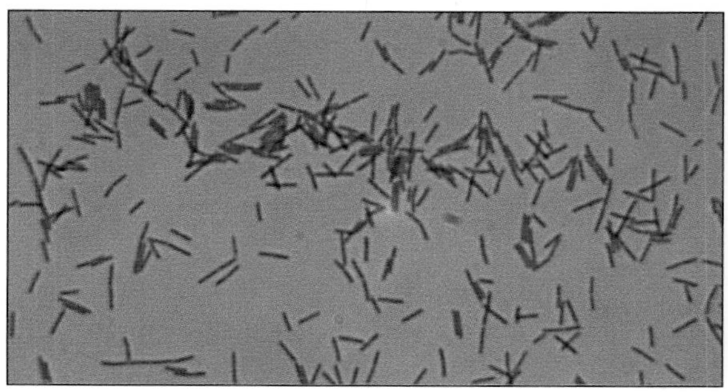

I.2 Gram-negative rods.
Escherichia coli (Exercise 5).

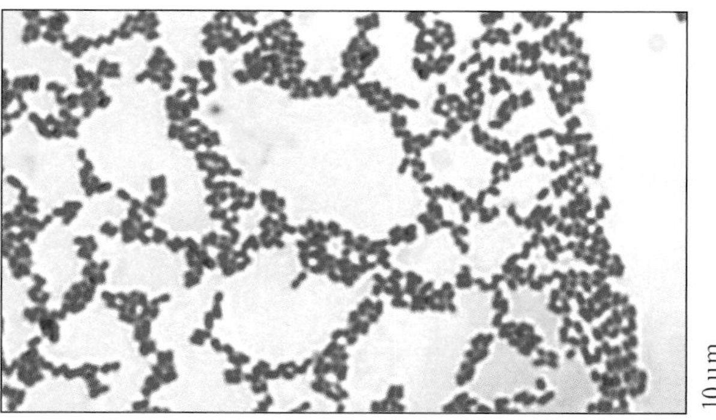

I.3 Gram-positive cocci. *Staphylococcus aureus* (Exercise 5).

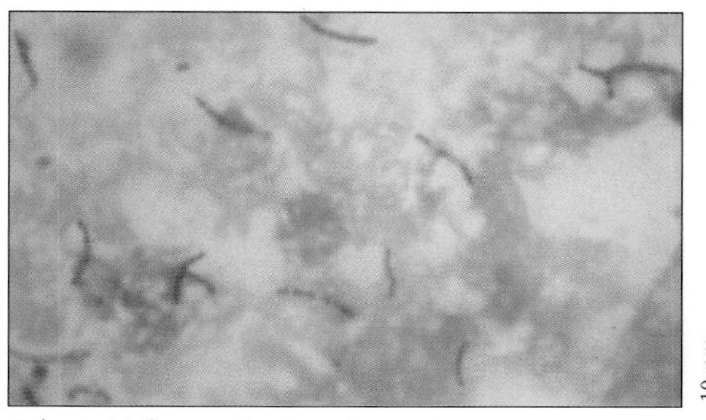

I.4 Acid-fast stain. *Mycobacterium tuberculosis* (Exercise 6).

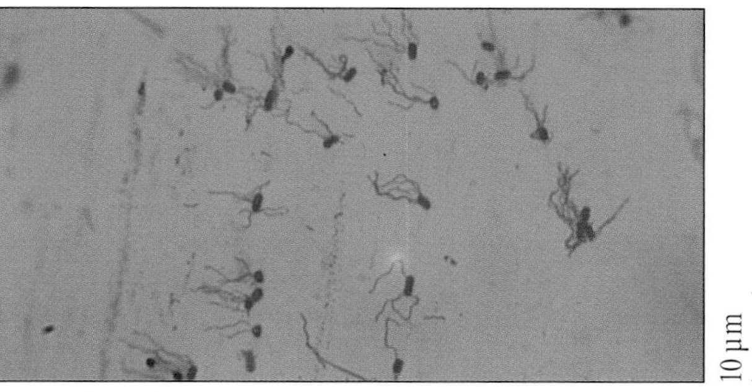

I.5 Flagella stain. Peritrichous flagella of *Proteus vulgaris* (Exercise 7).

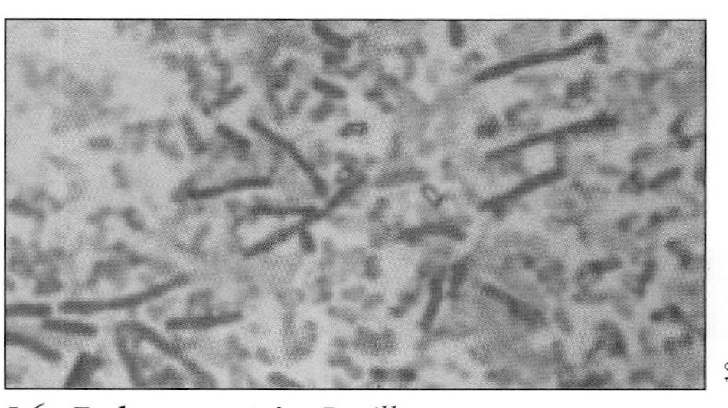

I.6 Endospore stain. *Bacillus* sp. (Exercise 7).

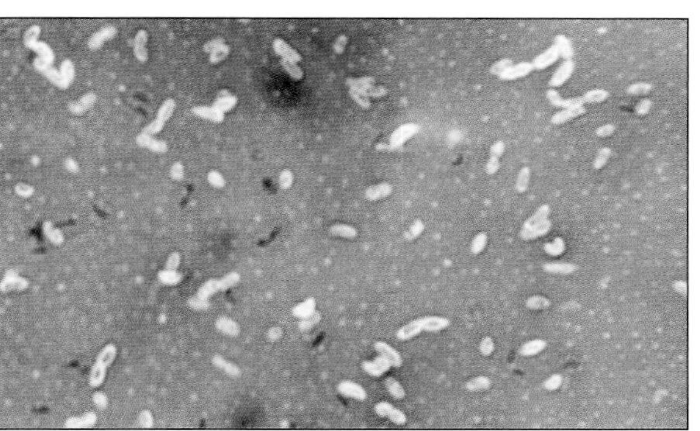

I.7 Capsule stain. *Streptococcus mutans* (Exercise 7).

PLATE II Biotechnology

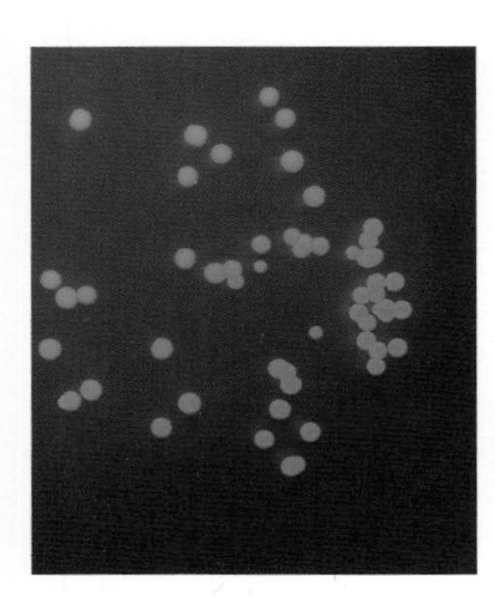

II.1 *E. coli.* The colonies fluoresce under UV light because the cells contain the jellyfish gene for green fluorescent protein (Exercise 27).

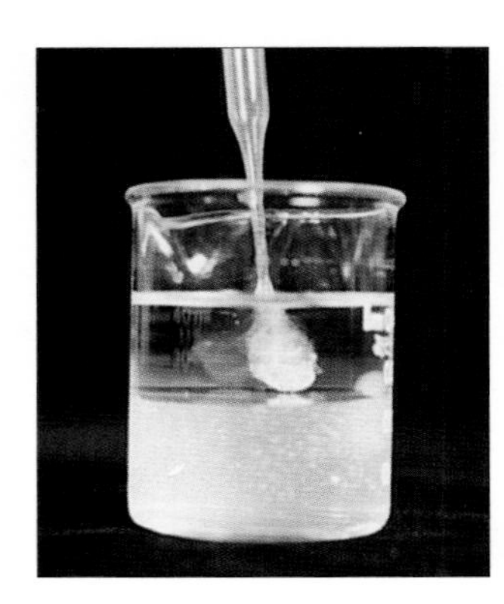

II.2 DNA is collected on a glass rod (Exercise 29).

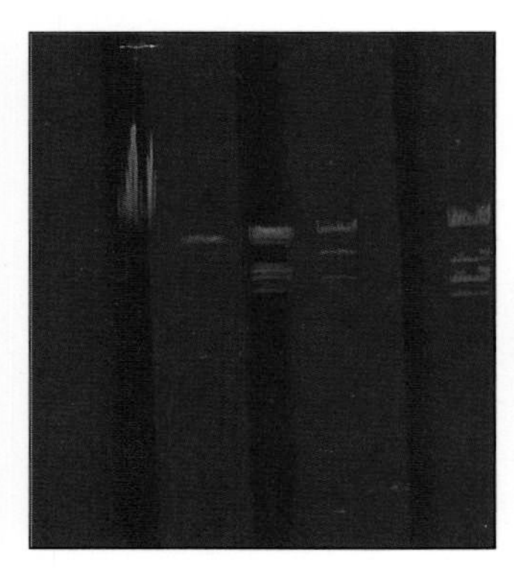

II.3 Restriction enzyme digests of small pieces of DNA. To see the bands, the DNA was stained with ethidium bromide and the gel was illuminated with UV light (Exercise 30).

II.4 Ames test. The chemical (ethidium bromide) diffused from the disk on the left and caused the his^- bacteria to revert to his^+ (colonies in the upper left). The his^- cells on the control plate (right) are unable to grow on glucose-minimal salts agar (Exercise 32).

PLATE III Microbial Metabolism

III.1 Reactions in OF-glucose medium. The species in tubes **1** and **2** is an oxidizer. The organism in tubes **3** and **4** does not use glucose. The culture in tubes **5** and **6** is a fermenter (Exercise 13).

III.2 Reactions in fermentation tubes. **1** is an uninoculated control. Growth and acid production from carbohydrate fermentation are seen in **2.** Acid and gas are produced from fermentation in **3** (Exercise 14).

III.3 Methyl red test. Red color (in **1**) after addition of methyl red indicates a positive test. **2** is methyl red–negative (Exercise 14).

III.4 Voges–Proskauer test. A positive Voges–Proskauer test develops a red color when exposed to oxygen (Exercise 14).

III.5 Citrate test. Utilization of citric acid as the sole carbon source in Simmons citrate agar causes the indicator to turn blue (**2**). **1** is citrate-negative (Exercise 14).

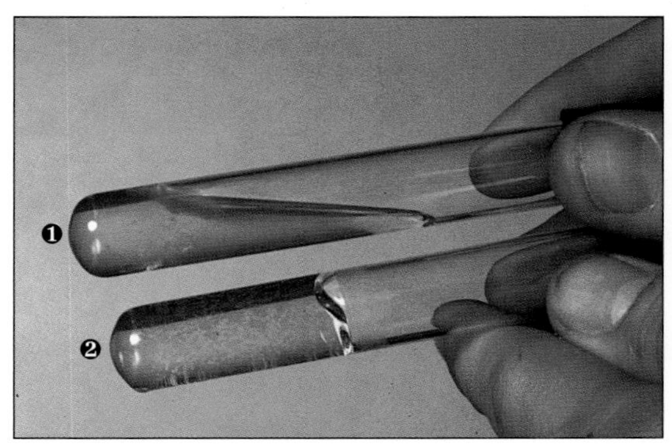

III.6 Gelatin hydrolysis. After hydrolysis (**1**), gelatin remains liquid. **2** is unhydrolyzed gelatin (Exercise 15).

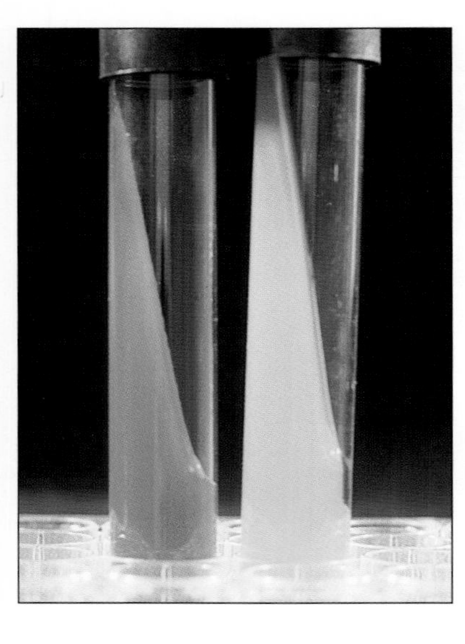

III.7 Urease production. Hydrolysis of urea produces ammonia, which turns the indicator fuchsia (Exercise 15).

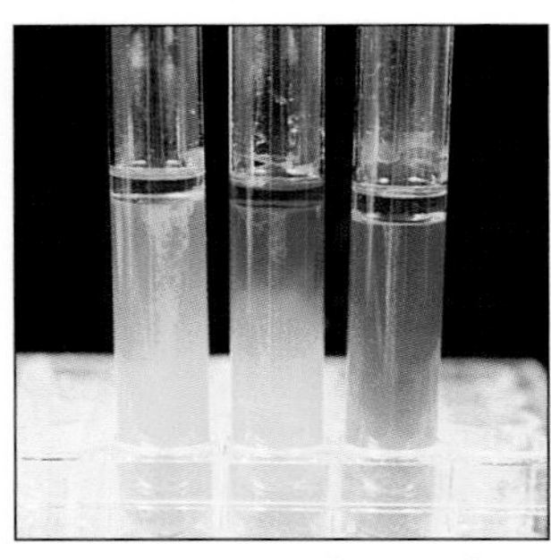

III.8 MIO medium. The culture in the tube on the left is *motile*. The red color of the Kovacs reagent added over the agar indicates *indole* production in the middle tube. Removal of CO_2 from *ornithine* turns the indicator purple, as seen in the tube on the right (Exercise 16).

III.9 Peptone iron agar is used to detect the production of H_2S. H_2S produced in the tube will precipitate as ferrous sulfide (Exercise 16).

III.10 Phenylalanine deaminase. Removal of the amino group produces an organic acid that forms a green complex with the ferric ion-containing indicator (Exercise 16).

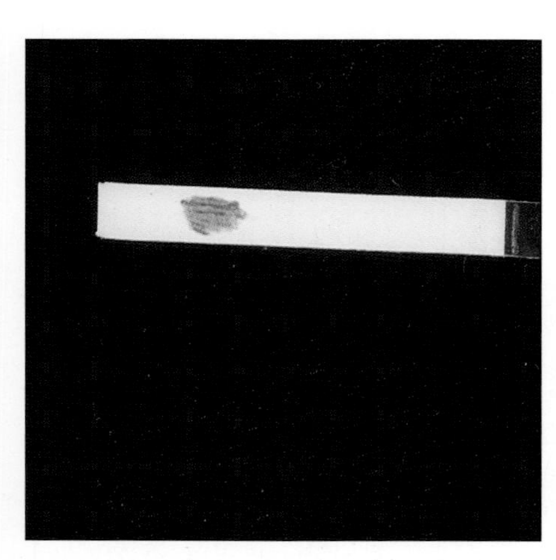

III.11 Oxidase test. Cytochrome oxidase-positive bacteria turn oxidase reagent pink to purple or black (Exercise 17).

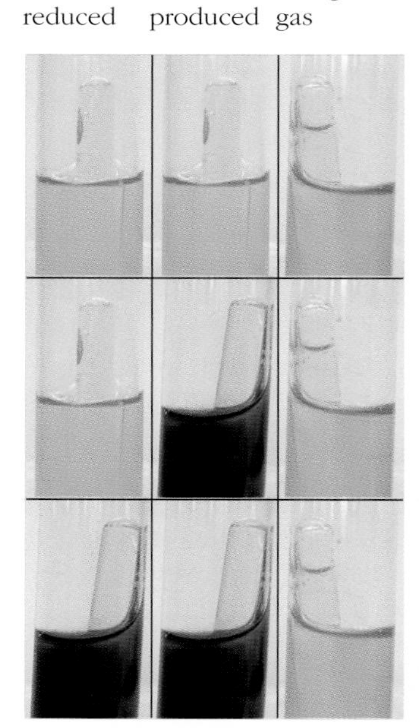

III.12 Nitrate reduction. Nitrate reagent A and nitrate reagent B are added to nitrate broth after incubation to determine nitrate redution. If the broth turns red after addition of nitrate reagents A and B, nitrate ion was reduced to nitrite ion. Zinc dust is added if no color change occurs. If the broth turns red, nitrate ions are present. If nitrates were reduced to nitrogen gas, no color change occurs and the gas should be visible in the Durham (inverted) tube (Exercises 17, 27, and 56).

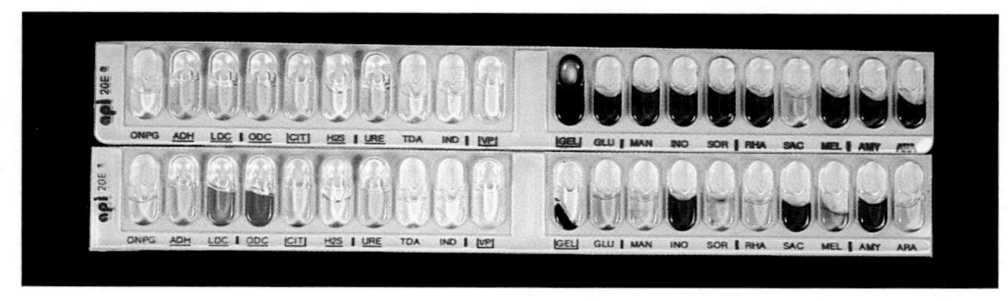

III.13 Twenty biochemical tests are performed in an API 20E® strip. The top strip is inoculated with *Proteus vulgaris* and the bottom strip with *Escherichia coli* (Exercise 18).

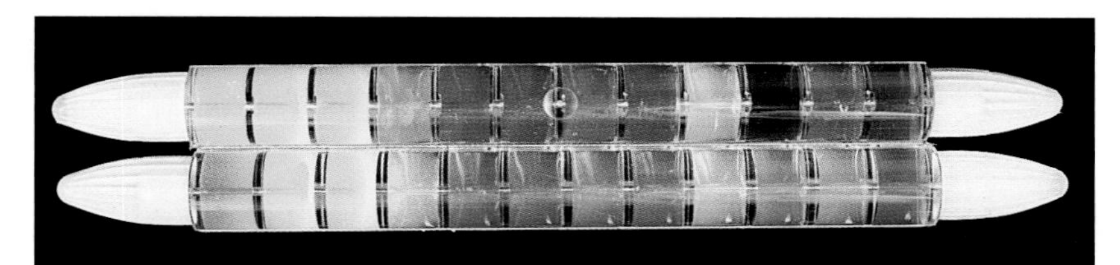

III.14 Fifteen biochemical tests are performed in Enterotube® II. The bottom tube is uninoculated (Exercise 18).

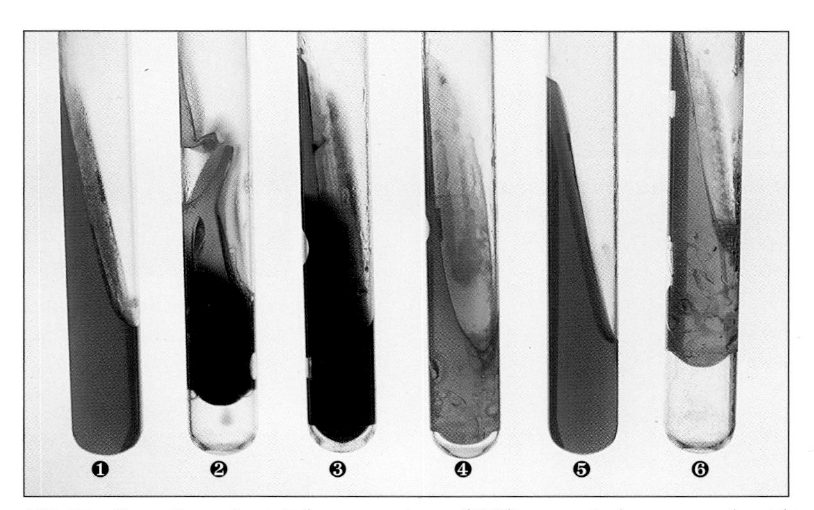

III.15 Reactions in triple sugar iron (TSI) agar. 1 shows growth with no fermentation or hydrogen sulfide (H_2S) production. **2** shows blackening due to H_2S and acid and gas from fermentation of glucose and sucrose and/or lactose. Sucrose and lactose were not fermented in **3**; H_2S production masks the glucose fermentation reaction although gas is produced. Acid and gas are produced from glucose in **4. 5** is uninoculated. **6** shows acid and gas production from glucose and sucrose and/or lactose (Exercise 49).

EUKARYOTES

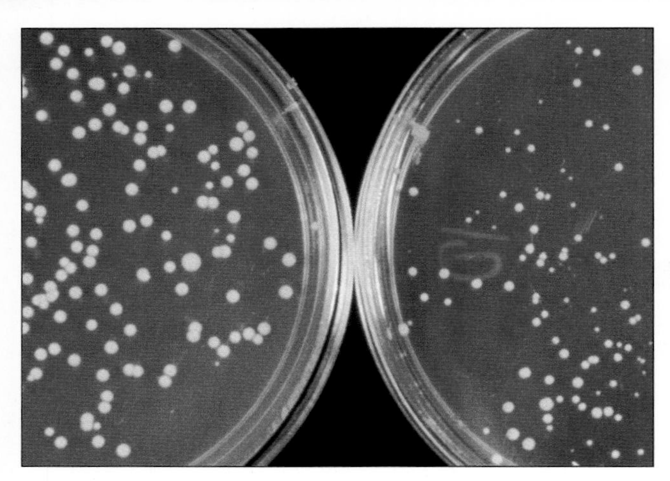

IV.1 The yeast *Saccharomyces cerevisiae* produces circular, convex, glistening colonies on Sabouraud agar (Exercise 34). The plate on the right was irradiated for 25 seconds (Exercise 23).

IV.2 *Rhizopus stolonifer* growing on Styrofoam impregnated with a nutrient medium. Dark sporangia are visible at the tips of the sporangiophores (Exercise 35).

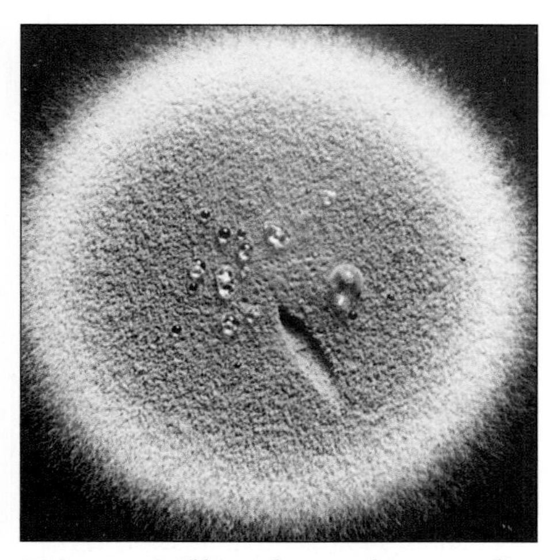

IV.3 *Penicillium* has a white mycelium and green conidiospores. Note the crystallized penicillin on top of the mold colony (Exercise 35).

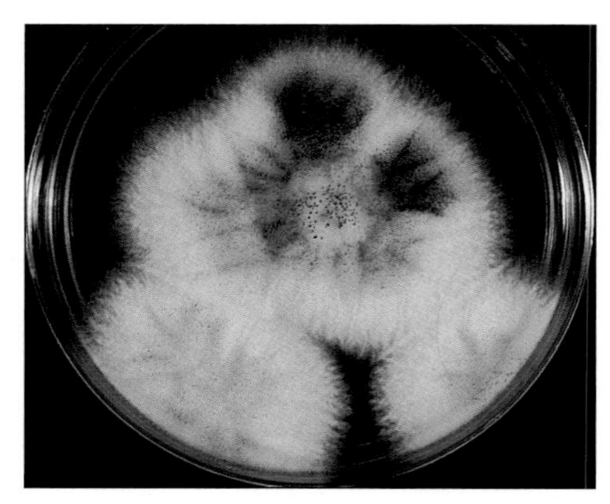

IV.4 Colonies of *Aspergillus niger* have white mycelia and black conidiospores (Exercise 35).

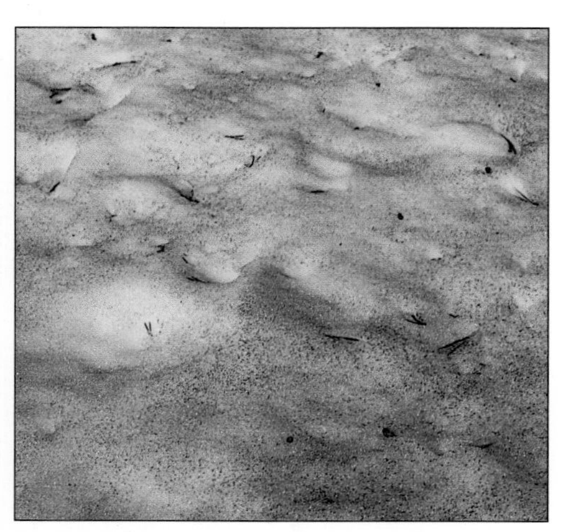

IV.5 Red snow at Tioga Pass (Yosemite National Park) caused by the green alga *Chlamydomonas nivalis.* The green color of chlorophyll is masked by red carotenoid pigments produced in the presence of intense light (Exercise 36).

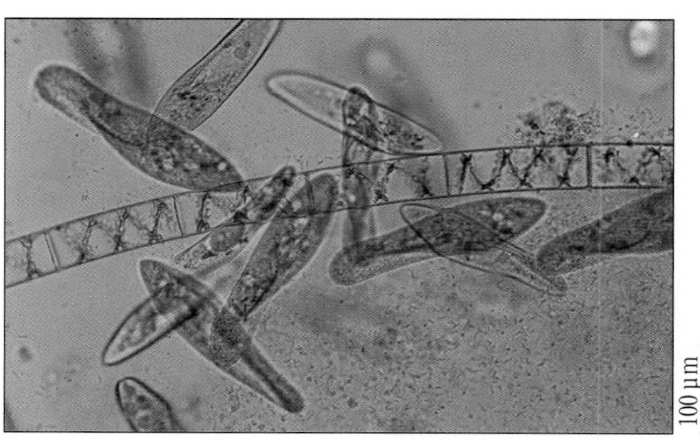

IV.6 The green alga *Spirogyra* and protozoan *Paramecium,* in pond water (Exercises 36 and 37).

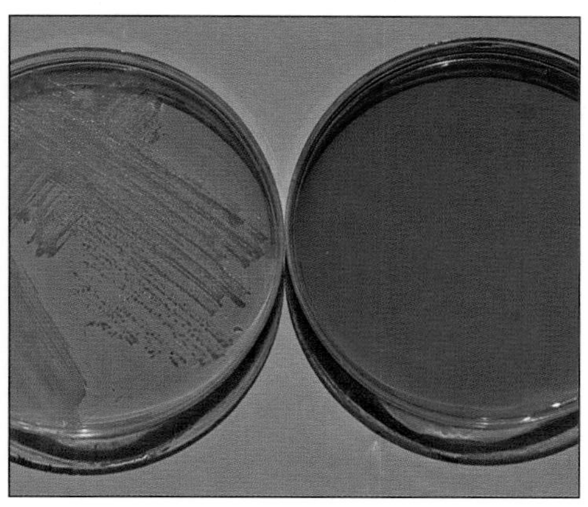

V.1 *Staphylococcus aureus* growing on mannitol salt agar. The yellow color indicates that mannitol is fermented. The red plate is uninoculated (Exercises 12 and 46).

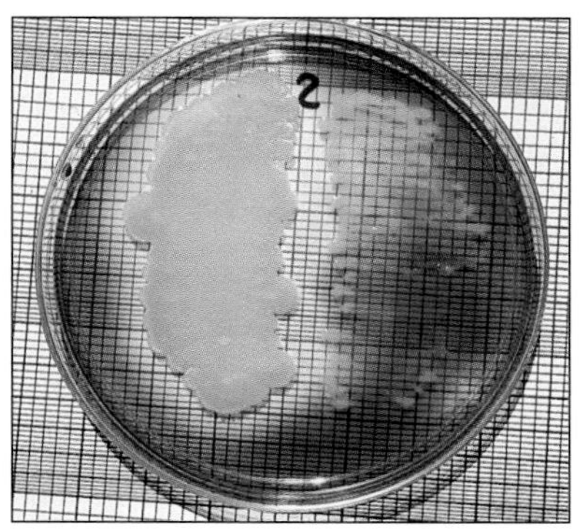

V.2 Starch hydrolysis. Iodine is used to detect the presence of starch on a starch agar plate. The clearing around the *Bacillus* indicates that the starch was hydrolyzed by an extracellular enzyme (Exercise 13).

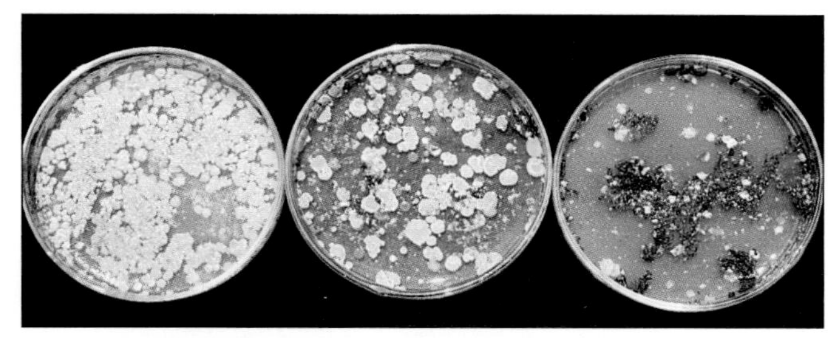

V.3 Compare the effects of dry heat and autoclaving. The control (left) is a pour plate inoculated with soil. The middle plate was inoculated with soil exposed to dry heat (121°C) for 1 hour. The soil on the right was autoclaved (121°C) and has no bacterial growth (Exercise 22).

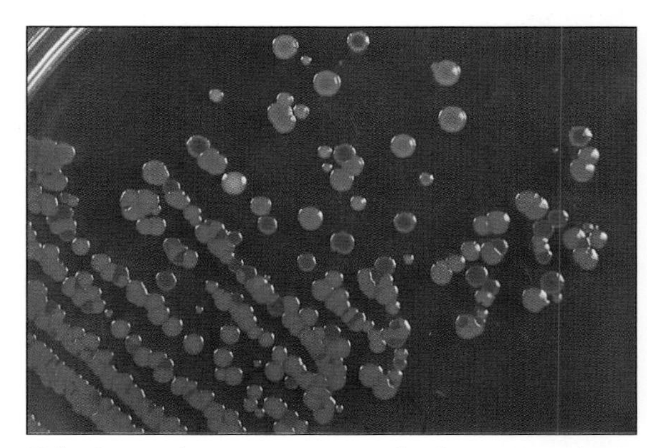

V.4 *Serratia marcescens* colonies on nutrient agar after incubation at 25°C. Note color variations caused by mutations in prodigiosin production (Exercise 28).

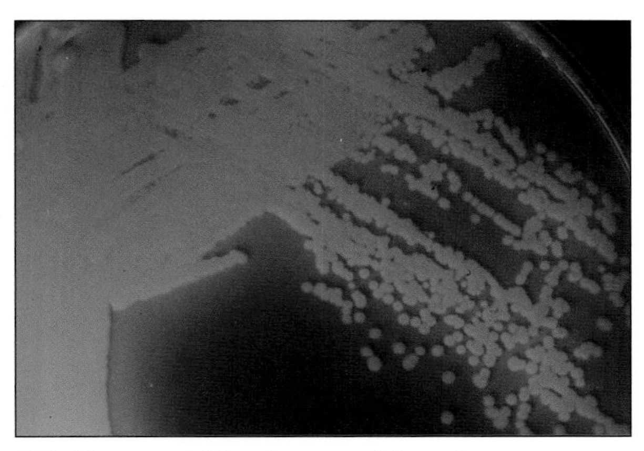

V.5 Water-soluble pigment of *Pseudomonas aeruginosa* fluoresces under ultraviolet light (Exercise 50).

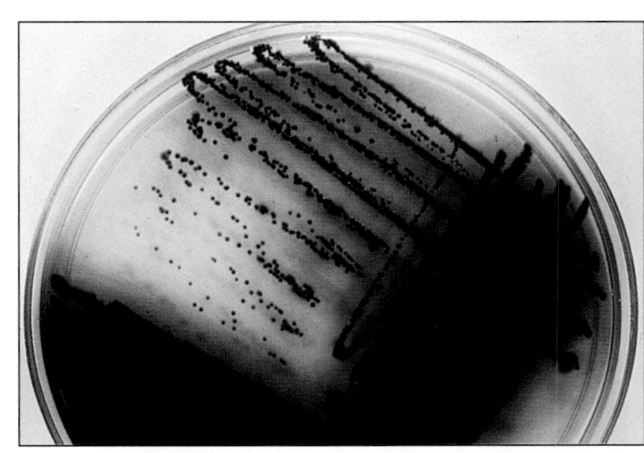

V.6 *Pseudomonas aeruginosa* produces a water-soluble blue pigment on Pseudomonas agar P (Exercise 50).

ENVIRONMENTAL MICROBIOLOGY

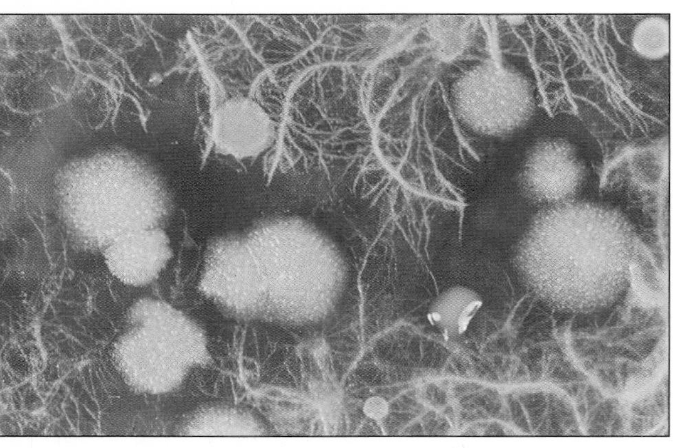

VI.1 New antibiotics are discovered by examining colonies of soil bacteria. In this sample, one species is producing an antibiotic that inhibits *Bacillus cereus* from growing near it (Exercise 9).

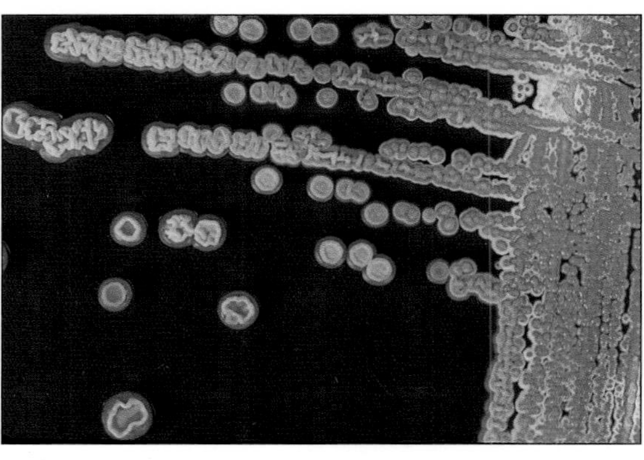

VI.2 Actinomycetes colonies. *Streptomyces griseus* colonies penetrate into the agar as well as extend above. The powdery appearance in the corner is caused by conidiospores (Exercise 11).

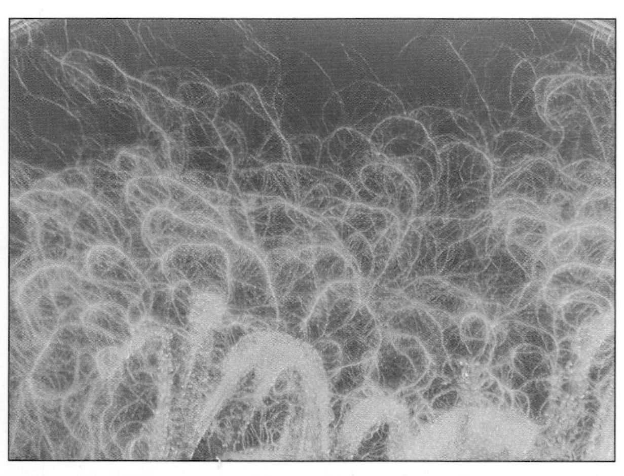

VI.3 *Bacillus mycoides* colonies. Easily recognized by their filamentous form. Bending of the filaments to the left or right is a strain characteristic (Exercise 11).

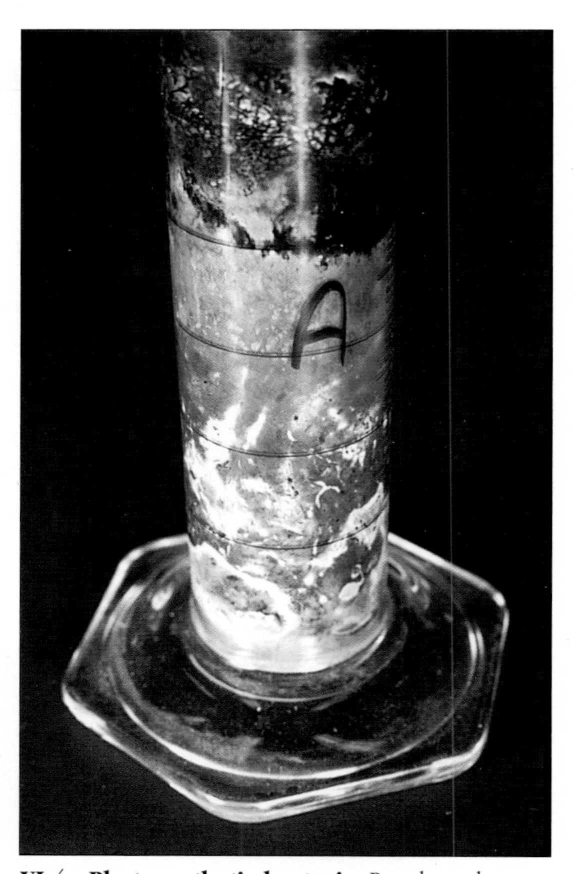

VI.4 Photosynthetic bacteria. Purple and green sulfur bacteria have colonized in this Winogradsky column (Exercise 56).

VI.5 Root nodules. *Rhizobium leguminosarum* infected the roots of this vetch plant and caused the production of pink nodules (Exercise 56).

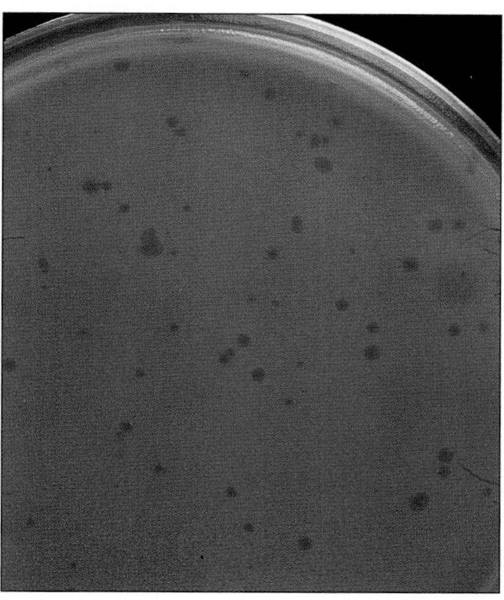

VII.1 Numerous plaques (clearings) in this *Escherichia coli* culture are caused by growth of a T-even bacteriophage (Exercise 38).

VII.2 Chlorosis (loss of green color) and plaques (spotting) in this tomato leaf are caused by a tobacco mosaic virus infection (Exercise 39).

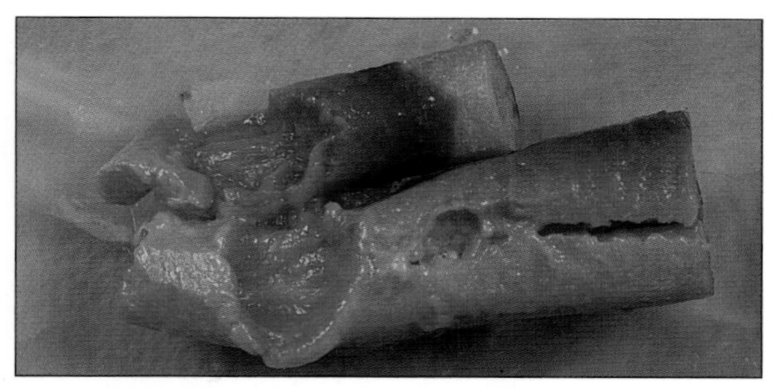

VII.3 Carrot soft rot caused by *Pectobacterium carotovorum* (Exercise 41).

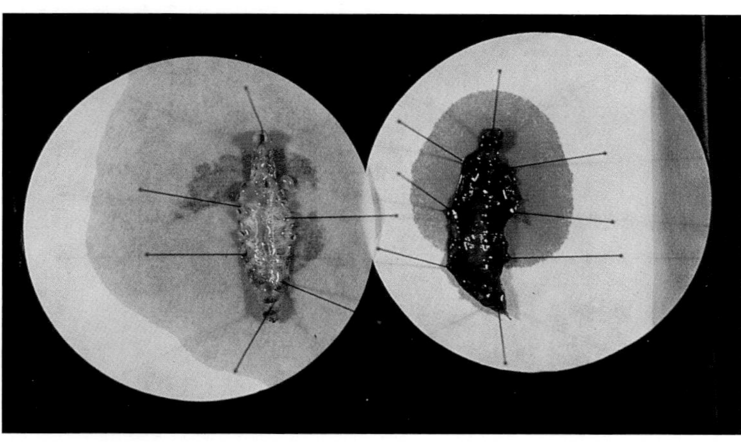

VII.4 Milky spore disease. The larva on the right ingested the toxin produced by *Bacillus thuringiensis* (Exercise 41).

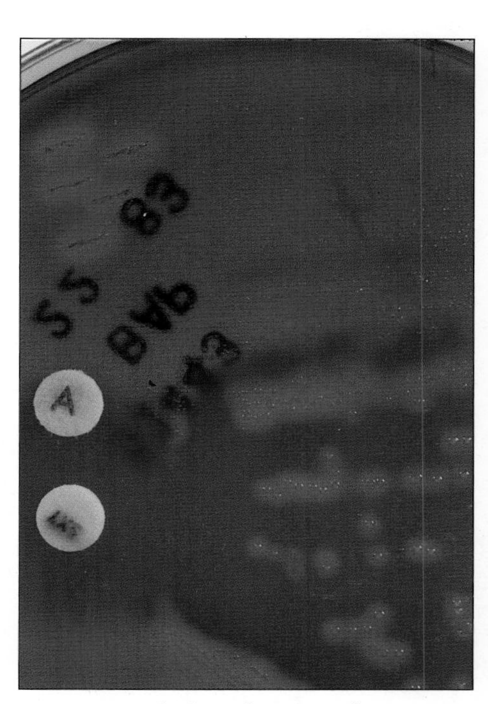

VII.5 Rapid identification of *Streptococcus pyogenes*. *S. pyogenes* is beta-hemolytic and inhibited by bacitracin (Exercise 47).

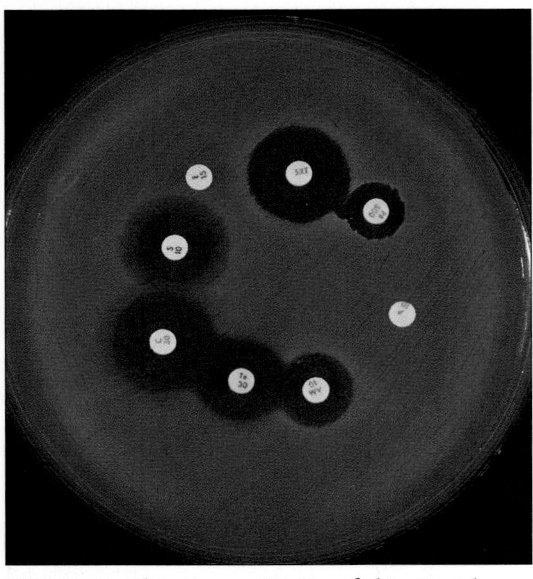

VIII.1 Antibiotic sensitivity of this *Staphylococcus aureus* culture is demonstrated by the Kirby–Bauer agar diffusion test (Exercise 25).

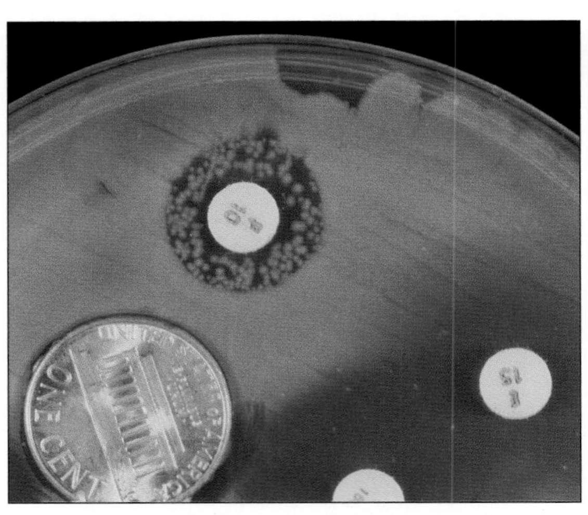

VIII.2 Some of the cells of this strain of *Staphylococcus aureus* are resistant to penicillin (Exercise 25).

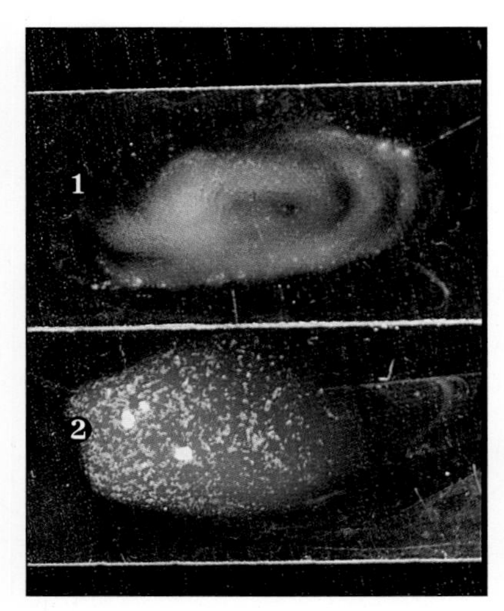

VIII.3 Coagulase test. Many pathogenic strains of *Staphylococcus aureus* can coagulate plasma. Bacteria are caught in fibrin clumps in **2** (Exercise 46).

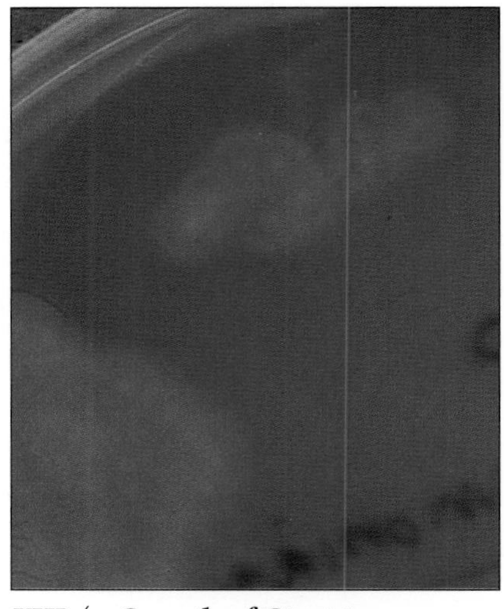

VIII.4 Growth of *Streptococcus pneumoniae* on blood agar demonstrates alpha-hemolysis. Note the greenish color around the colonies (Exercise 47).

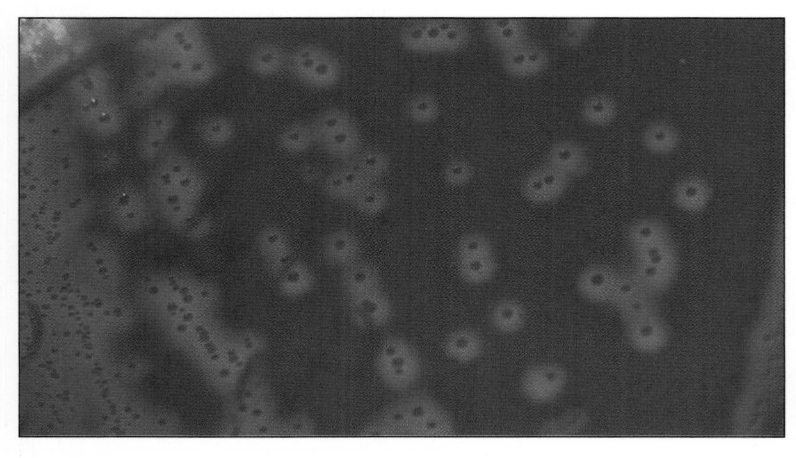

VIII.5 Beta-hemolysis produces a clear area around the colonies of *Streptococcus pyogenes* on blood agar (Exercise 47).

IMMUNOLOGY

PLATE IX

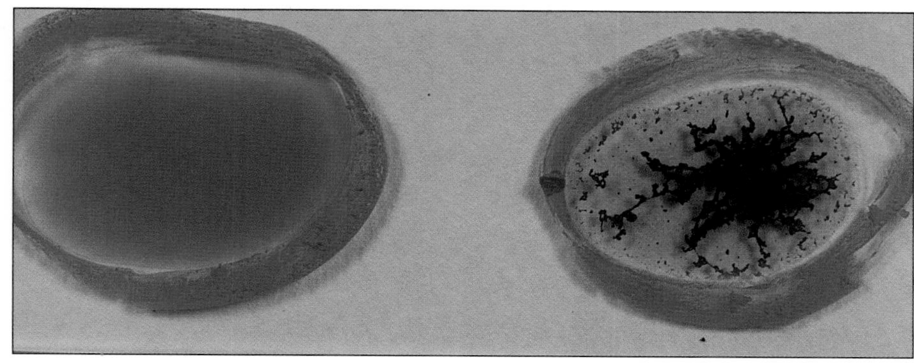

IX.1 Slide agglutination test for the ABO blood group. A negative reaction with anti-B (left) and a positive reaction with anti-A (indicated by agglutination) on the right (Exercise 43). What is this person's blood type?

GROWTH

X.1 Bacterial growth in broth. The nutrient broth in tube **1** is clear, indicating no bacterial growth. Bacterial growth is determined by **2** turbidity, **3** a pellicle on the surface, **4** flocculant suspended in the broth, and **5** a sediment at the bottom (Exercises 9 and 21).

BIOREMEDIATION

XI.1 In a wastewater treatment facility, sewage is aerated to encourage bacterial growth and decomposition of organic matter (Exercises 52, 53, and 57).

XI.2 After secondary sewage treatment, water is disinfected and discharged into a water course or onto land (Exercises 52, 53, and 57).

XI.3 Soil is excavated and layered with pipes at a former truck stop. The pipes are connected to supplies of nitrogen, phosphate, and oxygen to provide additional nutrients for bacteria that use hydrocarbons as their carbon source (Exercise 57).

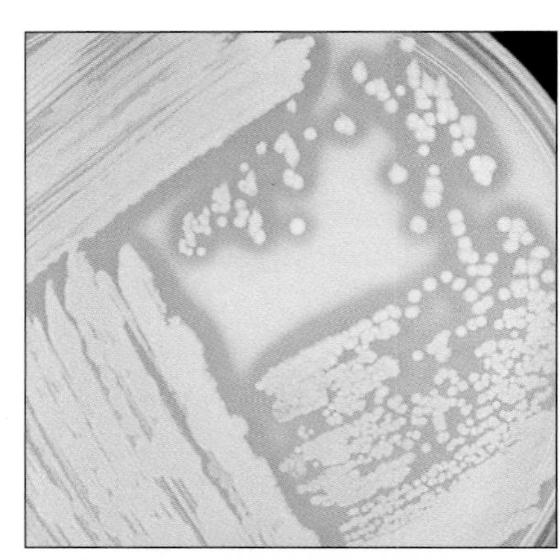

XI.4 Bacteria that degrade the hydrocarbon in the agar produce water-soluble products, resulting in clearing around the colonies (Exercise 57).

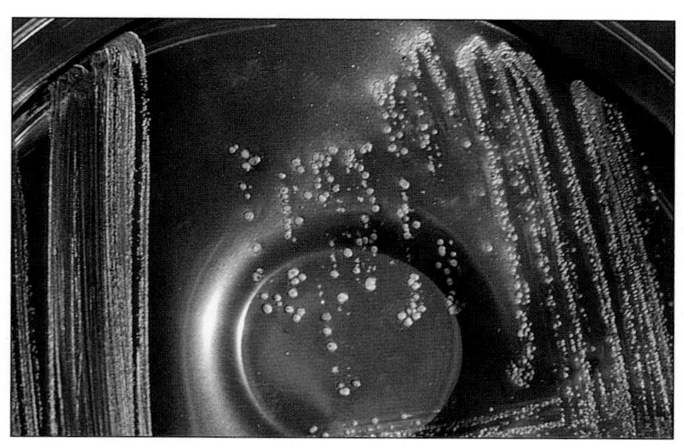

XII.1 Colonies of *Escherichia coli* (shown) and *Citrobacter* develop a metallic green sheen on EMB agar (Exercises 12 and 53).

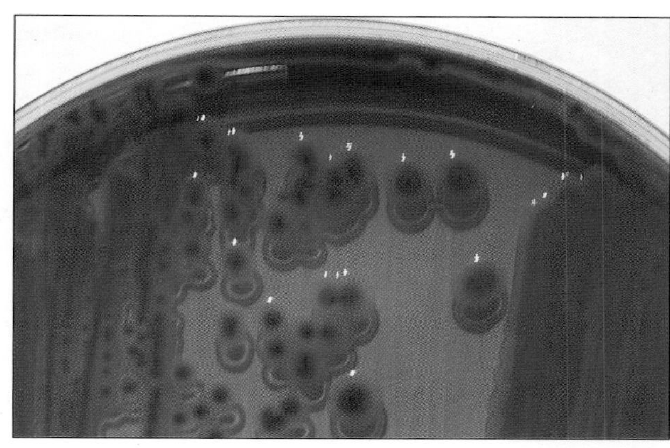

XII.2 Colonies of *Enterobacter* have a distinctive blue "fish-eye" appearance on EMB agar (Exercises 12 and 53).

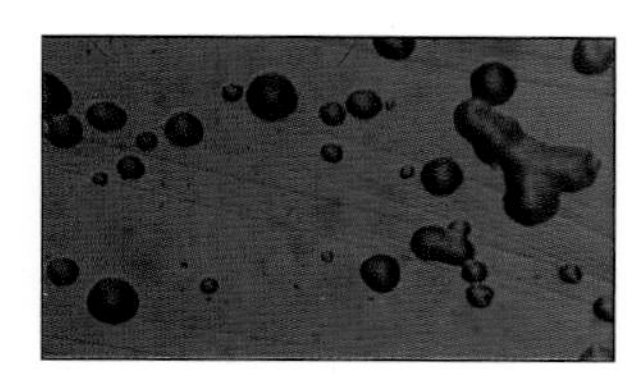

XII.3 Nonlactose fermenters such as *Shigella boydii* (shown) produce colorless colonies on EMB agar (Exercises 12 and 53).

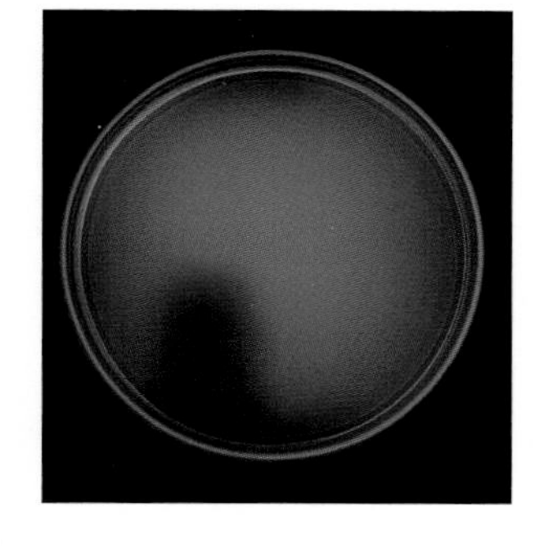

XII.4 *E. coli* is identified by its ability to convert MUG to a fluorescent compound that is visible with an ultraviolet lamp (Exercise 52).

XII.5 Coliforms trapped on a membrane filter grow into colonies when the filter is laid on EMB agar and incubated (Exercise 53).

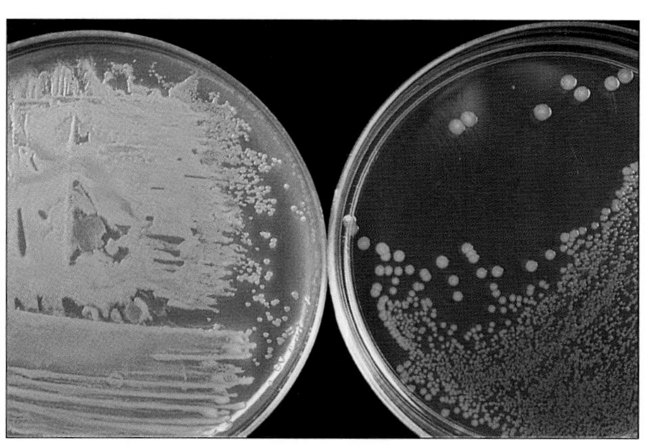

XII.6 MacConkey agar is selective for gram-negative enterics. Colonies of lactose-fermenting enterics are red and surrounded by an opaque precipitate (left). Lactose-negative bacteria produce colorless colonies (right) (Exercises 49 and 50).

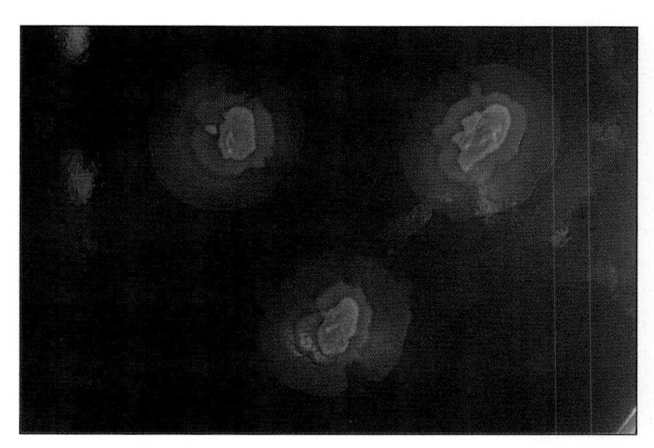

XII.7 Cells of *Proteus mirabilis* are highly motile. *P. mirabilis* colonies exhibit swarming over EMB agar as young cells move away from the colony's center (Exercises 50 and 53).

PHOTOTROPHIC MICROORGANISMS

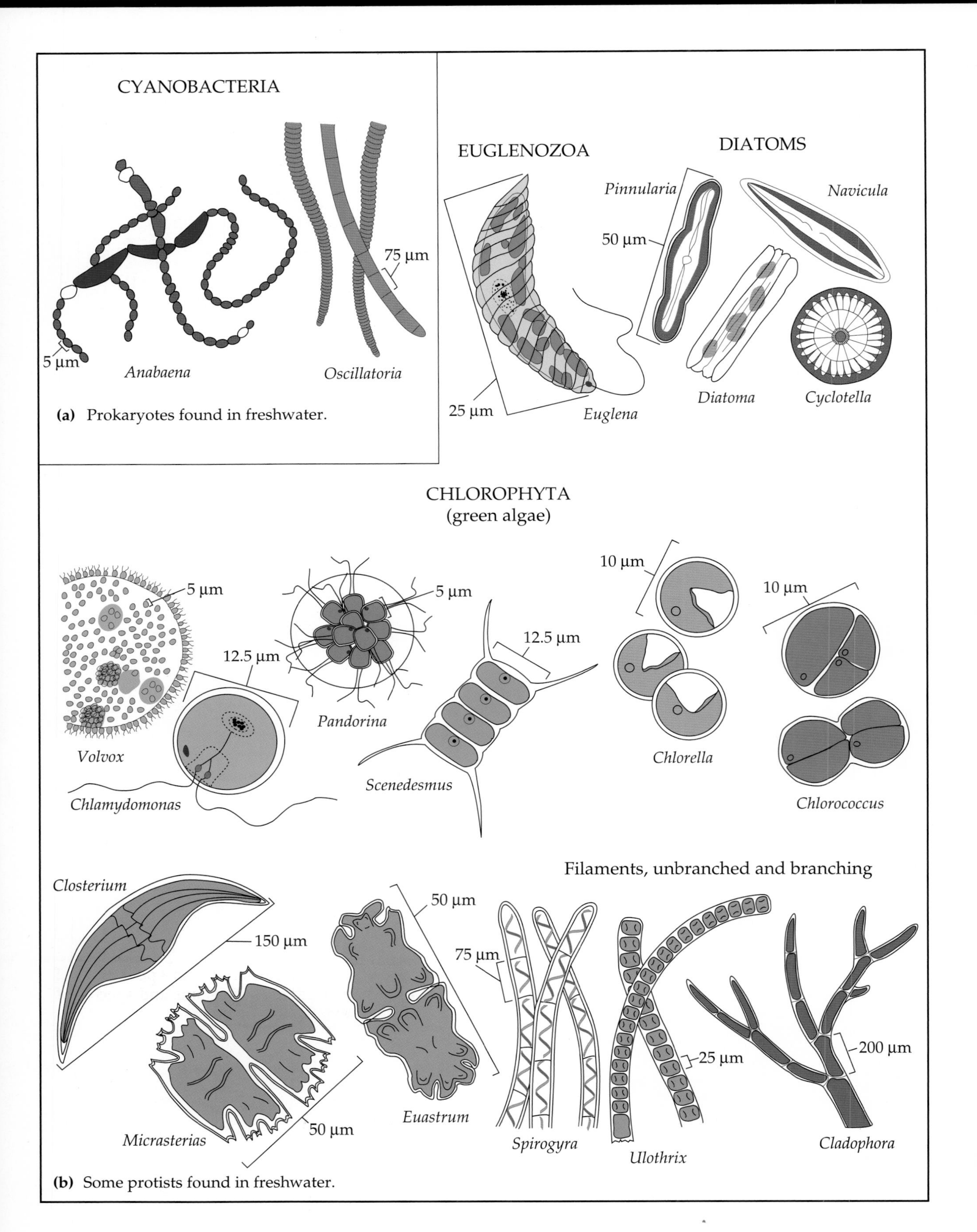

(a) Prokaryotes found in freshwater.

(b) Some protists found in freshwater.

UNKNOWN IDENTIFICATION: PHOTO QUIZ 1 PLATE XIV

This is an example of how to identify an unknown bacterium.

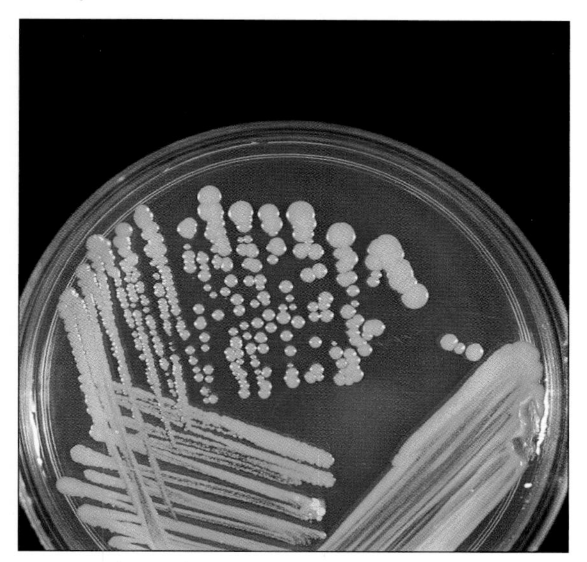

XIV.1 The unknown organism is isolated in pure culture (Exercise 11).

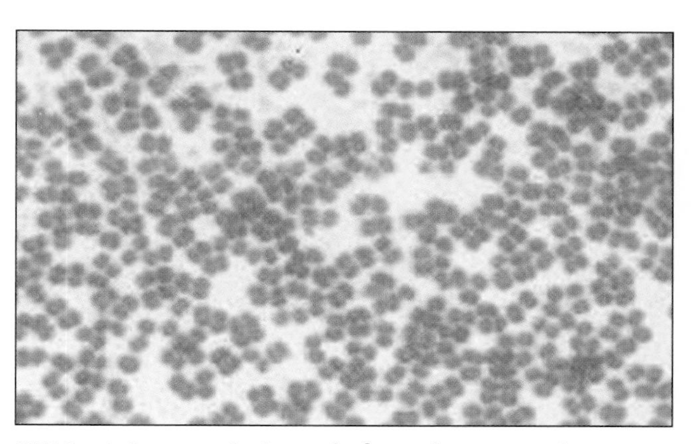

XIV.2 A Gram stain is made from the pure culture (Exercise 5).

XIV.3 Based on the result of the Gram stain, a catalase test is indicated (Exercise 17).

XIV.4 Results of the previous tests lead to the next series of tests. A glucose fermentation tube is inoculated with the unknown bacterium (Exercise 13).

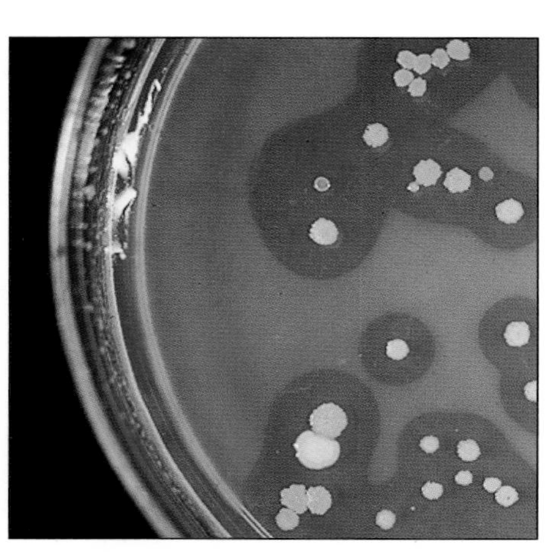

XIV.5 Lipid hydrolysis test. The clearing around the colonies indicates hydrolysis of the lipid in the medium (Exercise 57).

Using the identification keys in Appendix H of this manual, identify the genus and species of this bacterium.

Answer: *Micrococcus luteus*

Unknown Identification: Photo Quiz 2

The unknown is a gram-negative, oxidase-negative rod. An uninoculated control is shown on the left in plates XIV.7 through XIV.11.

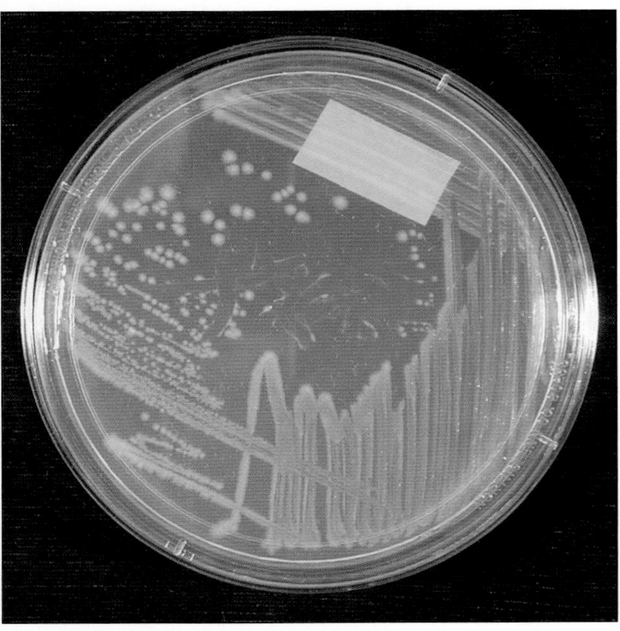

XIV.6 Colonies after incubation at 25°C.

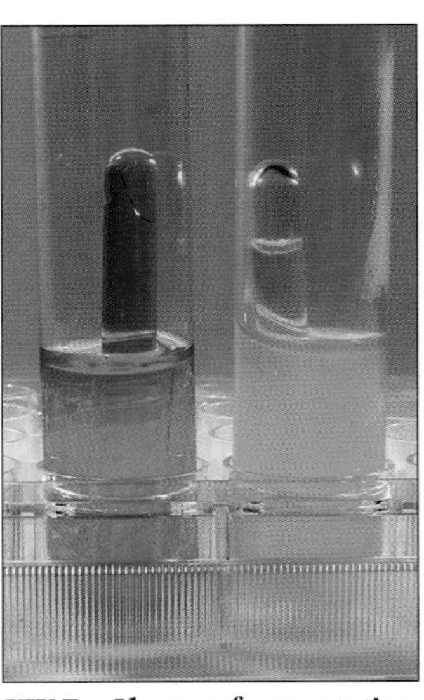

XIV.7 Glucose fermentation.

XIV.8 Lactose fermentation.

XIV.9 Peptone iron agar.

XIV.10 MIO. Kovacs reagent was added to the test (right).

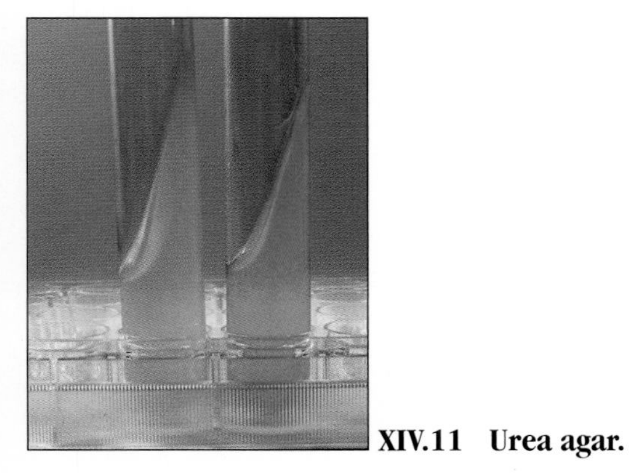

XIV.11 Urea agar.

Using the identification keys in Appendix H of this manual, identify the genus and species of this bacterium.

Answer: *Proteus mirabilis*

Part One Microscopy

EXERCISES

The microscope is a very important tool for a microbiologist. Microscopes and microscopy (microscope technique) are introduced in Exercises 1 and 2, which are designed to help you become familiar with and proficient in the use of the compound light microscope. This knowledge will be valuable in later exercises.

Beginning students frequently become impatient with the microscope and forgo this opportunity to practice and develop their observation skills. Simple observation is a critical part of any science. Making discoveries by observation requires *curiosity* and *patience*. We cannot provide procedures for observation, but we can offer this suggestion: Make *careful sketches* to enhance effective observation. You need not be an artist to draw what you see. In your drawings, pay special attention to the following:

1. *Size relationships*. For example, how big are bacteria relative to protozoa?
2. *Spatial relationships*. For example, where is one bacterium in relation to the others? Are they all together in chains?
3. *Behavior*. For example, are individual cells moving, or are they all flowing in the liquid medium?
4. *Sequence of events*. For example, were cells active when you first observed them?

Looking at objects through a microscope is not easy at first, but with a little practice you, like Antoni van Leeuwenhoek, will make discoveries in the microcosms of peppercorn infusions and raindrops (see the photograph on page 2). Van Leeuwenhoek wrote in 1684:

Tho my teeth are kept usually very clean, nevertheless when I view them in a Magnifying Glass, I find growing between them a little white matter as thick as wetted flour: In this substance tho I do not perceive any motion, I judged there might probably be living creatures.

*I therefore took some of this flour and mixed it either with pure rain water wherein were no Animals; or else with some of my Spittle (having no air bubbles to cause a motion in it) and then to my great surprise perceived that the aforesaid matter contained very many small living Animals, which moved themselves very extravagantly. Their motion was strong and nimble, and they darted themselves thro the water or spittle, as a Jack or Pike does thro the water.**

*Quoted in E. B. Fred. "Antoni van Leeuwenhoek." *Journal of Bacteriology* 25:1, 1933.

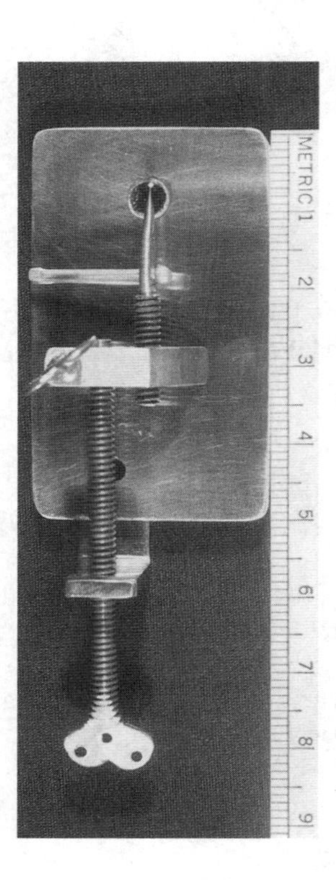

A replica of the simple microscope made by Antoni van Leeuwenhoek to observe living organisms too small to be seen with the naked eye. The specimen was placed on the tip of the adjustable point and viewed from the other side through the tiny round lens. The highest magnification with his lenses was about 300×.

Exercise 1

Use and Care of the Microscope

The most important discoveries of the laws, methods and progress of nature have nearly always sprung from the examination of the smallest objects which she contains.

JEAN BAPTISTE LAMARCK

Objectives

After completing this exercise, you should be able to:

1. Demonstrate the correct use of a compound light microscope.
2. Diagram the path of light through a compound microscope.
3. Name the major parts of a compound microscope.
4. Identify the three basic morphologies of bacteria.

Background

Virtually all organisms studied in microbiology cannot be seen with the naked eye but require the use of optical systems for magnification. The microscope was invented shortly before 1600 by Zacharias Janssen of the Netherlands. The microscope was not used to examine microorganisms until the 1680s, when a clerk in a dry-goods store, Antoni van Leeuwenhoek, examined scrapings of his teeth and any other substances he could find. The early microscopes, called **simple microscopes,** consisted of biconvex lenses and were essentially magnifying glasses. (See the photograph on p. 2.) To see microbes, a compound microscope, which has two lenses between the eye and the object, is required. This optical system magnifies the object, and an illumination system (sun and mirror or lamp) ensures that adequate light is available for viewing. A **brightfield compound microscope,** which shows dark objects in a bright field, is used most often.

You will be using a brightfield compound microscope similar to the one shown in Figure 1.1a. The basic frame of the microscope consists of a **base,** a **stage** to hold the slide, an **arm** for carrying the microscope, and a **body tube** for transmitting the magnified image. The stage may have two clips or a movable mechanical stage to hold the slide. The light source is in the base. Above the light source is a **condenser,** which consists of several lenses that concentrate light on the slide by focusing it into a cone, as shown in Figure 1.1b. The condenser has an **iris diaphragm,** which controls the angle and size of the cone of light. This ability to control the *amount* of light ensures that optimal light will reach the slide. Above the stage, on one end of the body tube, is a revolving nosepiece holding three or four **objective lenses.** At the upper end of the tube is an **ocular** or **eyepiece lens** (10× to 12.5×). If a microscope has only one ocular lens, it is called a **monocular** microscope; a **binocular** microscope has two ocular lenses.

By moving the tube closer to the slide or the stage closer to the objective lens, using the coarse- or fine-adjustment knobs, one can focus the image. The larger knob, the **coarse adjustment,** is used for focusing with the low-power objectives (4× and 10×), and the smaller knob, the **fine adjustment,** is used for focusing with the high-power and oil immersion lenses. The coarse-adjustment knob moves the lenses or the stage longer distances. The area seen through a microscope is called the **field of vision.**

The **magnification** of a microscope depends on the type of objective lens used with the ocular. Compound microscopes have three or four objective lenses mounted on a nosepiece: scanning (4×), low-power (10×), high-dry (40× to 45×), and oil immersion (97× to 100×). The magnification provided by each lens is stamped on the barrel. The total magnification of the object is calculated by multiplying the magnification of the ocular (usually 10×) by the magnification of the objective lens. The most important lens in microbiology is the **oil immersion lens;** it has the highest magnification (97× to 100×) and must be used with immersion oil. Optical systems could be built to magnify much more than the 1000× magnification of your microscope, but the resolution would be poor.

Resolution or **resolving power** refers to the ability of lenses to reveal fine detail or two points distinctly separated. An example of resolution involves a car approaching you at night. At first only one light appears, but as the car nears, you can distinguish two

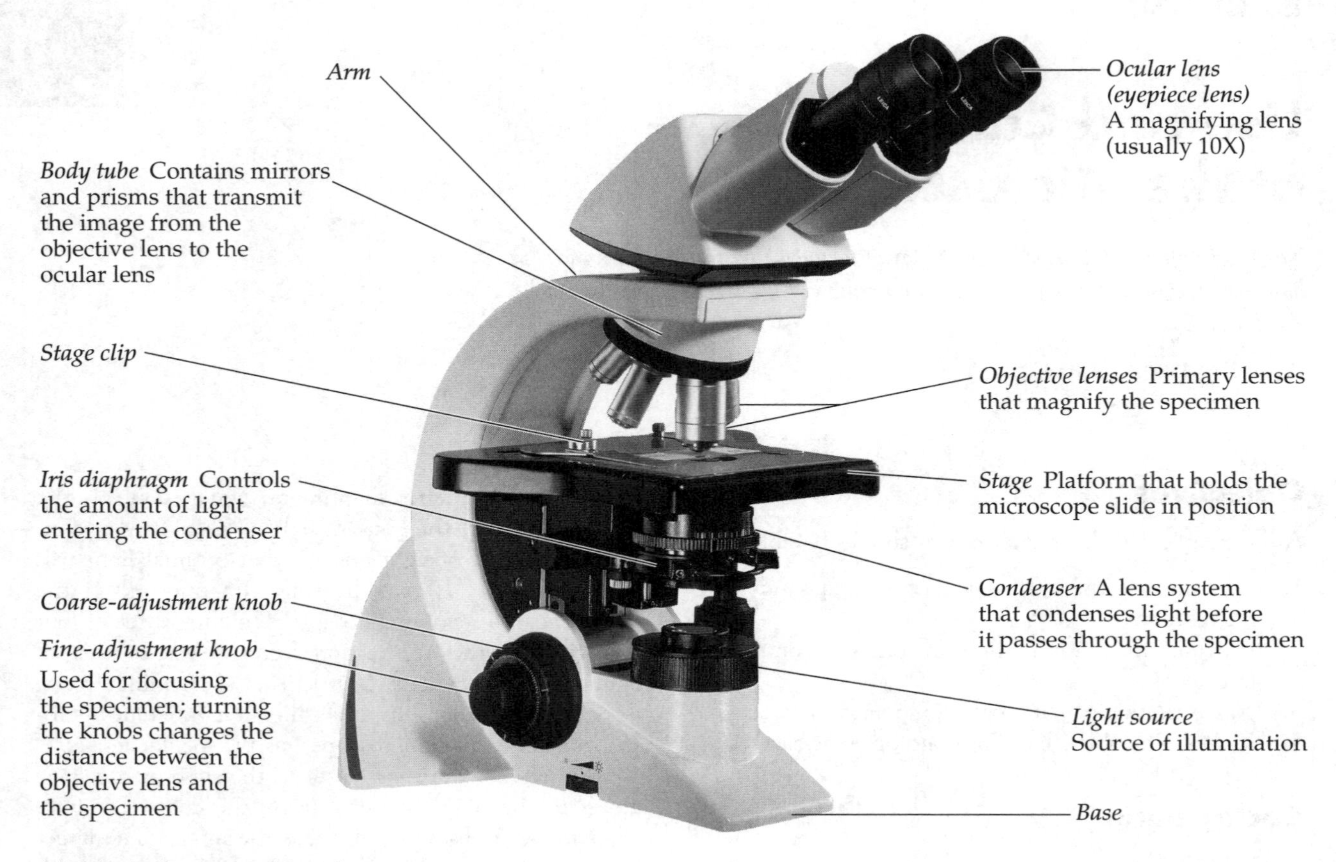

(a) Principal parts and functions

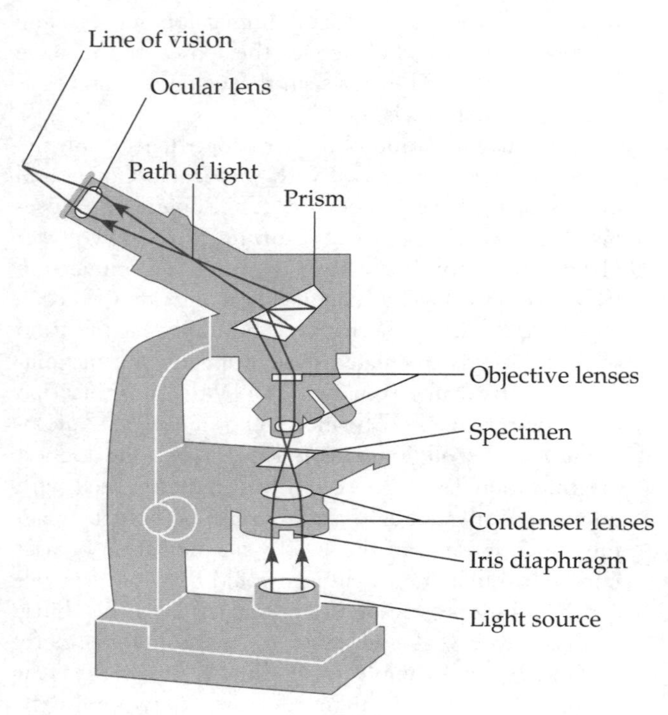

(b) Arrows show the path of light (bottom to top)

Figure 1.1

The compound light microscope. **(a)** Its principal parts and their functions. **(b)** Lines from the light source through the ocular lens illustrate the path of light.

headlights. The resolving power is a function of the wavelength of light used and a characteristic of the lens system called **numerical aperture.** Resolving power is best when two objects are seen as distinct even though they are very close together. Resolving power is expressed in units of length; the smaller the distance, the better the resolving power.

$$\text{Resolving power} = \frac{\text{Wavelength of light used}}{2 \times \text{numerical aperture}}$$

Smaller wavelengths of light improve resolving power. The effect of decreasing the wavelength can be seen in electron microscopes, which use electrons as a source of "light." The electrons have an extremely short wavelength and result in excellent resolving power. A light microscope has a resolving power of about 200 nanometers (nm), whereas an electron microscope has a resolving power of less than 0.2 nm. The numerical aperture is engraved on the side of each objective lens (usually abbreviated N.A.). If the numerical aperture increases—for example, from 0.65 to 1.25—the resolving power is improved. The numerical aperture is dependent on the maximum angle of the light entering the objective lens and on the **refractive index** (the amount the light bends) of the material (usually air) between the objective lens and the slide. This relationship is defined by the following:

$$\text{N.A.} = N \sin \theta$$

N = Refractive index of medium
θ = Angle between the most divergent light ray gathered by the lens and the center of the lens

As shown in Figure 1.2, light is refracted when it emerges from the slide because of the change in media as the light passes from glass to air. When immersion oil is placed between the slide and the oil immersion lens, the light ray continues without refraction because immersion oil has the same refractive index ($N = 1.52$) as glass ($N = 1.52$). This can be seen easily. When you look through a bottle of immersion oil, you cannot see the glass rod in it because of the identical N values of the glass and immersion oil. The result of using oil is that light loss is minimized, and the lens focuses very close to the slide.

As light rays pass through a lens, they are bent to converge at the **focal point,** where an image is formed (Figure 1.3a). When you bring the center of a microscope field into focus, the periphery may be fuzzy due to the curvature of the lens, resulting in multiple focal points. This is called **spherical aberration** (Figure 1.3b). Spherical aberrations can be minimized by the use of the iris diaphragm, which eliminates light rays to the periphery of the lens, or by a series of lenses resulting in essentially a flat optical system. Sometimes a multitude of colors, or **chromatic aberration,** is seen in the field (Figure 1.3c). This is caused by the prismlike effect of the lens as various wavelengths of white light pass through to a different focal point for each wavelength. Chromatic aberrations can be minimized by the use of filters (usually blue); or by lens systems corrected for red and blue light, called *achromatic lenses;* or by lenses corrected for red, blue, and other wavelengths, called *apochromatic lenses*. The most logical, but most expensive, method of eliminating chromatic aberrations is to use a light source of one wavelength, or **monochromatic light.**

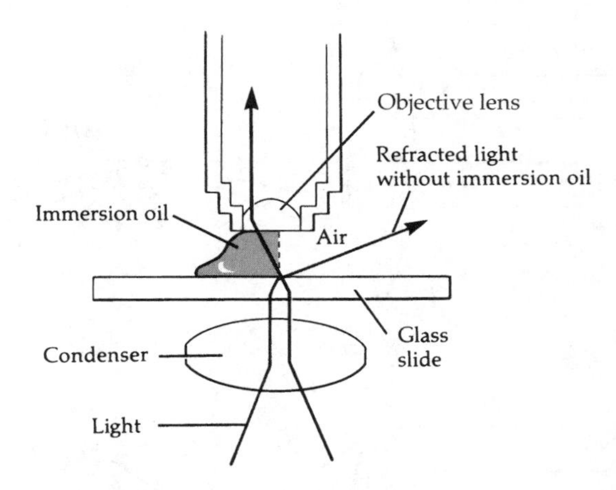

Figure 1.2

Refractive index. Because the refractive indexes of the glass microscope slide and immersion oil are the same, the oil keeps the light rays from refracting.

Compound microscopes require a light source. The light may be reflected to the condenser by a mirror under the stage. If your microscope has a mirror, the sun or a lamp may be used as the light source. Most compound microscopes have a built-in illuminator in the base. The *intensity* of the light can often be adjusted with a rheostat.

The microscope is a very important tool in microbiology, and it must be used carefully and correctly. Follow these guidelines *every* time you use a microscope.

General Guidelines

1. Carry the microscope with both hands: one hand beneath the base and one hand on the arm.
2. Do not tilt the microscope; instead, adjust your stool so you can comfortably use the instrument.
3. Observe the slide with both eyes open, to avoid eyestrain.
4. Always focus by moving the lens away from the slide.
5. Always focus slowly and carefully.

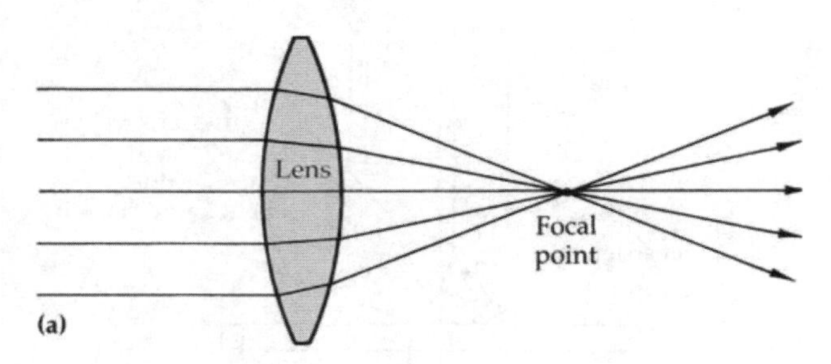

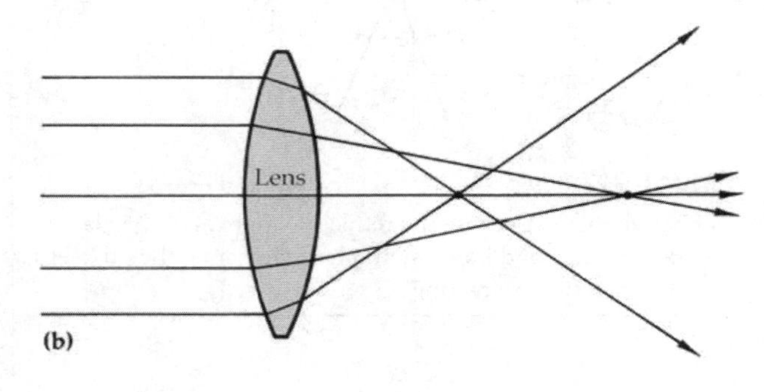

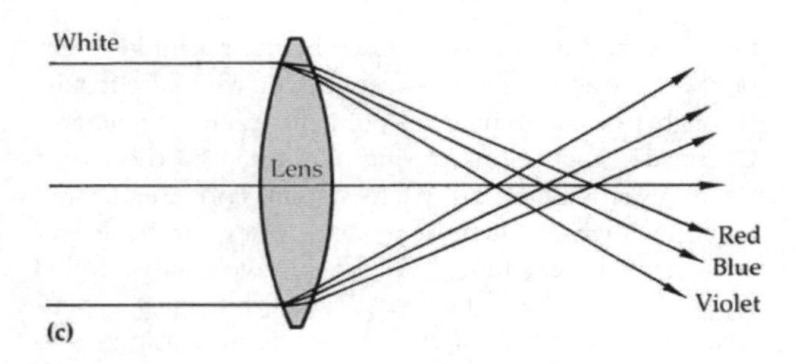

Figure 1.3

Focal point. **(a)** An image is formed when light converges at one point, called the focal point. **(b)** Spherical aberration. Curved lenses result in light passing through one region of the lens having a different focal point than light passing through another part of the lens. **(c)** Chromatic aberration. Each wavelength of light may be given a different focal point by the lens.

6. When using the low-power lens, the iris diaphragm should be barely open so that good contrast is achieved. More light is needed with higher magnification.
7. Before using the oil immersion lens, have your slide in focus under high power. *Always focus with low power first*.
8. Keep the stage clean and free of oil. Keep all lenses except the oil immersion lens free of oil.
9. Keep all lenses clean. Use *only* lens paper to clean them. Wipe oil off the oil immersion lens before putting your microscope away. Do not touch the lenses with your hands.
10. Clean the ocular lens carefully with lens paper. If dust is present, it will rotate as you turn the lens.
11. After use, remove the slide, wipe oil off it, put the dust cover on the microscope, and return it to the designated area.
12. When a problem does arise with the microscope, obtain help from the instructor. Do not use another microscope unless yours is declared "out of action."

Materials

Compound light microscope

Immersion oil

Lens paper

Prepared slides of algae, fungi, protozoa, and bacteria

Procedure

1. Place the microscope on the bench squarely in front of you.
2. Obtain a slide of algae or fungi and place it in the stage clips on the stage.
3. Adjust the eyepieces on a binocular microscope to your own personal measurements.
   a. Look through the eyepieces and, using the thumb wheel, adjust the distance between the eyepieces until one circle of light appears.
   b. With the low-power (10×) objective in place, cover the left eyepiece with a small card and focus the microscope on the slide. When the right eyepiece has been focused, remove your hand from the focusing knobs and cover the right eyepiece. Looking through the microscope with your left eye, focus the left eyepiece by turning the eyepiece adjustment. Make a note of the number at which you focused the left eyepiece so you can adjust any binocular microscope for your eyes.
4. Raise the condenser up to the stage. On some microscopes, the condenser can be focused by the following procedure:
   a. Focus with the 10× objective.
   b. Close the iris diaphragm so only a minimum of light enters the objective lens.

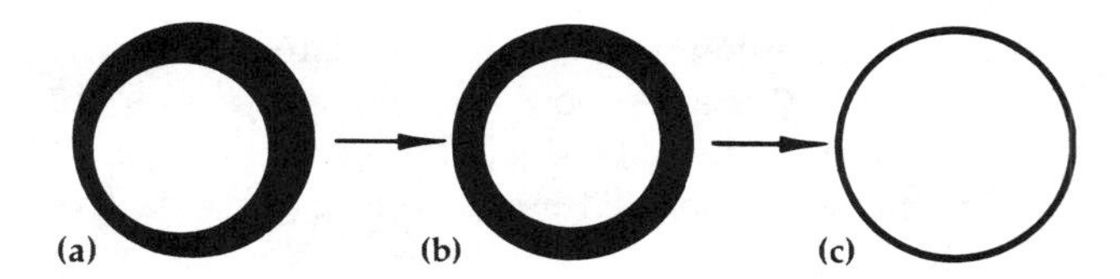

Figure 1.4

Using low power, lower the condenser until a distinct circle of light is visible **(a).** Center the circle of light using the centering screws **(b).** Open the iris diaphragm until the light just fills the field **(c).**

c. Lower the condenser until the light is seen as a circle in the center of the field. On some microscopes the circle of light may be centered (Figure 1.4) using the centering screws found on the condenser.

d. Raise the condenser up to the slide, lower it, and stop when the color on the periphery changes from pink to blue (usually 1 or 2 mm below the stage).

e. Open the iris diaphragm until the light just fills the field.

5. Diagram some of the cells on the slide under low power. Use a minimum of light by adjusting the ______________________.

6. When an image has been brought into focus with low power, rotate the turret to the next lens, and the subject will remain almost in focus. All of the objectives (with the possible exception of the 4×) are **parfocal;** that is, when a subject is in focus with one lens, it will be in focus with all of the lenses. When you have completed your observations under low power, swing the high-dry objective into position and focus. Use the fine adjustment. Only a slight adjustment should be required. Why? ______
____

More light is usually needed. Again, draw the general size and shape of some cells.

7. Move the high-dry lens out of position, and place a drop of immersion oil on the area of the slide you are observing. Carefully click the oil immersion lens into position. It should now be immersed in the oil (Figure 1.5). Careful use of the fine-adjustment knob should bring the object into focus. Note the shape and size of the cells. Did the color of the cells change with the different lenses? __________ Did the size of the field change? ______________

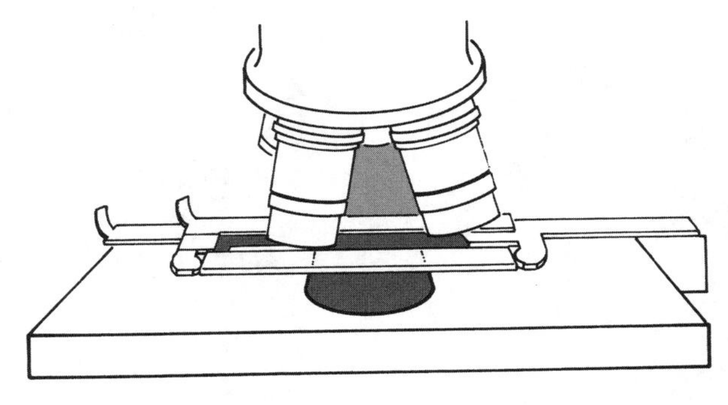

(a) Move the high-dry lens out of position.

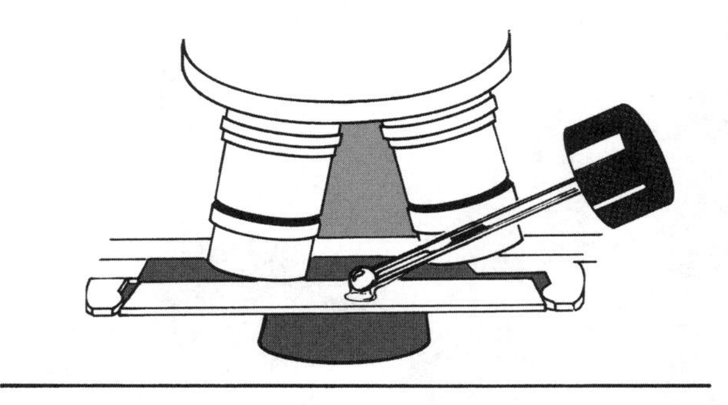

(b) Place a drop of immersion oil in the center of the slide.

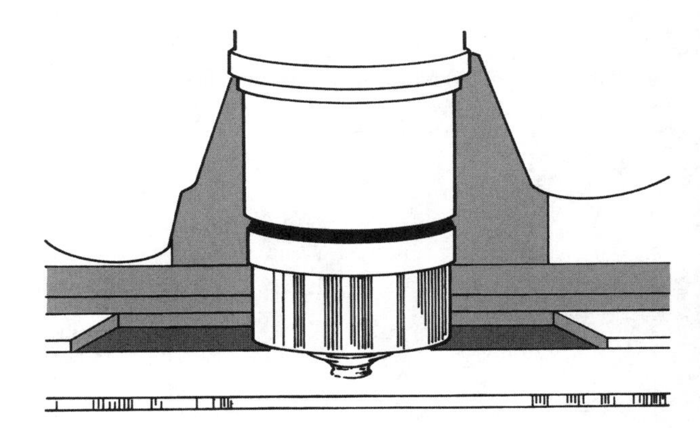

(c) Move the oil immersion lens into position.

Figure 1.5

Using the oil immersion lens.

8. Record your observations and note the magnifications. The following figures may be used as references:
Algae: Color Plate XIII
Protozoa: Figure 37.1
Fungi: Figure 35.1

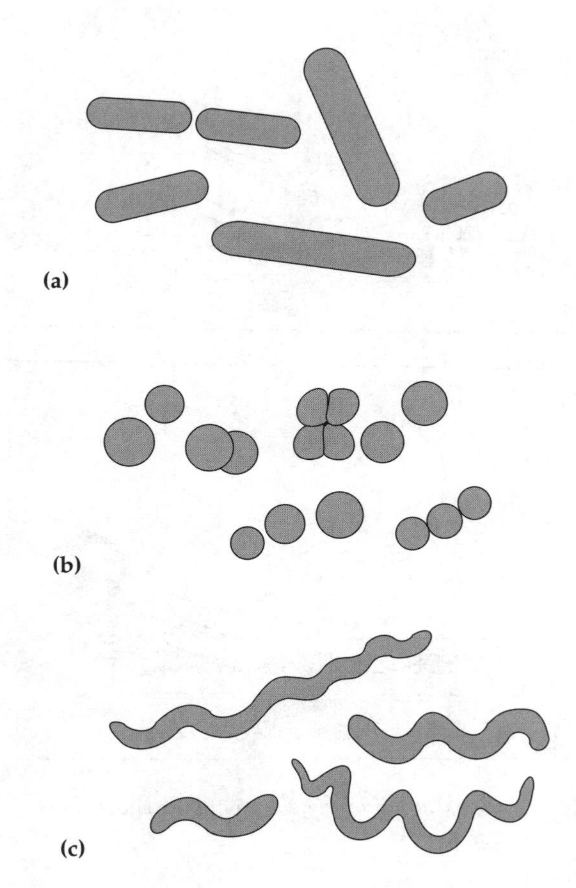

Figure 1.6

Basic shapes of bacteria. **(a)** Bacillus (plural: bacilli), or rod. **(b)** Coccus (plural: cocci). **(c)** Spiral.

9. When your observations are completed, move the turret to bring a low-power objective into position. *Do not* rotate the high-dry (40×) objective through the immersion oil. Clean the oil off the objective lens with lens paper, and clean off the slide with tissue paper or a paper towel. Remove the slide. Repeat this procedure with all the available slides. When observing the bacteria, note the three different morphologies, or shapes, shown in Figure 1.6.

Exercise 1 **Laboratory Report**

Use and Care of the Microscope

Name ______________________

Date ______________________

Lab Section ______________________

Purpose ______________________

____

____

Data

Microscope number: ________ Monocular or binocular: ______________________

Eyepiece adjustment notes: ______________________

Draw a few representative cells from each slide and show how they appeared at each magnification. Note the differences in size at each magnification.

Algae

Slide of ______________

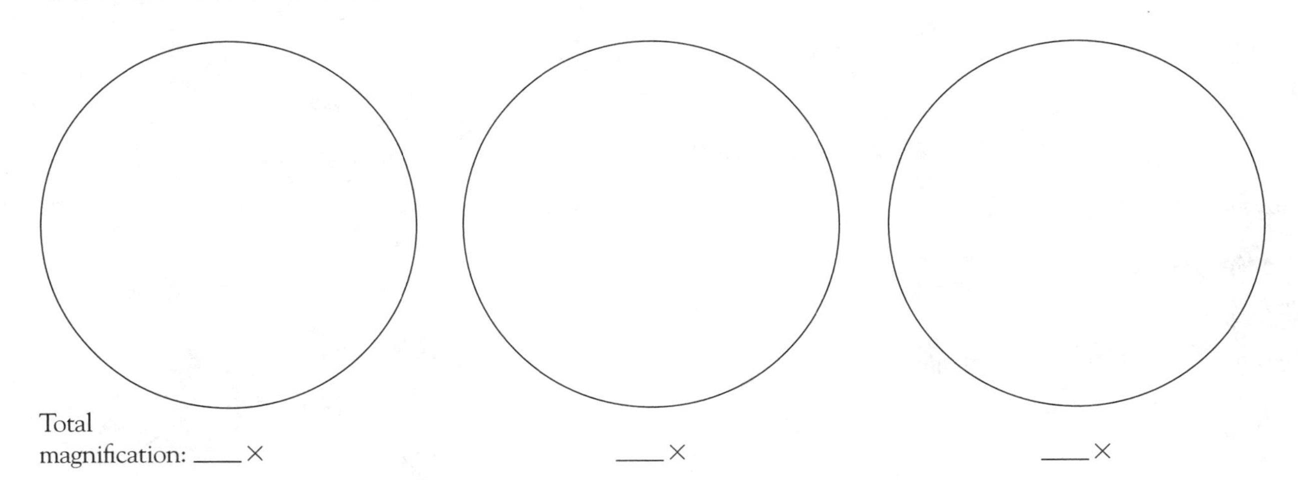

Total magnification: ___× ___× ___×

Fungi

Slide of ____________________

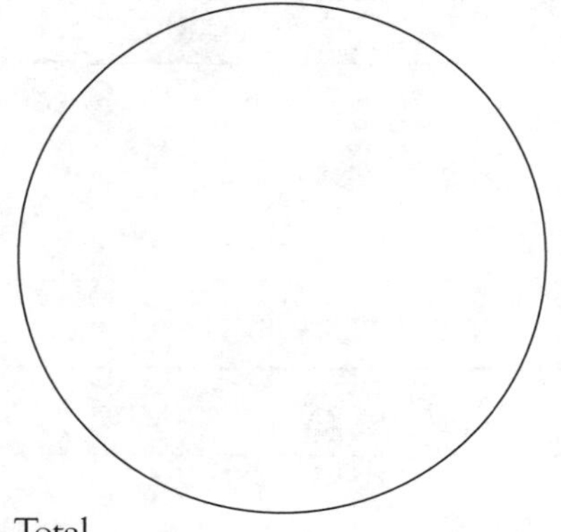

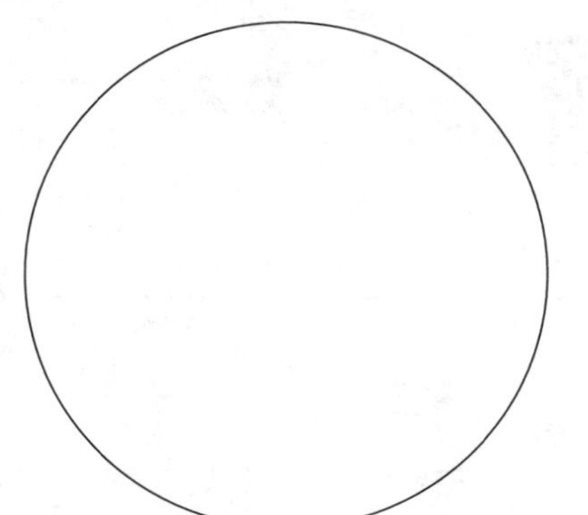

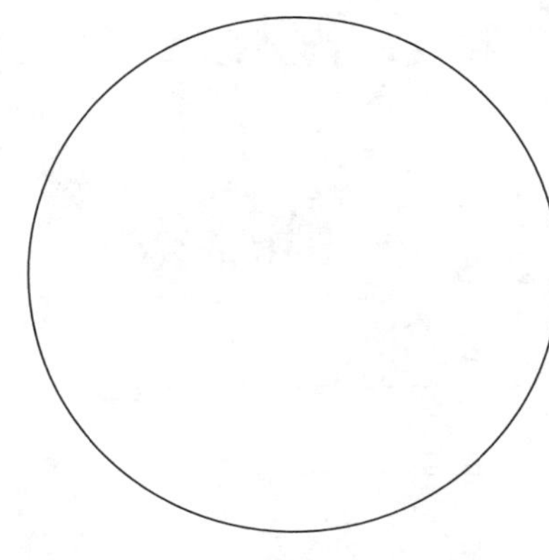

Total magnification: ____×

____×

____×

Protozoa

Slide of ____________________

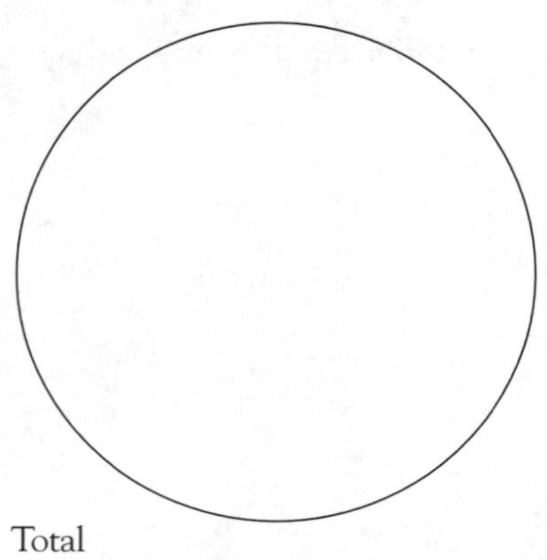

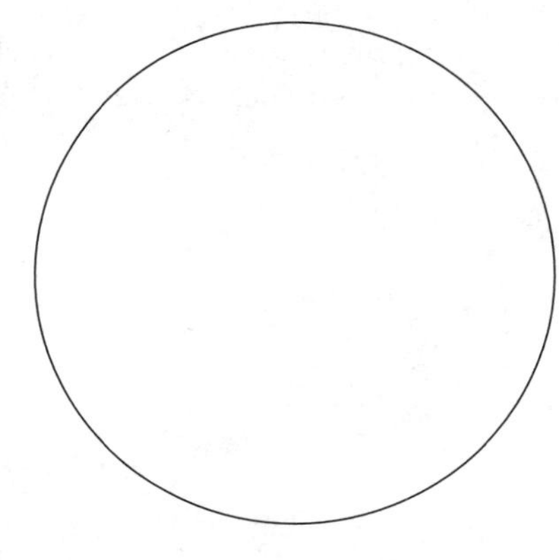

Total magnification: ____×

____×

____×

Bacteria

Slide of ____________________

Be sure to sketch all bacterial shapes observed.

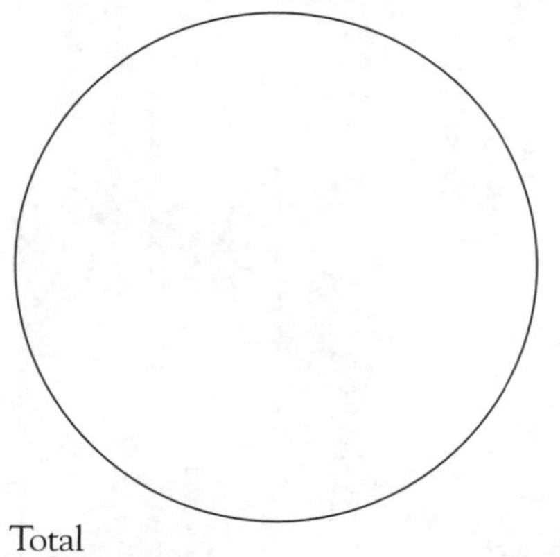

Total magnification: ____×

____×

____×

Questions

1. Why is it desirable that microscope objectives be parfocal? ______________________________

2. Which objective focuses closest to the slide? ______________________________

3. What controls the amount of light reaching the ocular lens? ______________________________

4. Is your lens corrected for chromatic aberrations? ______________________________

5. What three bacterial shapes did you observe? ______________________________

Critical Thinking

1. Assume the diameter of the field of vision in your microscope is 2 mm under low power. If one *Bacillus* cell is 2 μm, how many *Bacillus* cells could fit end to end across the field? How many 10 μm yeast cells could fit across the field?

2. What effect does increased magnification have on the field of vision?

3. Name two ways in which you can enhance the resolving power.

4. What advantages does the low-power objective have over the oil immersion objective for viewing fungi?

5. What would occur if water were accidentally used in place of immersion oil?

Exercise 2

Examination of Living Microorganisms

Objectives

After completing this exercise, you should be able to:

1. Prepare and observe wet-mount slides and hanging-drop slides.
2. Distinguish between true motility and Brownian movement.
3. Use a phase-contrast microscope.
4. Explain how phase-contrast and darkfield microscopy differ from brightfield microscopy.

Background

Antoni van Leeuwenhoek was the first known individual to observe living microbes in a suspension. Unfortunately, he was very protective of his homemade microscopes and left no descriptions of how to make them. During his lifetime he kept "for himself alone" his microscopes and his method of observing "animalcules." Directions for making a replica of van Leeuwenhoek's microscope can be found in *American Biology Teacher.** Fortunately, you will not have to make your own microscope.

In **brightfield microscopy,** objects are dark and the field is light. Brightfield microscopy can be used to observe unstained microorganisms. However, because the optical properties of the organisms and their aqueous environment are similar, very little contrast can be seen. Two other types of compound microscopes, however, are useful for observing living organisms: darkfield and phase-contrast microscopes. These microscopes optically increase contrasts between the organism and background by using special condensers.

In **darkfield microscopy,** the objects are light and the field is dark. In brightfield microscopy, light rays that strike the specimen are reflected away from the lens (Figure 2.1a). The darkfield condenser concentrates the light into a hollow cone of light at such an angle that none of the light rays reach the objective lens unless they pass through an object such as a cell to change their direction (Figure 2.1b). An opaque darkfield disk eliminates all central light rays. Thus, the objects appear brightly illuminated against a dark background. Darkfield microscopy allows the investigator to observe the shape and motility of unstained live organisms. Darkfield microscopy is valuable for observing the spirochete (*Treponema pallidum*) that causes syphilis. This bacterium is not stainable with conventional stains but can be observed in direct smears with darkfield microscopy.

In **phase-contrast microscopy,** small differences in the refractive properties of the objects and the aqueous environment are transformed into corresponding variations of brightness. In phase-contrast microscopy, a ring of light passes through the object, and light rays are **diffracted** (retarded) and out of phase with the light rays not hitting the object. The phase-contrast microscope enhances these phase differences so that the eye detects the difference as contrast (Figure 2.2) between the organisms and background and between structures within a cell. In phase-contrast microscopy, the organisms appear as degrees of brightness against a darker background. The advantage of phase-contrast microscopy is that structural detail within live cells can be studied.

In this exercise, you will examine, using wet-mount techniques, different environments to help you become aware of the numbers and varieties of microbes found in nature. The microbes will exhibit either Brownian movement or true motility. **Brownian movement** is not true motility but rather movement caused by the molecules in the liquid striking an object and causing the object to shake or bounce. In Brownian movement, the particles and microorganisms all vibrate at about the same rate and maintain their relative positions. Motile microorganisms move from one position to another. Their movement appears more directed than Brownian movement, and occasionally the cells may spin or roll.

Many kinds of microbes, such as protozoa, algae, fungi, and bacteria, can be found in pond water and in infusions of organic matter. Van Leeuwenhoek made some of his discoveries using a peppercorn infusion similar to the one you will see in this exercise. Direct examination of living microorganisms is very useful in determining size, shape, and movement. A wet mount is a fast way to observe bacteria. Motility and larger

*W. G. Walter and H. Via. "Making a Leeuwenhoek Microscope Replica." *American Biology Teacher* 30(6):537–539, 1968.

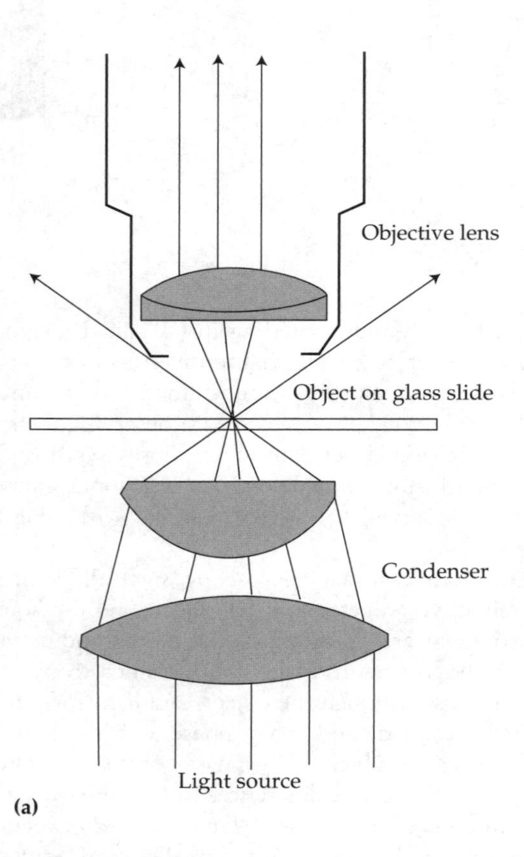

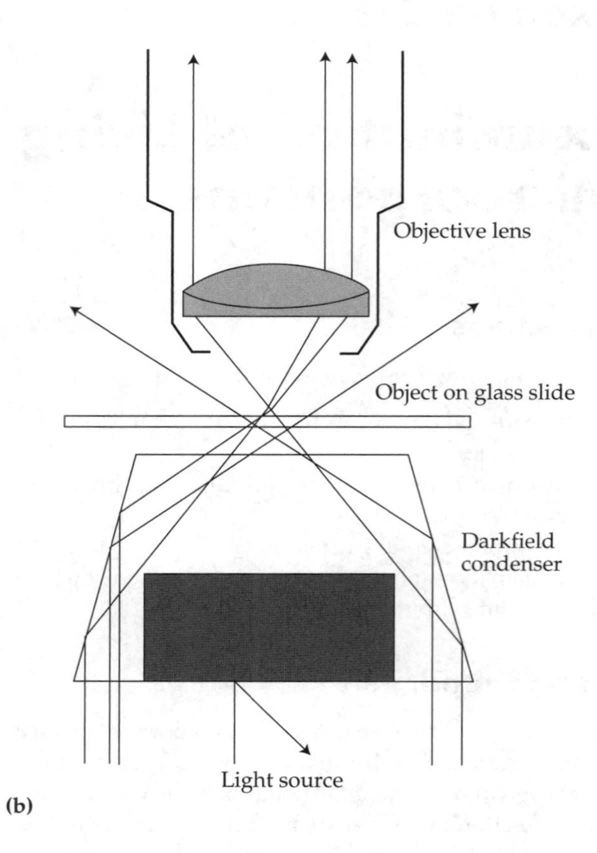

Figure 2.1
A comparison of brightfield and darkfield microscopy. **(a)** In brightfield microscopy, light is reflected away from the objective lens by the specimen. **(b)** In darkfield microscopy, only light rays that go through the object reach the objective lens.

microbes are more easily observed in the greater depth provided by a hanging drop. Evaporation of the suspended drop of fluid is reduced by using a petroleum jelly seal.

In this exercise, you will examine living microorganisms by brightfield and by phase-contrast microscopy.

Materials

Slides

Coverslips

Hanging-drop (depression) slide

Petroleum jelly

Toothpick

Pasteur pipettes

Alcohol

Gram's iodine

Phase-contrast microscope and centering telescope

Inoculating loop

Cultures

Hay infusion, incubated 1 week in light

Hay infusion, incubated 1 week in dark

Peppercorn infusion

Pond water with algae

18- to 24-hour-old broth culture of *Bacillus*

Techniques Required

Compound light microscopy, Exercise 1

Procedure

Wet-Mount Technique

1. Suspend the infusions by stirring or shaking them carefully. Using a Pasteur pipette, transfer a small drop of one hay infusion to a slide or transfer a loopful using the inoculating loop, as demonstrated by your instructor.
2. Handle the coverslip carefully by its edges and place it on the drop.

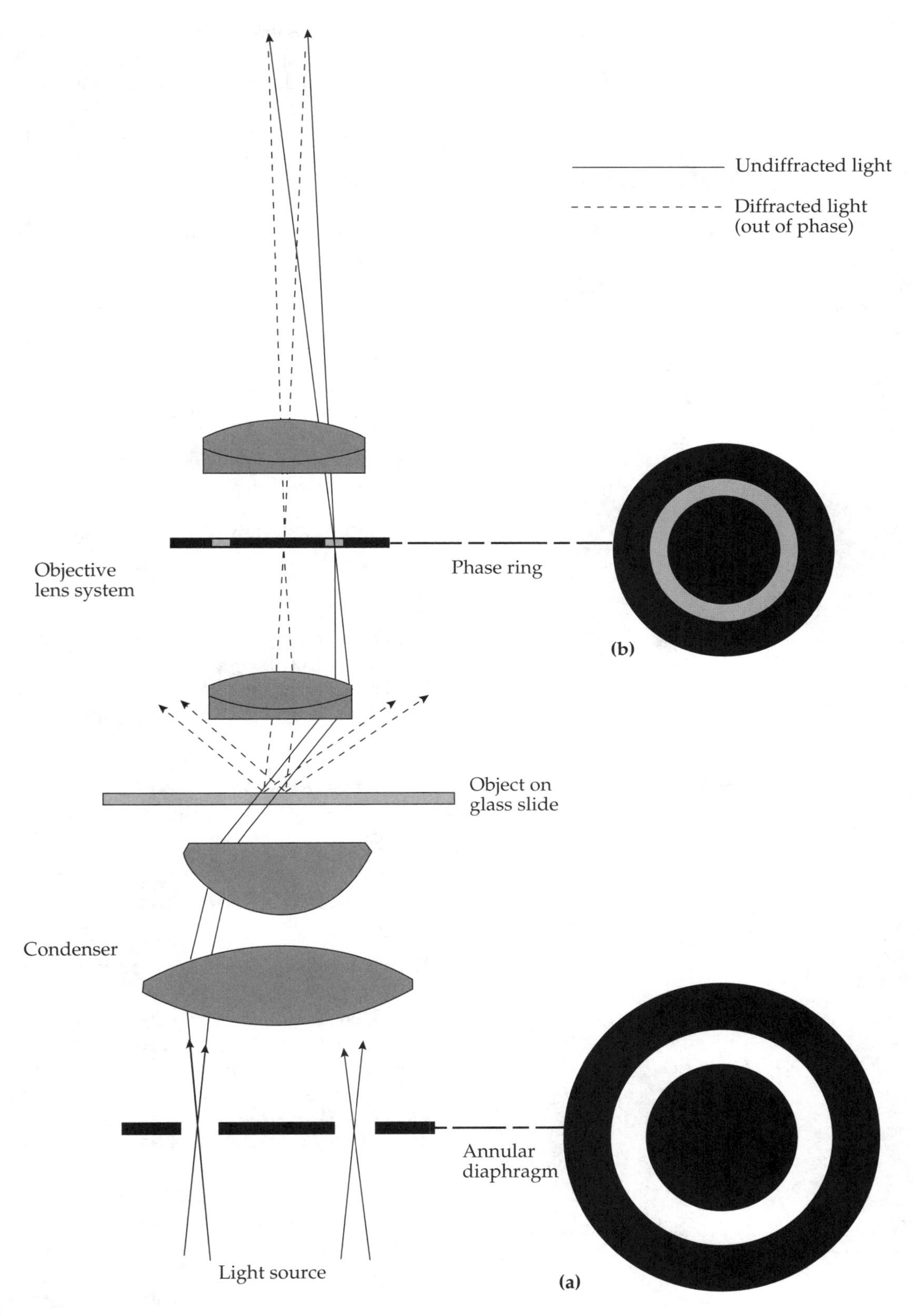

Figure 2.2

Phase-contrast microscopy. **(a)** A hollow cone of light is formed by the annular diaphragm. Diffracted rays are further retarded by the phase ring **(b)** in the objective lens.

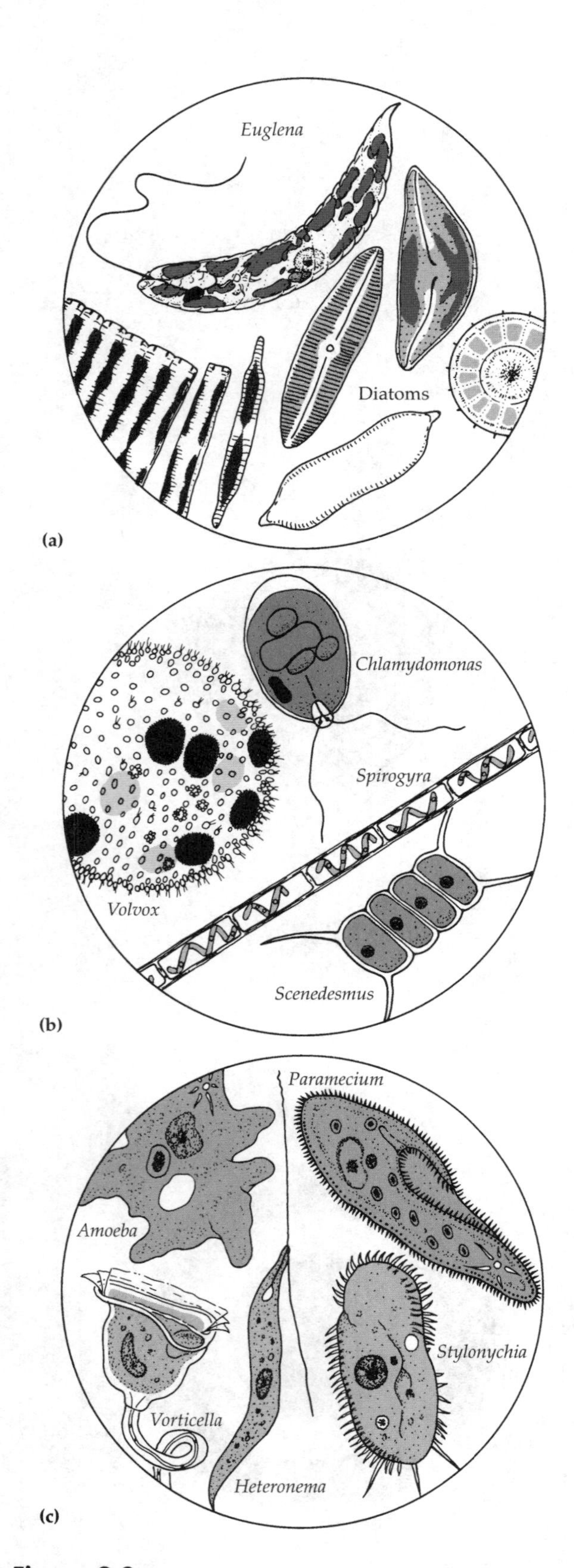

Figure 2.3

Some common protozoa and algae that can be found in infusions: **(a and b)** algae and **(c)** protozoa.

3. Gently press on the coverslip with the end of a pencil or loop handle.
4. Place the slide on the microscope stage and observe it with low power (10×). Adjust the iris diaphragm so a small amount of light is admitted. Concentrate your observations on the larger, more rapidly moving organisms. At this magnification, bacteria are barely discernible as tiny dots. Figure 2.3 and Color Plate XIII may help you to identify some of the microorganisms.
5. Examine the slide with the high-dry lens (40×); then increase the light and focus carefully. Bacteria should now be magnified sufficiently to be seen.
6. After recording your observations, examine the slide with the oil immersion lens. Some microorganisms are motile, while others exhibit Brownian movement.
7. If you want to observe the motile organisms further, place a drop of alcohol or Gram's iodine at the edge of the coverslip and allow it to run under and mix with the infusion. What does the alcohol or iodine do to these organisms? ______________________

   ____

   They can now be observed more carefully.
8. Record your observations, noting the relative size and shape of the organisms.
9. Make a wet mount from the other hay infusion, and observe it, using the low and high-dry objectives. Record your observations.
10. Clean all the slides and return them to the slide box. Coverslips can be discarded in the disinfectant jar. Wipe the oil from the objective lens with lens paper.

Hanging-Drop Procedure

1. Obtain a clean hanging-drop (depression) slide.
2. Pick up a small amount of petroleum jelly on a toothpick.
3. Pick up a coverslip (by its edges) and carefully touch the petroleum jelly with an edge of the coverslip to get a small rim of petroleum jelly. Repeat with the other three edges (Figure 2.4a), keeping the petroleum jelly on the same side of the coverslip.
4. Place the coverslip on a paper towel, with the petroleum jelly-side up.
5. Transfer a small drop or loopful of the peppercorn infusion to the coverslip (Figure 2.4b).
6. Place a depression slide over the drop and quickly invert it so the drop is suspended (Figure 2.4c). Why should the drop be hanging? ____________

   ____
7. Examine the drop under low power (Figure 2.4d) by locating the edge of the drop and moving the slide so the edge of the drop crosses the center of the field.

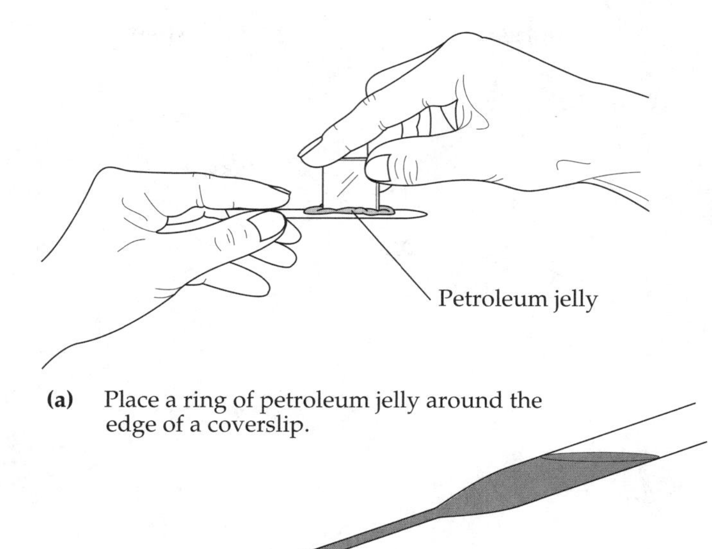

(a) Place a ring of petroleum jelly around the edge of a coverslip.

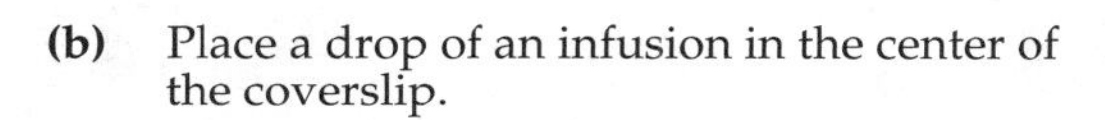

(b) Place a drop of an infusion in the center of the coverslip.

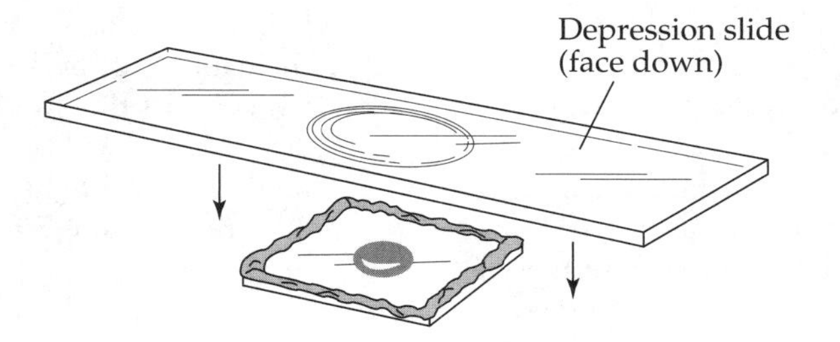

(c) Place the depression slide on the coverslip.

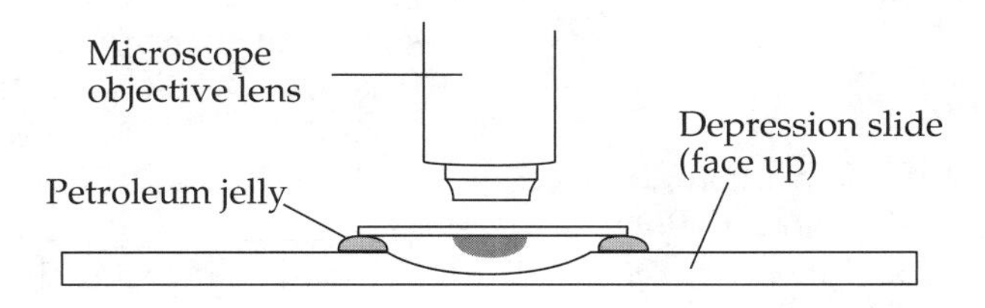

(d) Turn the slide over; place the slide, coverslip up, on the microscope stage; and observe it under the low and high-dry objectives.

Figure 2.4
Hanging-drop preparation.

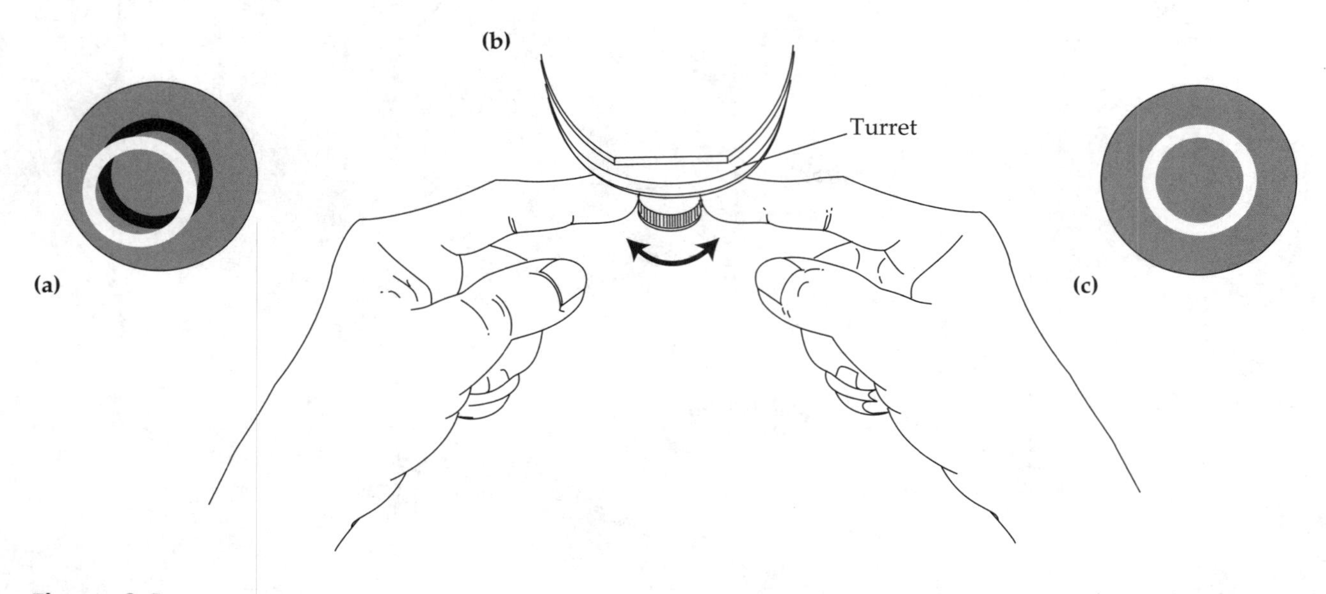

Figure 2.5
Adjustment of phase rings. **(a)** The two images seen before adjustment of the phase rings. By moving the wheel under the condenser turret **(b),** the images are made to coincide **(c).**

8. Reduce the light with the iris diaphragm and focus. Observe the different sizes, shapes, and types of movement.
9. Switch to high-dry and record your observations. Do not focus down. Why not? ______________

____

Do not attempt to use the oil immersion lens.

10. When finished, clean your slide and, using a new coverslip, repeat the procedure with the culture of *Bacillus*. Record your observations.
11. Return your microscope to its proper location. Clean your slides well and return them. Discard the coverslips in the disinfectant jar.

Phase-Contrast Microscopy

1. Make a wet mount of any of the suspensions.
2. Place the slide on the stage and turn on the light.
3. Start with the 10× objective and move the condenser diaphragm to match setting "10."
4. Focus on an obvious clump of material with the coarse and fine adjustments.
5. Close the iris diaphragm, and move the condenser up and down until a light octagon comes into focus. Then open the diaphragm until light just fills the field.
6. Adjust the phase rings.
   a. Replace the eyepiece (screw out) with the centering telescope, and focus the telescope on the phase plate ring by revolving its head.
   b. You will see two rings (Figure 2.5a): one, a bright image of the phase ring, and the other, the dark image of the phase plate in the objective. Adjust the knurled wheel under the condenser turret with your fingertips (Figure 2.5b) to make the bright image coincide with the dark ring (Figure 2.5c). Do not touch the iris diaphragm.
   c. When the ring has been centered (Figure 2.5c), replace the telescope with the ocular lens.
7. Observe.
8. Focus the slide with the 40× objective and the 40× condenser turret.
9. Readjust the phase rings using the telescope, as done in step 6. Record your observations. Can you distinguish any of the organelles in the organisms?

____

10. Focus the slide with the 100× objective and the 100× condenser turret, and then readjust the phase rings using the telescope. Diagram your observations.
11. Clean the slide. Make a wet mount of one other sample. Observe. Are motile organisms present?

____

Compare your observations with those made with the brightfield microscope.

Exercise 2

LABORATORY REPORT

Examination of Living Microorganisms

NAME ____________

DATE ____________

LAB SECTION ____________

Purpose

____

Data

Wet-Mount Technique

Draw the types of protozoa, algae, fungi, and bacteria observed. Indicate their relative sizes and shapes. Record the magnification.

Sample:	Hay infusion, light	Hay infusion, dark
Total magnification:	____×	____×

Compare the size and shape of organisms observed in the "light" and "dark" hay infusion.

____

Hanging-Drop Procedure

Draw the types of microorganisms seen under high-dry magnification. Indicate their relative sizes and shapes.

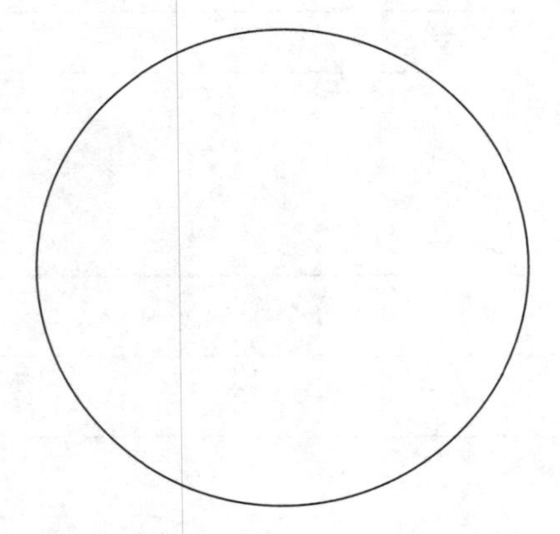

Sample:	Peppercorn infusion	*Bacillus* culture
Total magnification:	___×	___×

Phase-Contrast Microscopy

Carefully draw the organisms and their internal structure.

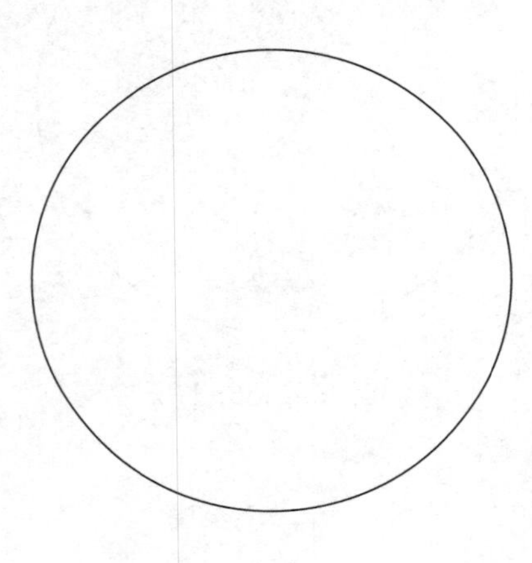

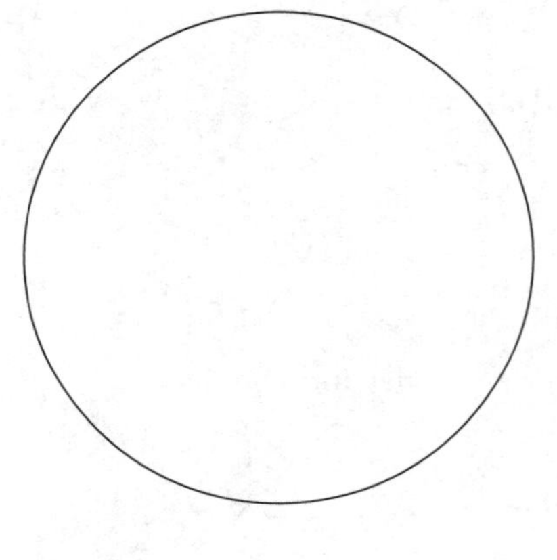

Sample:	______________	______________
Total magnification:	___×	___×

Bacteria

In the following table, record the relative numbers of each bacterial shape observed. Record your data as 4+ (most abundant), 3+, 2+, +, – (not seen).

Culture	Shape		
	Bacilli (rods)	Cocci	Spiral
Hay infusion, dark			
Hay infusion, light			
Peppercorn infusion			
Bacillus culture			
Pond water			

Questions

1. What, if any, practical value do these techniques have? ______________________

____

2. Did any of the bacteria exhibit true motility? __________ How do you distinguish true motility from Brownian movement or motion of the fluid? ______________________

____

3. What is the advantage of the hanging-drop procedure over the wet mount? ______________________

____

4. Compare the appearance of microorganisms observed using phase-contrast microscopy versus brightfield microscopy. ______________________

____

____

5. Why is petroleum jelly used in the hanging-drop procedure? ______________________

____

____

6. Which infusion had the most bacteria? ______________________ Briefly explain why. __________

____

____

____

Critical Thinking

1. Why are microorganisms hard to see in wet preparations?

2. Can you distinguish the prokaryotic organisms from the eukaryotic organisms? Explain.

3. Why isn't the oil immersion lens used in the hanging-drop procedure?

4. From where did the organisms in the infusions come?

Part Two Staining Methods

EXERCISES

In 1877, Robert Koch wrote:

> *How many incomplete and false observations might have remained unpublished instead of swelling the bacterial literature into a turbid stream, if investigators had checked their preparations with each other?**

To solve this problem, he introduced into microbiology the procedures of air drying, chemical fixation, and staining with aniline dyes.

In addition to making a lasting preparation, staining bacteria enhances the contrast between bacteria and the surrounding material and permits observation of greater detail and resolution than wet-mount procedures. Microorganisms are prepared for staining by smearing them onto a microscope slide.

In these exercises, bacteria will be transferred from growth or culture media to microscope slides using an inoculating loop. An **inoculating loop** is a Nichrome wire held with an insulated handle. Before and after it is used, the inoculating loop is sterilized by heating or **flaming**—that is, the loop is held in the flame of a burner (photograph a) or electric incinerator (photograph b) until it is red-hot.

---

*Quoted in H. A. Lechevalier and M. Solotorovsky. *Three Centuries of Microbiology*. New York: Dover Publications, 1974, p. 79.

Staining techniques may involve **simple stains,** in which only one reagent is used and all bacteria are usually stained similarly (Exercises 3 and 4), or **differential stains,** in which multiple reagents are used and bacteria react to the reagents differently (Exercises 5 and 6). Structural stains are used to identify specific parts of microorganisms (Exercise 7).

Most of the bacteria that are grown in laboratories are rods or cocci (see Color Plates I.2 and I.3). Rods and cocci will be used throughout this part. Staining and microscopic examination are usually the first steps in identifying microorganisms. In Exercise 8, Morphologic Unknown, you will be asked to characterize selected bacteria by their microscopic appearance.

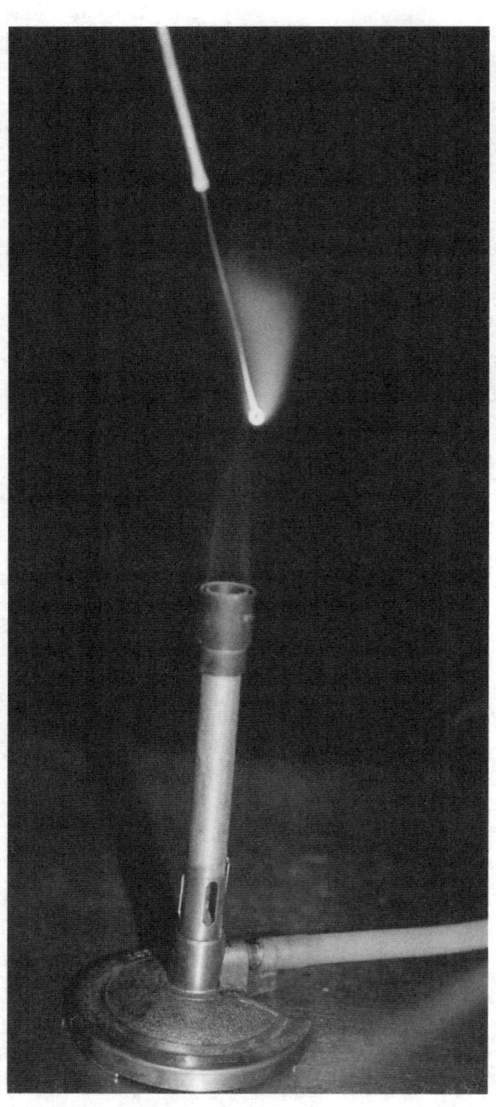

(a) Flaming a loop.

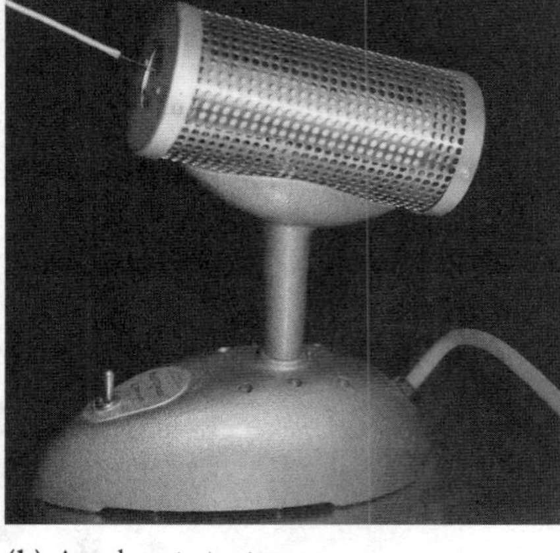

(b) An electric incinerator.

Sterilizing an inoculating loop. **(a)** Before and after use, sterilize the loop with a flame until it is red-hot. Place the wire in the hottest part of the flame, at the edge of the inner blue area. **(b)** The loop can be sterilized in an electric incinerator.

Preparation of Smears and Simple Staining

Objectives

After completing this exercise, you should be able to:

1. Prepare and fix a smear.
2. List the advantages of staining microorganisms.
3. Explain the basic mechanism of staining.
4. Perform a simple direct stain.

Background

Most stains used in microbiology are synthetic aniline (coal tar derivative) dyes derived from benzene. The dyes are usually salts, although a few are acids or bases, composed of charged colored ions. The ion that is colored is referred to as a **chromophore.** For example,

$$\text{Methylene blue chloride} \leftrightharpoons \underset{\text{(Chromophore)}}{\text{Methylene blue}^{+}} + \text{Cl}^{-}$$

If the chromophore is a positive ion like the methylene blue in the equation shown, the stain is considered a **basic stain;** if it is a negative ion, it is an **acidic stain.** Most bacteria are stained when a basic stain permeates the cell wall and adheres by weak ionic bonds to the negative charges of the bacterial cell.

Staining procedures that use only one stain are called **simple stains.** A simple stain that stains the bacteria is a **direct stain,** and a simple stain that stains the background but leaves the bacteria unstained is a **negative stain.** Simple stains can be used to determine cell morphology, size, and arrangement.

Before bacteria can be stained, a thin film of bacterial cells, called a **smear,** must be placed on a slide. A smear is made by spreading a bacterial suspension on a clean slide and allowing it to air-dry. The smear must be **fixed** to kill the bacteria; coagulated proteins from the cells will cause cells to stick to the slide. The dry smear is passed through a Bunsen burner flame several times to **heat-fix** the bacteria. Heat fixing may not kill all the bacteria. Alternatively, the dry smear can be placed on a 60°C slide warmer for 10 minutes or until chemically fixed. To **chemically fix** the bacteria, cover the smear with 95% methyl alcohol for 1 minute. Fixing denatures bacterial enzymes, preventing them from digesting cell parts, which causes the cell to break, a process called *autolysis*. Fixing also enhances the adherence of bacterial cells to the microscope slide.

Materials

Methylene blue

Wash bottle of distilled water

Slide

Inoculating loop

Cultures

Staphylococcus epidermidis slant

Bacillus megaterium broth

Techniques Required

Compound light microscopy, Exercise 1

Inoculating loop

Procedure

1. Clean your slide well with abrasive soap or cleanser; rinse and dry. Handle clean slides by the end or edge. Use a marker to make two dime-sized circles on the bottom of each slide so they will not wash off. Label each circle according to the bacterial culture used.
2. Sterilize your inoculating loop by holding it in the hottest part of the flame (at the edge of the inner blue area) or the electric incinerator until it is red-hot (see the figure on page 24). The entire wire should get red. Allow the loop to cool so that bacteria picked up with the loop won't be killed. Allow the loop to cool without touching it or setting it down. Cooling takes about 30 seconds. You will determine the appropriate time with a little practice.

The loop must be cool before inserting it into a medium. A hot loop will spatter the medium and move bacteria into the air.

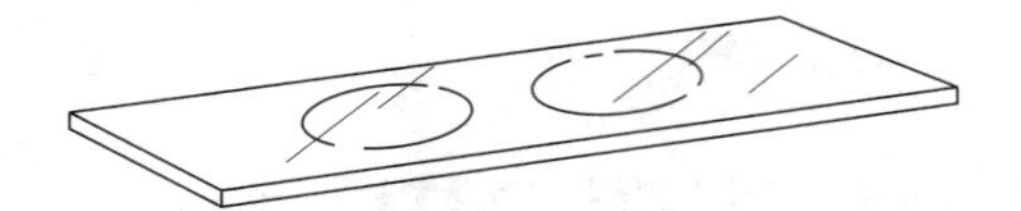

(a) Mark the smear areas with a marking pencil on the underside of a clean slide.

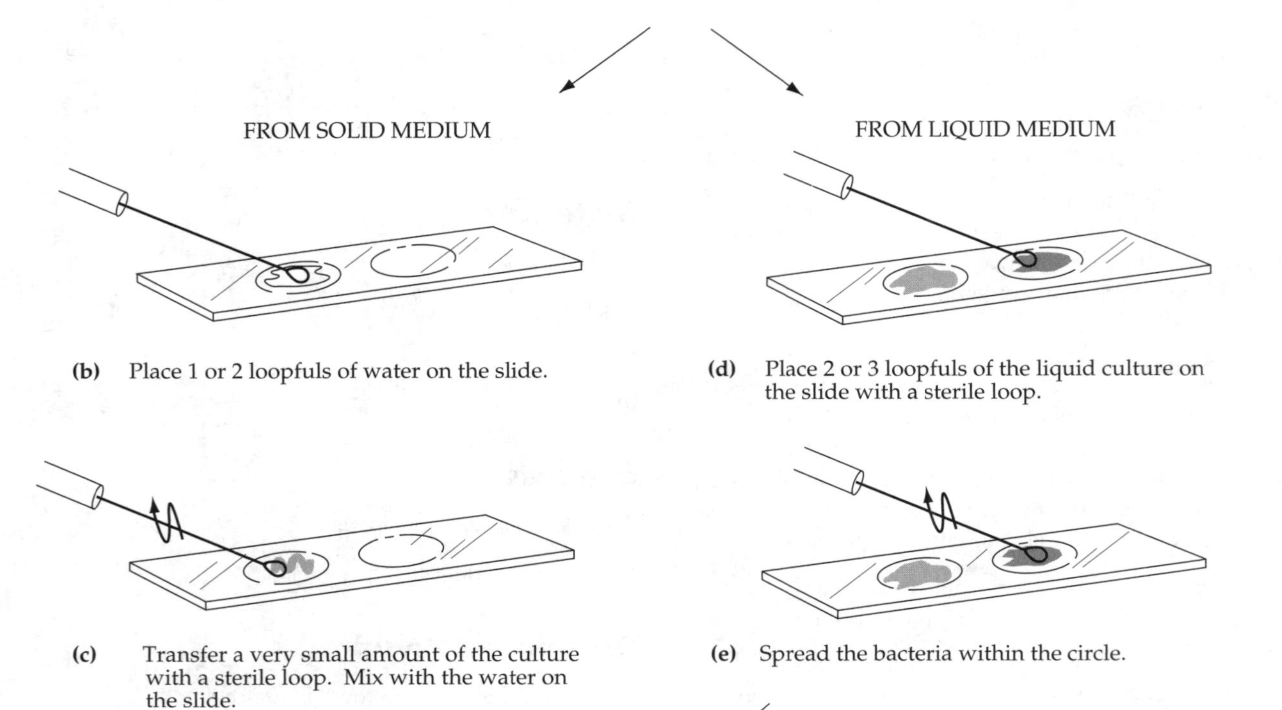

(b) Place 1 or 2 loopfuls of water on the slide.

(d) Place 2 or 3 loopfuls of the liquid culture on the slide with a sterile loop.

(c) Transfer a very small amount of the culture with a sterile loop. Mix with the water on the slide.

(e) Spread the bacteria within the circle.

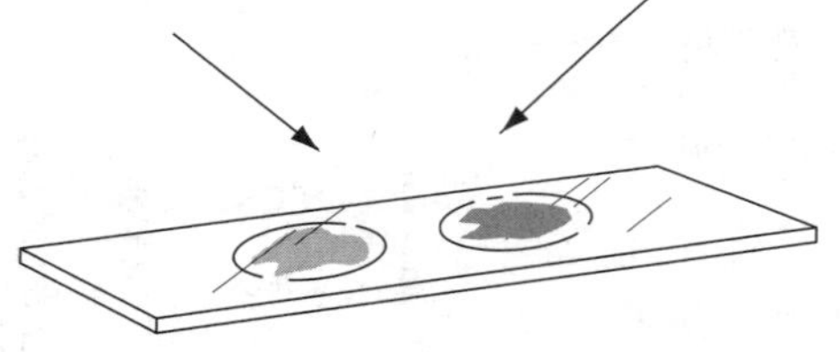

(f) Allow the smears to air-dry at room temperature.

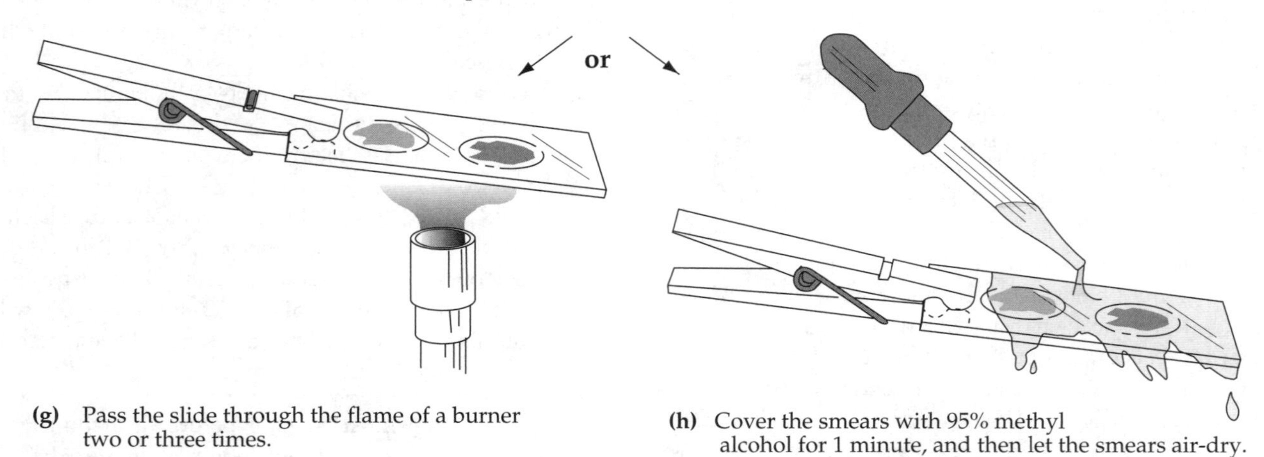

(g) Pass the slide through the flame of a burner two or three times.

(h) Cover the smears with 95% methyl alcohol for 1 minute, and then let the smears air-dry.

Figure 3.1

Preparing a bacterial smear.

3. Prepare smears (Figure 3.1).
   a. Make a smear of bacteria from the broth culture in the center of one circle. Flick the tube of broth culture lightly with your finger to resuspend sedimented bacteria, and place 2 or 3 loopfuls of the culture in the circle. Sterilize your loop between each loopful. Spread the culture within the circle.
   b. Sterilize your loop.

Always sterilize your loop after using it and before setting it down.

   c. For the bacterial culture on solid media, place 1 or 2 loopfuls of distilled water in the center of the other circle, using the sterile inoculating loop. Which bacterium is on a solid medium? ______________________
   Sterilize your loop.
   d. Using the cooled loop, scrape a *small* amount of the culture off the slant—do not take the agar (Figure 3.2). If you hear the sizzle of boiling water when you touch the agar with the loop, resterilize your loop and begin again. Why? ______________________
   Try not to gouge the agar. Emulsify (to a milky suspension) the cells in the drop of water, and spread the suspension to fill a majority of the circle. The smear should look like diluted skim milk. Sterilize your loop again.
   e. Let the smears dry. *Do not* blow on the slide because this will move the bacterial suspension. *Do not* flame the slide because flaming will distort the cells' shapes.
   f. Hold the slide with a clothespin and fix the smears by one of the following methods (Figure 3.1g or h):
   (1) Pass the slide quickly through the blue flame two or three times or place it on a 60° slide warmer for 10 minutes.
   (2) Cover the smear with 95% methyl alcohol for 1 minute. Tip the slide to let the alcohol run off, and let the slide air-dry before staining. Do not fix until the smears are completely dry. Why? ______________________

____

5. Stain smears (Figure 3.3).
   a. Use a clothespin to hold the slide, or place it on a staining rack.
   b. Cover the smear with methylene blue and leave it for 30 to 60 seconds (Figure 3.3a).

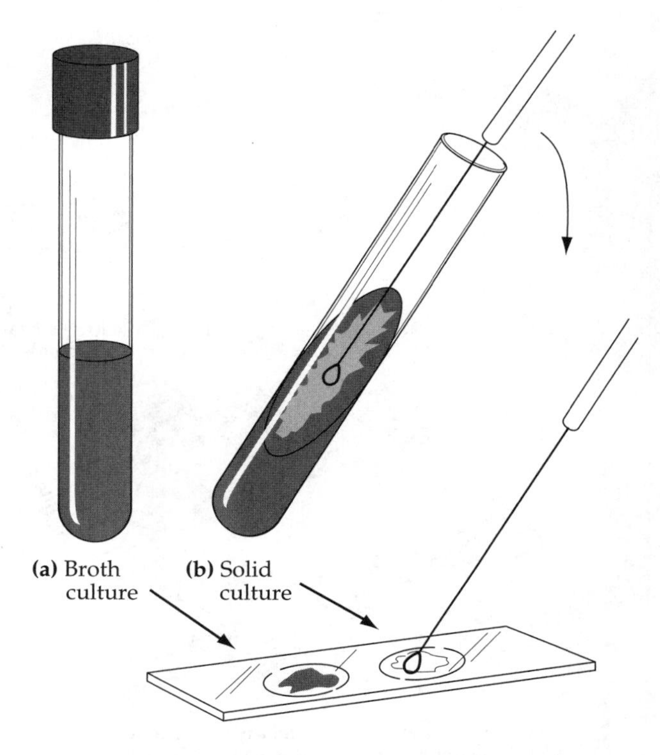

Figure 3.2

Transferring bacteria. **(a)** Transfer 2 or 3 loopfuls of microbial suspension to a slide. **(b)** Gently scrape bacteria from the agar surface and transfer the bacteria to a loopful of water on a slide. Be careful to avoid gouging into the agar.

   c. Carefully wash the excess stain off with distilled water from a wash bottle. Let the water run down the tilted slide (Figure 3.3b).
   d. Gently blot the smear with a paper towel or absorbent paper and let it dry (Figure 3.3c).
6. Examine your stained smears microscopically using the low, high-dry, and oil immersion objectives. Put the oil *directly* on the smear; coverslips are not needed. Record your observations with labeled drawings.
7. Blot the oil from the objective lens with lens paper, and return your microscope to its proper location. Clean your slides well, or save them as described in step 8.
8. Stained bacterial slides can be stored in a slide box. Remove the oil from the slide by blotting it with a paper towel. Any residual oil won't matter.

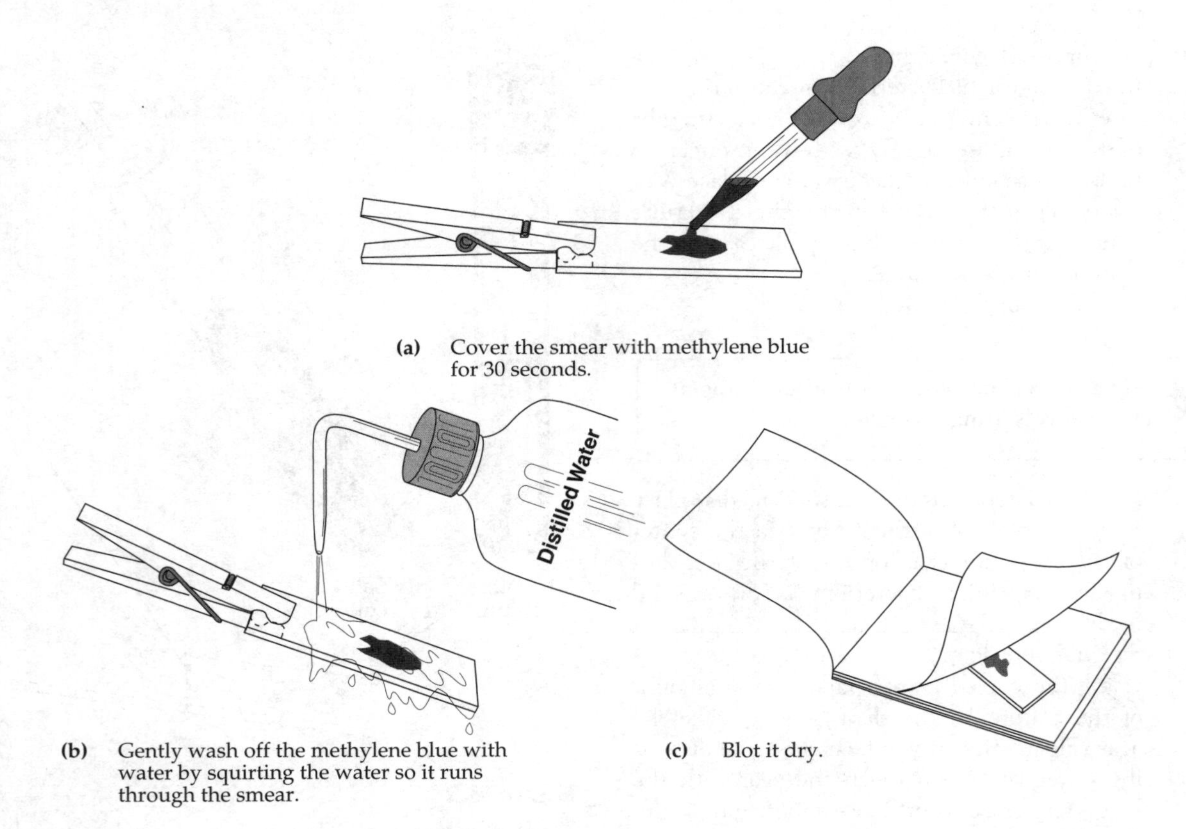

Figure 3.3

Simple staining.

Exercise 3

LABORATORY REPORT

Preparation of Smears and Simple Staining

NAME ______________________

DATE ______________________

LAB SECTION ______________________

Purpose

____

____

____

Data

Sketch a few bacteria viewed with the oil immersion objective lens.

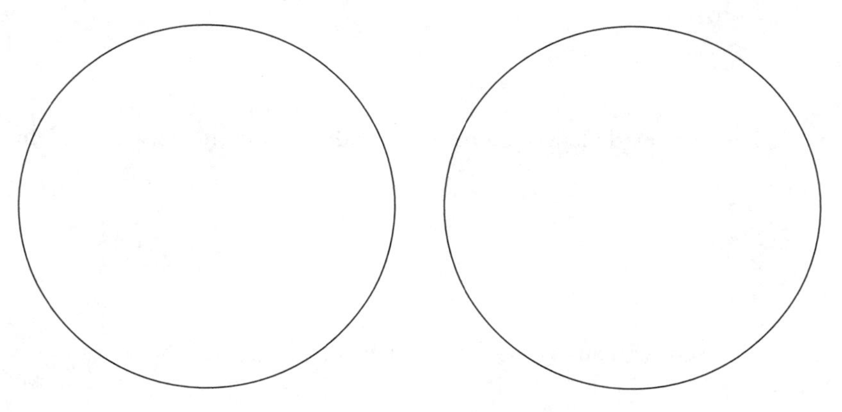

Bacteria:	*Staphylococcus epidermidis*	*Bacillus megaterium*
Total magnification:	____×	____×
Morphology (shape):	______________	______________
Arrangement of cells relative to one another:	______________	______________

Questions

1. Which bacterium is a rod? ______________________

2. Of what value is a simple stain? ______________________

____

____

3. What is the purpose of fixing the smear? ______________________________

____

4. What are the two methods of fixing a smear? ______________________________

   Which one did you use? ______________________________

5. In heat fixing, what would happen if too much heat were applied? ______________________________

____

Critical Thinking

1. Methylene blue can be prepared as a basic stain or an acidic stain. How would the pH of the stain affect the staining of bacteria?

2. Can dyes other than methylene blue be used for direct staining? Briefly explain.

3. Bacteria can be seen without staining. Why then was Koch's recommendation for fixing and staining important for microbiology?

EXERCISE 4

Exercise 4

Negative Staining

Objectives

After completing this exercise, you should be able to:

1. Explain the application and mechanism of the negative stain technique.
2. Prepare a negative stain.

Background

The **negative stain** technique does not stain the bacteria but stains the background. The bacteria will appear clear against a stained background. The stain does not stain the bacteria because of ionic repulsion: The bacteria and the acidic stain both have negative charges.

No heat fixing or strong chemicals are used so the bacteria are less distorted than in other staining procedures. The negative stain technique is useful in situations where other staining techniques don't clearly indicate cell morphology or size.

Materials

Nigrosin

Clean slides (6)

Distilled water

Sterile toothpicks

Cultures

Bacillus subtilis

Staphylococcus epidermidis

Techniques Required

Compound light microscopy, Exercise 1

Smear preparation, Exercise 3

Procedure (Figure 4.1)

1. Slides must be clean and grease-free.
2. a. Place a *small* drop of nigrosin at the end of the slide. For cultures on solid media, add a loopful of distilled water and emulsify a small amount of the culture in the nigrosin-water drop. For broth cultures, mix a loopful of the culture into the drop of nigrosin (Figure 4.1a). Do not spread the drop or let it dry.
   b. Using the end edge of another slide, spread out the drop (Figure 4.1b and c) to produce a smear varying from opaque black to gray. The angle of the spreading slide will determine the thickness of the smear.
   c. Let the smear air-dry (Figure 4.1d). *Do not heat-fix it.*
   d. Prepare a negative stain of the other culture.
3. Examine the stained slides microscopically, using the low, high-dry, and oil immersion objectives (Figure 4.2). As a general rule, if a few short rods are seen with small cocci, the morphology is rod shaped. The apparent cocci are short rods viewed from the end or products of the cell division of small rods. (See Color Plate I.1.)
4. a. Place a small drop of nigrosin and a loopful of water at the end of a slide.
   b. Scrape the base of your teeth and gums with a sterile toothpick.
   c. Mix the nigrosin-water with the toothpick to get an emulsion of bacteria from your mouth.

Discard the toothpick in disinfectant.

   d. Follow steps 2b and 2c to complete the negative stain. Observe the stain and describe your results.

Discard your slides in disinfectant.

5. Wipe the oil off your microscope and return it.

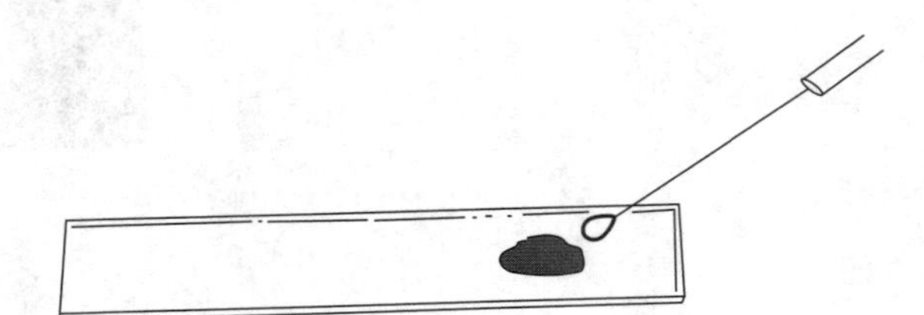

(a) Place a small drop of nigrosin near one end of a slide. Mix a loopful of broth culture in the drop. When the organisms are taken from a solid medium, mix a loopful of water in the nigrosin.

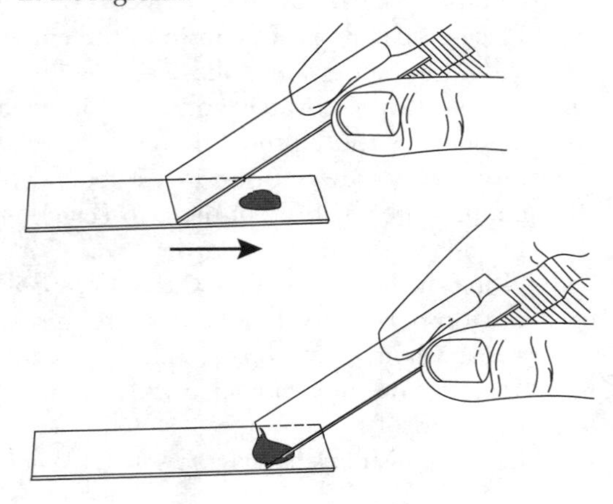

(b) Gently draw a second slide across the surface of the first until it contacts the drop. The drop will spread across the edge of the top slide.

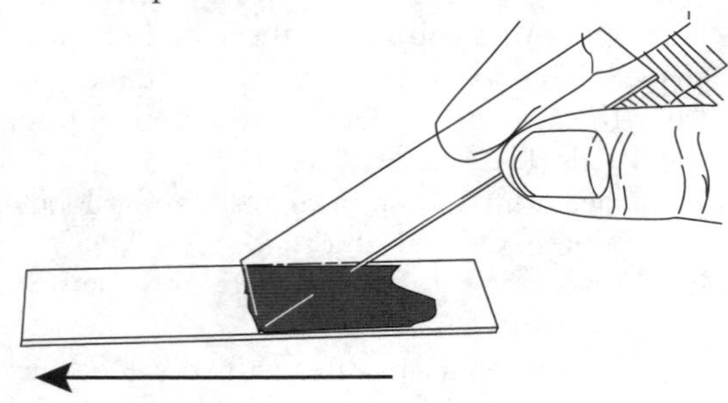

(c) Push the top slide to the left along the entire surface of the bottom slide.

(d) Let the smear air-dry.

Figure 4.1

Preparing a negative stain.

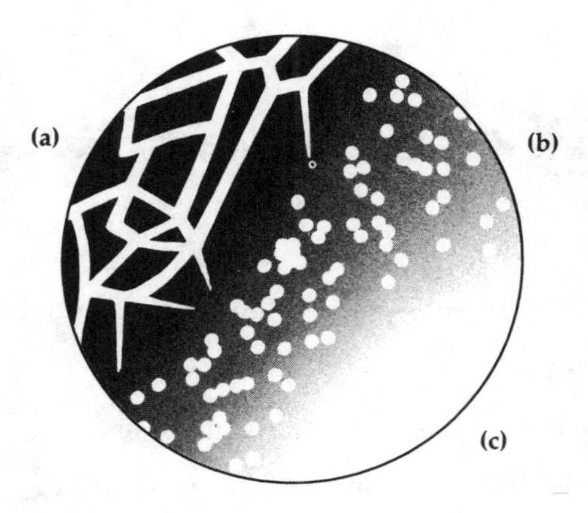

Figure 4.2

A negative stain viewed under a microscope. **(a)** This part of the smear is too heavy: Cracks in the stain can be seen. **(b)** Colorless cells are visible here. **(c)** Too little stain is in this area of the smear.

Exercise 4

LABORATORY REPORT

Negative Staining

NAME ______________________

DATE ______________________

LAB SECTION ______________________

Purpose

____

____

____

Data

Sketch a few bacteria (oil immersion objective lens).

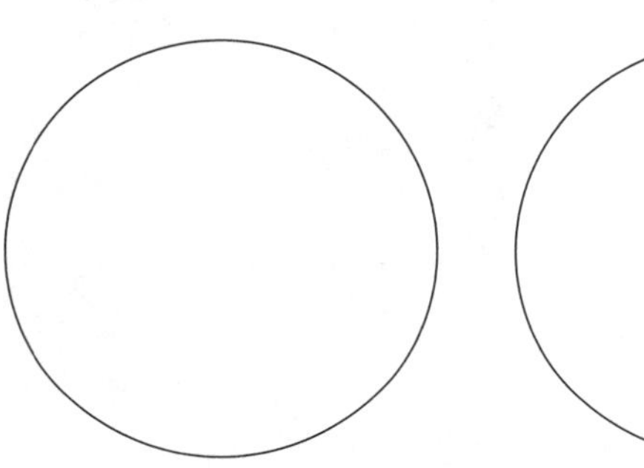

Specimen:	*Bacillus subtilis*	*Staphylococcus epidermidis*	Tooth and gum scraping
Total magnification:	____×	____×	____×
Morphology and arrangement:	______________	______________	______________

Questions

1. Which cell is a rod? ______________________ How does its appearance differ from the rod you stained with methylene blue (simple stain)? ______________________

2. Why is the size more accurate in a negative stain than in a simple stain? ______________________

   ____

3. Could any dye be used in place of nigrosin for negative staining? ________ What types of dyes are used for negative staining? ________ Briefly explain. ______________________

   ____

Critical Thinking

1. What microscopic technique gives a field similar in appearance to that seen in the negative stain?

2. Carbolfuchsin can be used as a simple stain and as a negative stain. As a simple stain, the pH is ________.

3. India ink can give the appearance of a negative stain when used in a wet mount. What is the basis (e.g., pH) for this stain? (India ink is a suspension of 0.5–1.0 μm carbon particles in water.)

Exercise 5

Gram Staining

Objectives

After completing this exercise, you should be able to:

1. Explain the rationale and procedure for the Gram stain.
2. Perform and interpret Gram stains.

Background

The Gram stain is a useful stain for identifying and classifying bacteria. The **Gram stain** is a differential stain that allows you to classify bacteria as either gram-positive or gram-negative. The Gram-staining technique was discovered by Hans Christian Gram in 1884, when he attempted to stain cells and found that some lost their color when excess stain was washed off.

The staining technique consists of the following steps:

1. Apply **primary stain** (crystal violet). All bacteria are stained purple by this basic dye.
2. Apply **mordant** (Gram's iodine). The iodine combines with the crystal violet in the cell to form a crystal violet–iodine complex (CV–I).
3. Apply **decolorizing agent** (ethyl alcohol or ethyl alcohol–acetone). The primary stain is washed out (decolorized) of some bacteria, while others are unaffected.
4. Apply **secondary stain** or **counterstain** (safranin). This basic dye stains the decolorized bacteria red.

The most important determining factor in the procedure is that bacteria differ in their *rate* of decolorization. Those that decolorize easily are referred to as **gram-negative,** whereas those that decolorize slowly and retain the primary stain are called **gram-positive.**

Bacteria stain differently because of chemical and physical differences in their cell walls. Crystal violet is picked up by the cell. Iodine reacts with the dye in the cytoplasm to form a CV–I that is larger than the crystal violet that entered the cell. The CV–I cannot be washed out of gram-positive cells. In gram-negative cells, the decolorizing agent dissolves the outer lipopolysaccharide layer, and the CV–I washes out through the thin layer of peptidoglycan.

The Gram stain is most consistent when done on young cultures of bacteria (less than 24 hours old). When bacteria die, their cell walls degrade and may not retain the primary stain, giving inaccurate results. Because Gram staining is usually the first step in identifying bacteria, the procedure should be memorized.

Materials

Gram-staining reagents:

- Crystal violet
- Gram's iodine
- Ethyl alcohol
- Safranin

Wash bottle of distilled water

Slides (3)

Cultures

Staphylococcus epidermidis

Escherichia coli

Bacillus subtilis

Techniques Required

Compound light microscopy, Exercise 1

Smear preparation, Exercise 3

Simple staining, Exercise 3

Procedure (Figure 5.1)

1. Prepare and fix smears (Figure 3.1). Clean the slides well, and make a circle on each slide with a marker. Label each slide for one of the cultures.
2. Prepare a Gram stain of one smear. Use a clothespin or slide rack to hold the slides.
   a. Cover the smear with crystal violet and leave it for 30 seconds (Figure 5.1a).
   b. Wash the slide carefully with distilled water from a wash bottle. Do not squirt water directly onto the smear (Figure 5.1b).
   c. Cover the smear with Gram's iodine for 10 seconds (Figure 5.1c).

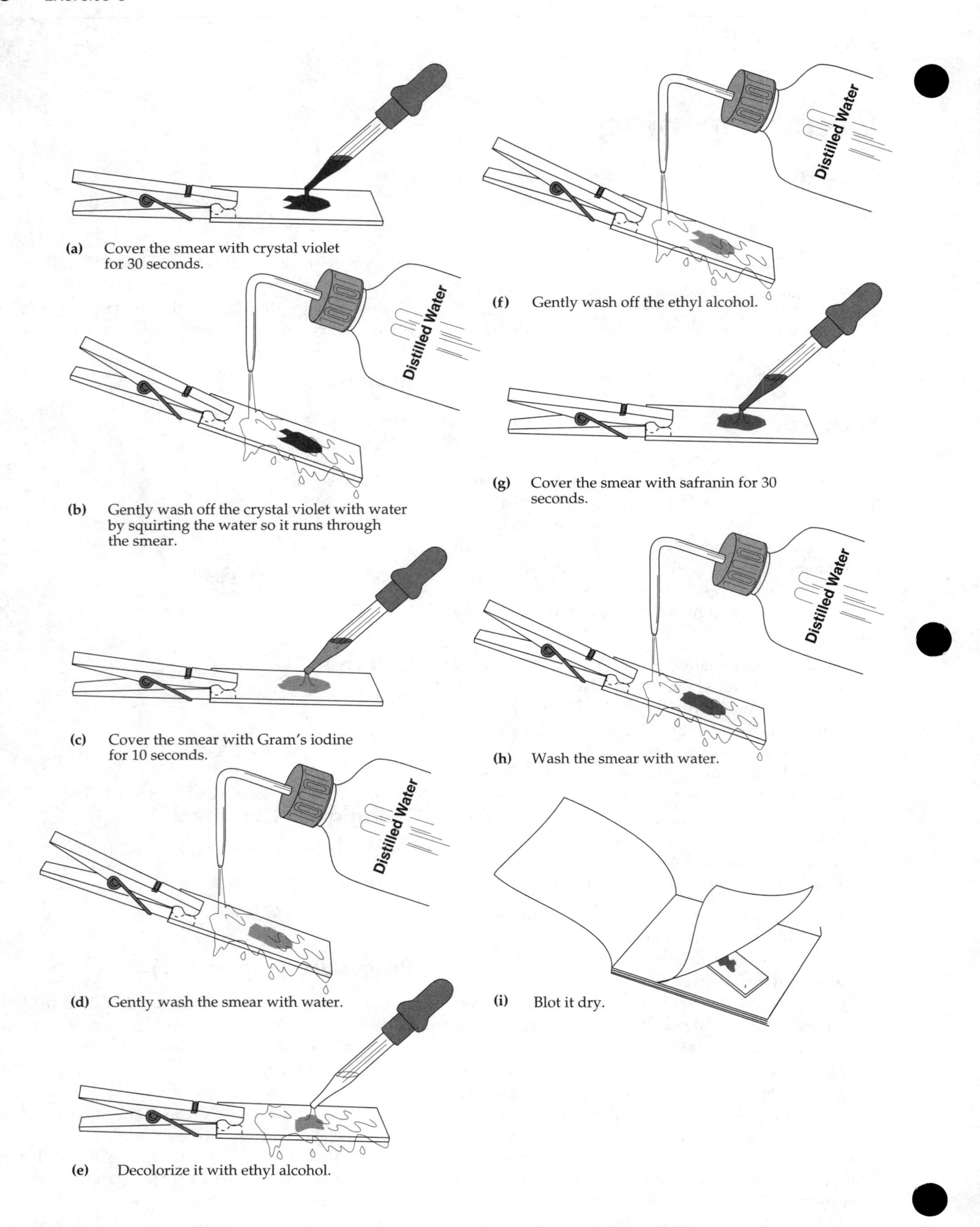

(a) Cover the smear with crystal violet for 30 seconds.

(b) Gently wash off the crystal violet with water by squirting the water so it runs through the smear.

(c) Cover the smear with Gram's iodine for 10 seconds.

(d) Gently wash the smear with water.

(e) Decolorize it with ethyl alcohol.

(f) Gently wash off the ethyl alcohol.

(g) Cover the smear with safranin for 30 seconds.

(h) Wash the smear with water.

(i) Blot it dry.

Figure 5.1

The Gram stain.

d. Wash off the iodine by tilting the slide and squirting water above the smear so that the water runs over the smear (Figure 5.1d).
e. Decolorize it with 95% ethyl alcohol (Figure 5.1e). Let the alcohol run through the smear until no large amounts of purple wash out (usually 10 to 20 seconds). The degree of decolorizing depends on the thickness of the smear. This is a critical step. *Do not overdecolorize*. However, experience is the only way you will be able to determine how long to decolorize. Very thick smears will give inaccurate results. Why? ______

____

f. Immediately wash gently with distilled water (Figure 5.1f). Why? ______________
g. Add safranin for 30 seconds (Figure 5.1g).
h. Wash the slide with distilled water and blot it dry with a paper towel or absorbent paper (Figure 5.1h and i).

3. Repeat step 2 to stain your remaining slides.
4. Examine the stained slides microscopically, using the low, high-dry, and oil immersion objectives. Put the oil directly on the smear. Record your observations. (See Color Plates I.2 and I.3.) Do they agree with those given in your textbook? __________
If not, try to determine why. Some common sources of Gram-staining errors are the following:
a. The loop was too hot.
b. Excessive heat was applied during heat fixing.
c. The decolorizing agent (ethyl alcohol) was left on the smear too long.
d. The culture was too old.
e. The smear was too thick.

Exercise 5

Laboratory Report

Gram Staining

Name ____________________

Date ____________________

Lab Section ____________________

Purpose

____

Data

Sketch a few bacteria (oil immersion objective lens).

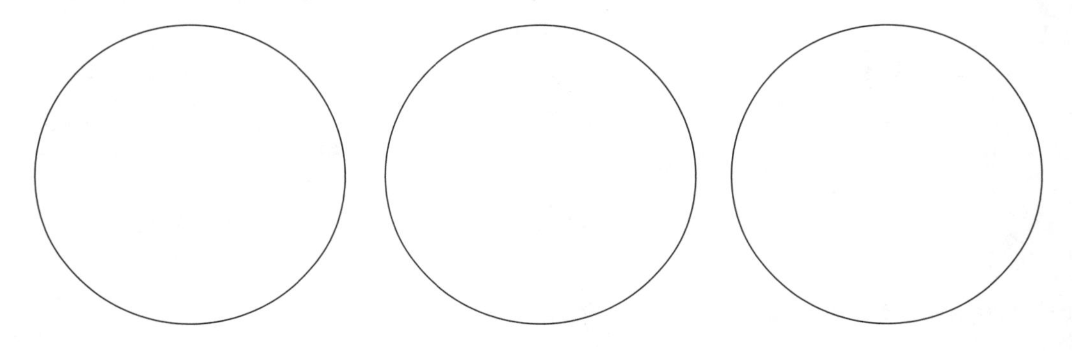

Bacteria:	*Staphylococcus epidermidis*	*Bacillus subtilis*	*Escherichia coli*
Total magnification:	___×	___×	___×
Morphology and arrangement:	______	______	______
Color:	______	______	______
Gram reaction:	______	______	______

Which organism is the largest? ______________ The smallest? ______________

Questions

1. Did your results agree with the information in your textbook? ______________ If not, why not?

2. Why will gram-positive cells more than 24 hours old stain gram-negative? ______________

3. Can iodine be added before the primary stain in a Gram stain? ____________________

____

____

4. List the steps of the Gram-staining procedure in order (omit washings), and fill in the color of gram-positive cells and gram-negative cells after each step.

Step	Chemical	Appearance	
		Gram-Positive Cells	Gram-Negative Cells
1			
2			
3			
4			

5. Which step can be omitted without affecting determination of the Gram reaction? ____________________

Critical Thinking

1. Suppose you performed a Gram stain on a sample from a pure culture of bacteria and observed a field of red and purple cocci. Adjacent cells were not always the same color. What do you conclude?

2. Suppose you are viewing a Gram-stained field of red rods and purple cocci through the microscope. What do you conclude?

3. Considering you can't identify bacteria from a Gram stain, why might a physician perform a Gram stain on a sample before prescribing an antibiotic?

4. If you performed a Gram stain on human cells, what would happen?

Exercise 6

Acid-Fast Staining

Objectives

After completing this exercise, you should be able to:

1. Apply the acid-fast procedure.
2. Explain what is occurring during the acid-fast staining procedure.
3. Perform and interpret an acid-fast stain.

Background

The **acid-fast stain** is a differential stain. In 1882, Paul Ehrlich discovered that *Mycobacterium tuberculosis* (the causative agent of tuberculosis) retained the primary stain even after washing with an acid-alcohol mixture. (We hope you can appreciate the phenomenal strides that were made in microbiology in the 1880s. Most of the staining and culturing techniques used today originated during that time.) Most bacteria are decolorized by acid-alcohol, with only the families Mycobacteriaceae, Nocardiaceae, Gordoniaceae, Dietziaceae, and Tsukamurellaceae (*Bergey's Manual**) being acid-fast. The acid-fast technique has great value as a diagnostic procedure because both *Mycobacterium* and *Nocardia* contain **pathogenic** (disease-causing) species.

The cell walls of acid-fast organisms contain a waxlike lipid called **mycolic acid,** which renders the cell wall impermeable to most stains. The cell wall is so impermeable, in fact, that a clinical specimen is usually treated with strong sodium hydroxide to remove debris and contaminating bacteria prior to culturing mycobacteria. The mycobacteria are not killed by this procedure.

Today, the techniques developed by Franz Ziehl and Friedrich Neelsen and by Joseph J. Kinyoun are the most widely used acid-fast stains. In the **Ziehl–Neelsen procedure,** the smear is flooded with carbolfuchsin (a dark red dye containing 5% phenol), which has a high affinity for a chemical component of the bacterial cell. The smear is heated to facilitate penetration of the stain into the bacteria. The stained smears are washed with an acid-alcohol mixture that easily decolorizes most bacteria except the acid-fast microbes. Methylene blue is then used as a counterstain to enable you to observe the non-acid-fast organisms. In the **Kinyoun modification,** called a *cold stain,* the concentrations of phenol and carbolfuchsin are increased so heating isn't necessary.

The mechanism of the acid-fast stain is probably the result of the relative solubility of carbolfuchsin and the impermeability of the cell wall. Fuchsin is more soluble in carbolic acid (phenol) than in water, and carbolic acid solubilizes more easily in lipids than in acid-alcohol. Therefore, carbolfuchsin has a higher affinity for lipids than for acid-alcohol and will remain with the cell wall when washed with acid-alcohol.

Bergey's Manual of Systematic Bacteriology, 2nd ed., 5 vols. (2005), is the standard reference for classification of prokaryotic organisms. *Bergey's Manual of Determinative Bacteriology*, 9th ed. (1994), is the standard reference for identification of culturable bacteria and archaea.

Materials

Acid-fast staining reagents:

- Kinyoun's carbolfuchsin
- Acid-alcohol
- Methylene blue

Wash bottle of distilled water

Slide

Cultures

Mycobacterium phlei

Escherichia coli

Demonstration Slides

Acid-fast sputum slides

Techniques Required

Compound light microscopy, Exercise 1

Smear preparation, Exercise 3

Simple staining, Exercise 3

Procedure (Figure 6.1)

1. Prepare and fix a smear of each culture (Figure 3.1).
2. Cover the smears with carbolfuchsin and let the smears stand for 5 minutes (Figure 6.1a).
3. Gently wash the slide with distilled water from a wash bottle. Do not squirt water directly onto the smear (Figure 6.1b).

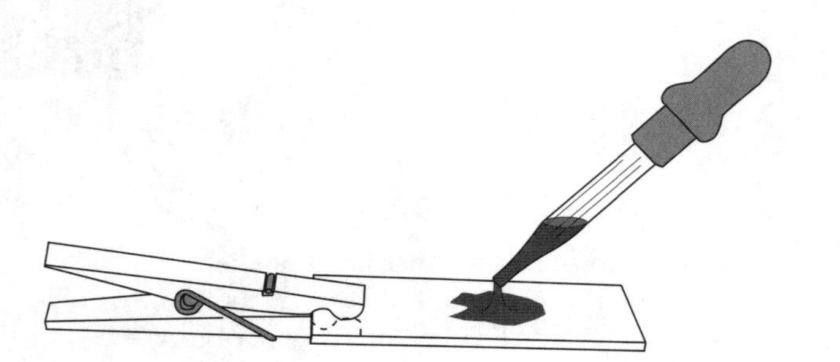

(a) Cover the smear with carbolfuchsin for 5 minutes.

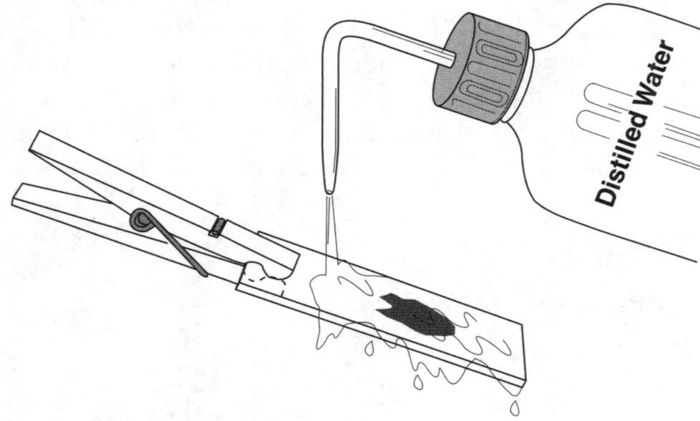

(b) Gently wash off the carbolfuchsin with water by squirting the water so it runs through the smear.

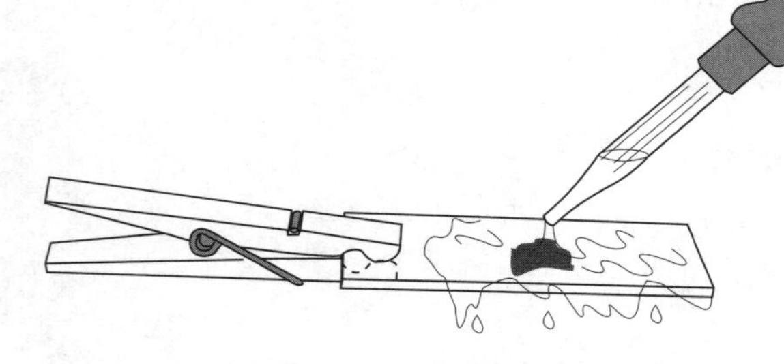

(c) Wash the smear with acid-alcohol for 1 minute.

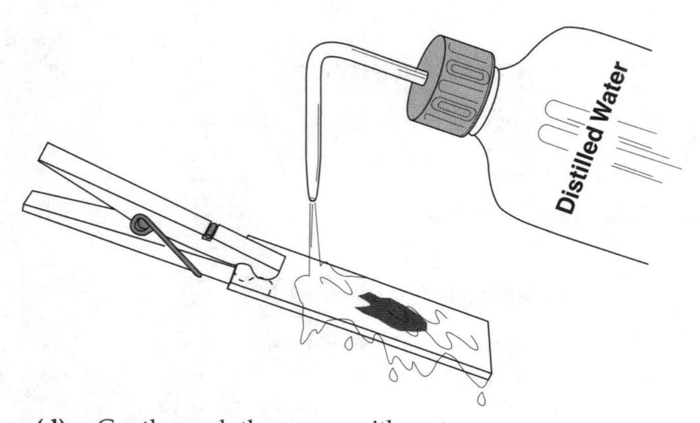

(d) Gently wash the smear with water.

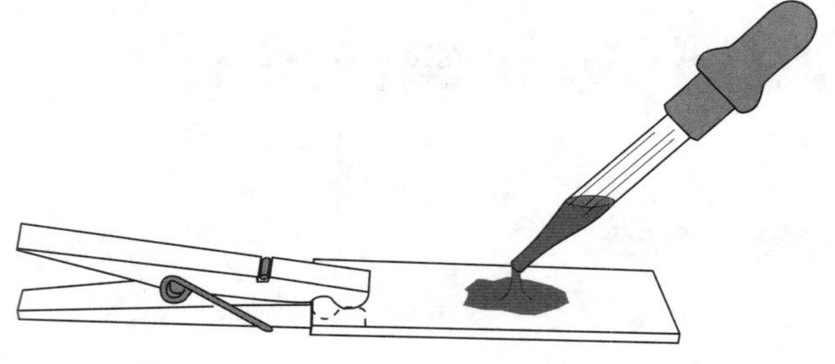

(e) Cover the smear with methylene blue for 1 minute.

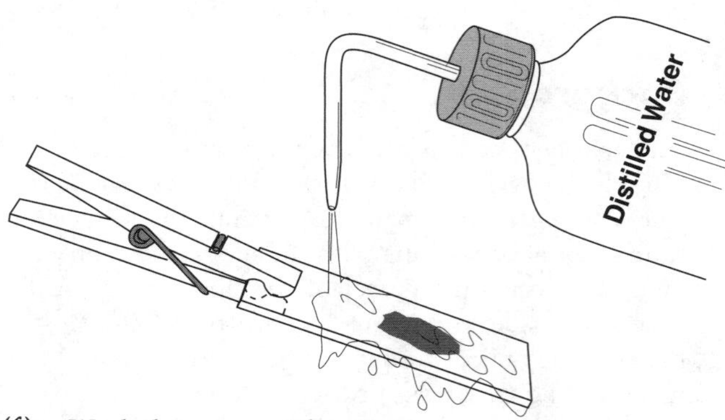

(f) Wash the smear with water.

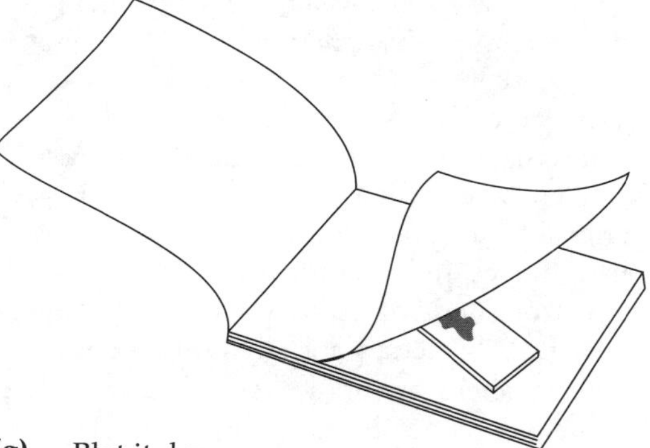

(g) Blot it dry.

Figure 6.1

The acid-fast stain.

4. Without drying it, wash the smear with decolorizer (acid-alcohol) for 1 minute or until no more red color runs off when the slide is tipped (Figure 6.1c).
5. Wash the smear carefully with distilled water (Figure 6.1d).
6. Counterstain the smear for about 1 minute with methylene blue (Figure 6.1e).
7. Wash the smear with distilled water and blot it dry (Figure 6.1f and g).
8. Examine the acid-fast stained slide microscopically and record your observations.
9. Observe the demonstration slides. (See Color Plate I.4.)

Exercise 6

LABORATORY REPORT

Acid-Fast Staining

NAME ______________________

DATE ______________________

LAB SECTION ______________________

Purpose

____

____

____

Data

Sketch a few bacteria (oil immersion objective lens).

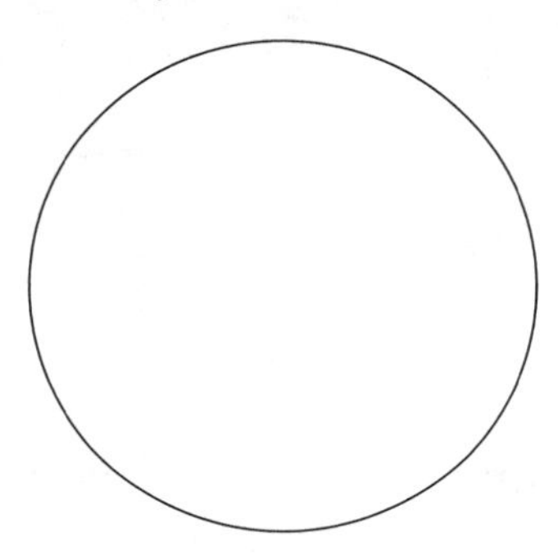

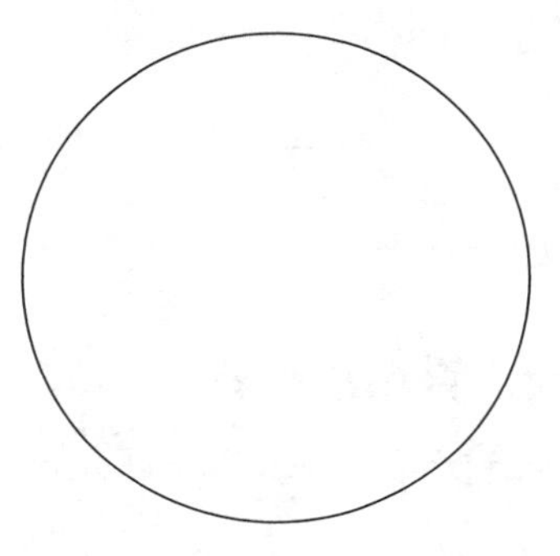

Bacteria:	*Mycobacterium phlei*	*Escherichia coli*
Total magnification:	____×	____×
Morphology:	____________	____________
Acid-fast reaction:	____________	____________

Demonstration Slide 1

Demonstration Slide 2

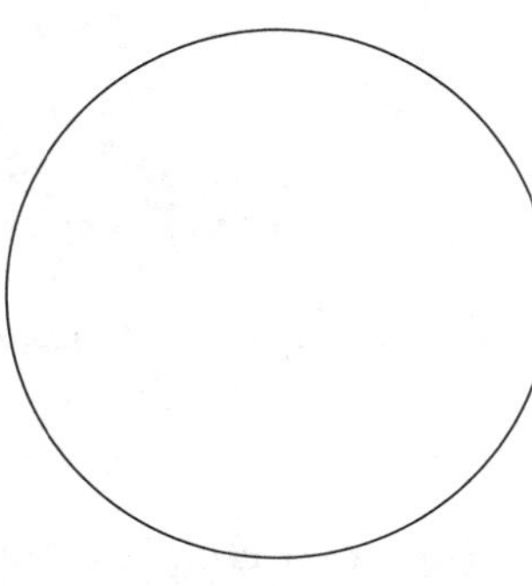

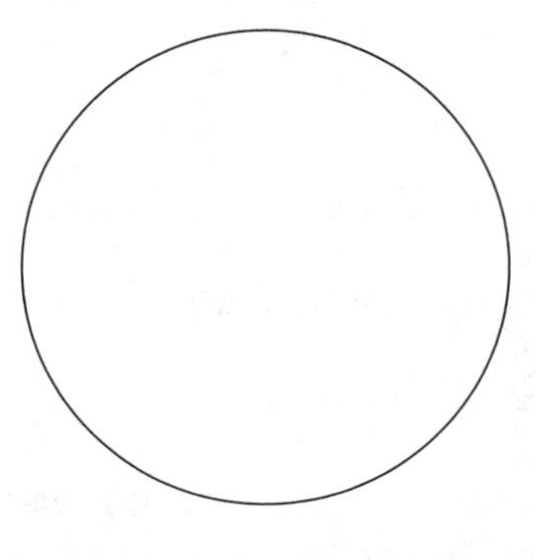

Specimen:	____________	____________
Total magnification:	____×	____×
Morphology:	____________	____________
Acid-fast reaction:	____________	____________

Questions

1. What are the large stained areas on the sputum slide? ______

2. What is the decolorizing agent in the Gram stain? ______

   In the acid-fast stain? ______

3. What diseases are diagnosed using the acid-fast procedure? ______

4. What is phenol (carbolic acid), and what is its *usual* application? ______

Critical Thinking

1. How might the acid-fast characteristic of *Mycobacterium* enhance the organism's ability to cause disease?

2. Clinical specimens suspected of containing *Mycobacterium* are digested with sodium hydroxide (NaOH) for 30 minutes prior to staining. Why is this technique used? Why isn't this technique used for staining other bacteria?

3. The acid-fast stain is used to detect *Cryptosporidium* protozoa in fecal samples. Which of the following would you expect to be a major component of their cell walls: carbohydrates, lipids, or proteins? ______

   What disease is caused by *Cryptosporidium*? ______

4. In 1882, after experimenting with staining *Mycobacterium*, Paul Ehrlich wrote that only alkaline disinfectants would be effective against *Mycobacterium*. How did he reach this conclusion without testing the disinfectants?

Exercise 7

Structural Stains (Endospore, Capsule, and Flagella)

Objectives

After completing this exercise, you should be able to:

1. Prepare and interpret endospore, capsule, and flagella stains.
2. Recognize the different types of flagellar arrangements.
3. Identify functions of endospores, capsules, and flagella.

Background

Structural stains can be used to identify and study the structure of bacteria. Currently, most of the fine structural details are examined using an electron microscope, but historically, staining techniques have given much insight into bacterial fine structure. We will examine a few structural stains that are still useful today. These stains are used to observe endospores, capsules, and flagella.

Endospores

Endospores are formed by several genera in the orders Bacillales and Clostridiales.* *Bacillus* and *Clostridium* are the most familiar genera. Endospores are called "resting bodies" because they do not metabolize and are resistant to heating, various chemicals, and many harsh environmental conditions. Endospores are not for reproduction; they are formed when essential nutrients or water are not available. Once an endospore forms in a cell, the cell will disintegrate (Figure 7.1a). Endospores can remain dormant for long periods of time. However, an endospore may return to its vegetative or growing state.

Taxonomically, it is helpful to know whether a bacterium is an endospore former and also the position of the endospores (Figure 7.1b through e). Endospores are impermeable to most stains so heat is usually applied to drive the stain into the endospore. Once stained, the endospores do not readily decolorize. We will use the **Schaeffer–Fulton** endospore stain.

**Bergey's Manual of Systematic Bacteriology*, 2nd ed. (2005).

Capsules

Many bacteria secrete chemicals that adhere to their surfaces, forming a viscous coat. This structure is called a **capsule** when it is round or oval in shape, and a **slime layer** when it is irregularly shaped and loosely bound to the bacterium. The ability to form capsules is genetically determined, but the size of the capsule is influenced by the medium on which the bacterium is growing. Most capsules are composed of polysaccharides, which are water soluble and uncharged. Because of the capsule's nonionic nature, simple stains will not adhere to it. Most capsule-staining techniques stain the bacteria and the background, leaving the capsules unstained—essentially, a "negative" capsule stain.

Capsules have an important role in the **virulence** (disease-causing ability) of some bacteria. For example, when bacteria such as *Streptococcus pneumoniae* have a capsule, the body's white blood cells cannot phagocytize the bacteria efficiently, and disease occurs. When *S. pneumoniae* lack a capsule, they are easily engulfed and are not virulent.

Flagella

Many bacteria are **motile,** which means they have the ability to move from one position to another in a directed manner. Most motile bacteria possess flagella, but other forms of motility occur. Myxobacteria exhibit gliding motion, and spirochetes undulate using axial filaments.

Flagella, the most common means of motility, are thin proteinaceous structures that originate in the cytoplasm and project out from the cell wall. They are very fragile and are not visible with a light microscope. They can be stained after carefully coating them using a mordant, which increases their diameter. The presence and location of flagella are helpful in the identification and classification of bacteria. Flagella are of two main types: **peritrichous** (all around the bacterium) and **polar** (at one or both ends of the cell) (Figure 7.2).

Motility may be determined by observing hanging-drop or wet-mount preparations of unstained bacteria, flagella stains, or inoculation of soft (or semisolid) agar deeps (Exercise 10). If time does not permit doing flagella stains, observe the demonstration slides.

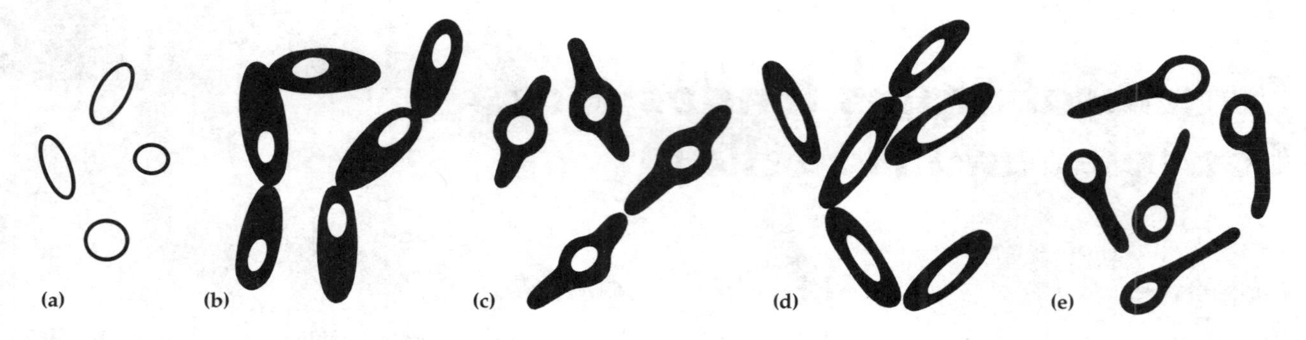

Figure 7.1

Some examples of bacterial endospores. **(a)** Free endospores after the cell has disintegrated. **(b)** Subterminal endospores (*Bacillus macerans*). **(c)** Central, swollen endospores (*Clostridium perfringens*). **(d)** Central endospores (*Bacillus polymyxa*). **(e)** Terminal, swollen endospores (*Clostridium tetani*).

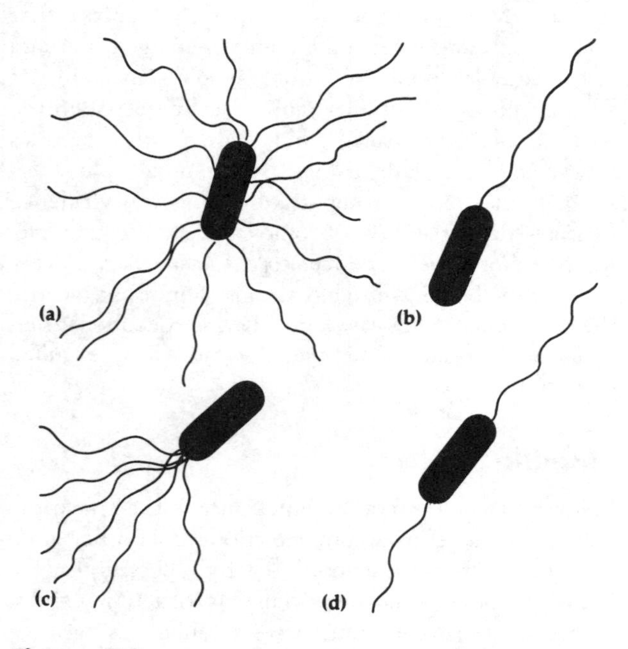

Figure 7.2

Flagellar arrangements. **(a)** Peritrichous flagella. Types of polar flagella: **(b)** monotrichous flagella, **(c)** lophotrichous flagella, and **(d)** amphitrichous flagella.

Materials

Slides

Coverslips

Paper towels

Wash bottle of distilled water

Forceps

Scalpel

Beaker with alcohol

Endospore stain reagents: malachite green and safranin

Capsule stain reagents: Congo red, acid-alcohol, and acid fuchsin

Flagella stain reagents: flagella mordant and carbolfuchsin

Cultures (as needed)

Endospore stain:

Bacillus megaterium (24-hour)

Bacillus subtilis (24-hour)

Bacillus subtilis (72-hour)

Capsule stain:

Streptococcus salivarius

Enterobacter aerogenes

Flagella stain:

Proteus vulgaris (18-hour)

Demonstration Slides

Endospore stain

Capsule stain

Flagella stain

Techniques Required

Compound light microscopy, Exercise 1

Smear preparation, Exercise 3

Simple staining, Exercise 3

Negative staining, Exercise 4

Procedure

Endospore Stain (Figure 7.3)

Be careful. The malachite green has a messy habit of ending up everywhere. But most likely you will end up with green fingers no matter how careful you are.

1. Make smears of the three *Bacillus* cultures on one or two slides, let them air-dry, and heat-fix them (Figure 3.1).
2. Tear out small pieces of paper towel and place them on each smear to reduce evaporation of the stain. The paper should be smaller than the slide (Figure 7.3a).
3. Cover the smears and paper with malachite green; steam the slide for 5 minutes. Add more stain as needed. *Keep it wet* (Figure 7.3b and c). What is the purpose of the paper? ______________________
4. Remove the towel and discard it carefully. *Do not put it in the sink.* Wash the stained smears well with distilled water (Figure 7.3d).
5. Counterstain with safranin for 30 seconds (Figure 7.3e).
6. Wash the smear with distilled water and blot it dry (Figure 7.3f).
7. Examine the slide microscopically and record your observations.
8. Observe the demonstration slides of the bacterial endospores. (See Color Plate I.6.)

Capsule Stain (Figure 7.4)

1. Draw two circles on a slide. Place a loopful of Congo red in each circle (Figure 7.4a).
2. Prepare a thick smear of *S. salivarius* in the Congo red in one circle. Prepare a thick smear of *E. aerogenes* in the other circle. Let the smears air-dry (Figure 7.4b and c). What color are the smears?

   ____
3. Fix the smears with acid-alcohol for 15 seconds (Figure 7.4d). What color are the smears? __________
4. Wash the smears with distilled water and cover them with acid fuchsin for 1 minute (Figure 7.4e and f).
5. Wash the smears with distilled water and blot them dry (Figure 7.4g).
6. Examine the slide microscopically. The bacteria will stain red, and the capsules will be colorless against a dark blue background. Record your observations.
7. Observe the demonstration slides of capsule stains. (See Color Plate I.7.)

Flagella Stain (Figure 7.5)

1. Flagella stains require special precautions to avoid damaging the flagella. Scrupulously clean slides are essential, and the culture must be handled carefully to prevent flagella from coming off the cells.
2. Without touching the bacterial culture, use a scalpel and forceps to cut out a piece of agar on which *Proteus* is growing (Figure 7.5a).

 **Do not touch the culture with your hands.**

   Gently place the agar, culture-side down, on a clean glass slide (Figure 7.5b). Then carefully remove the agar with the forceps. Place the piece of agar in a Petri plate. Place the forceps and scalpel in alcohol.
3. Allow the organisms adhering to the slide to air-dry. *Do not heat-fix the slide* (Figure 7.5c).
4. Cover the slide with flagella mordant and allow it to stand for 10 minutes (Figure 7.5d).
5. Gently rinse off the stain with distilled water (Figure 7.5e).
6. Cover the slide with carbolfuchsin for 5 minutes (Figure 7.5f). Rinse it gently with distilled water (Figure 7.5g). What color should the flagella be?

   ____

   The cells? ______________________________
7. Allow the rinsed smear to air-dry (*do not blot it*), and examine it microscopically for flagella.
8. Observe the demonstration slides illustrating various flagellar arrangements. (See Color Plate I.5.)

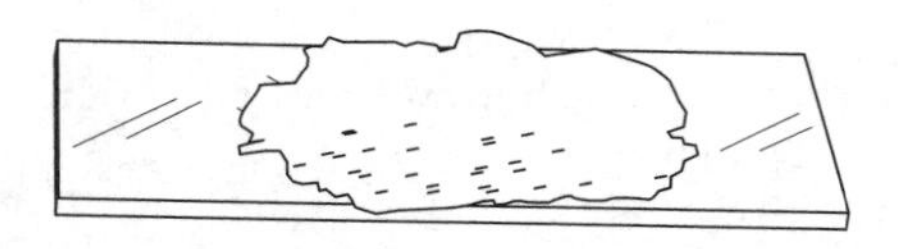

(a) Place a piece of absorbent paper over the smear.

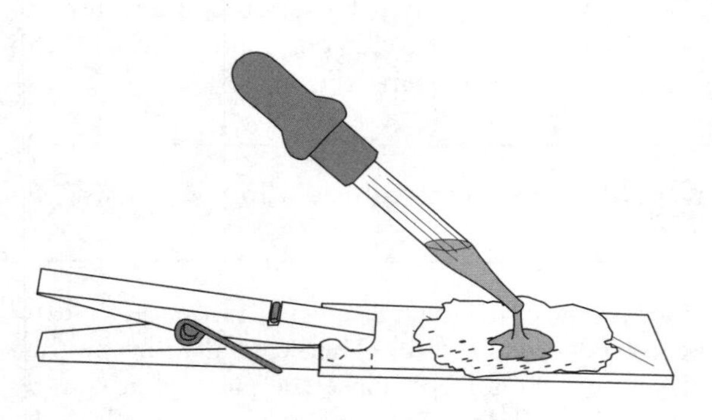

(b) Cover the paper with malachite green.

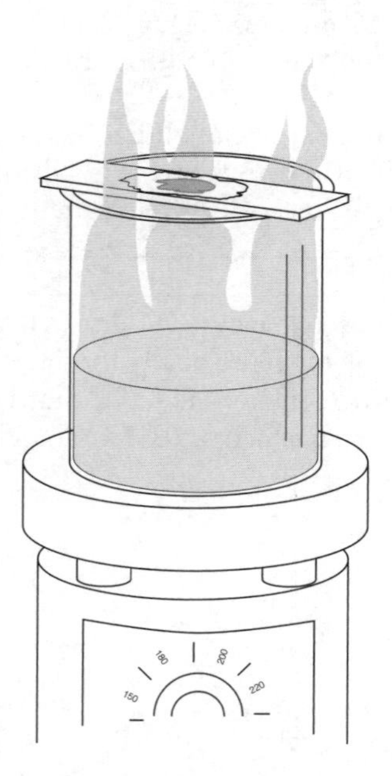

(c) Steam the slide for 5 minutes.

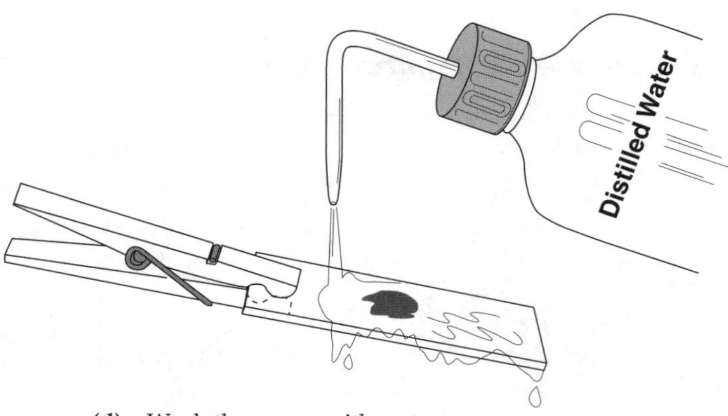

(d) Wash the smear with water.

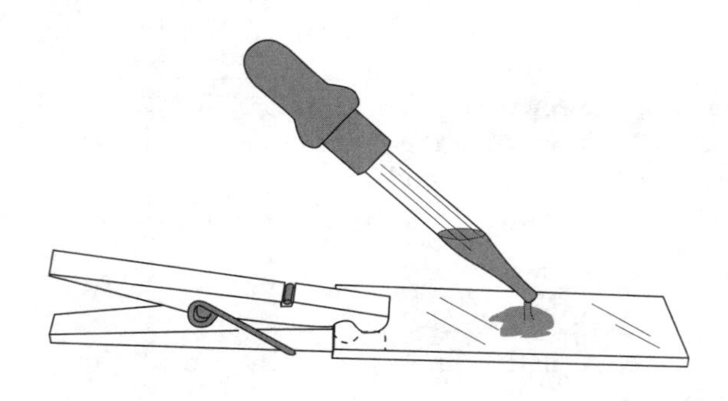

(e) Cover the smear with safranin for 30 seconds.

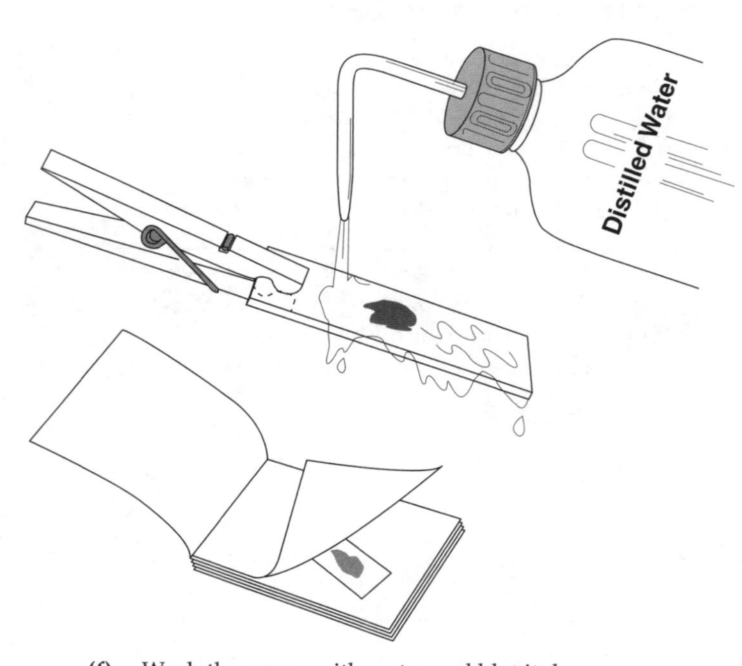

(f) Wash the smear with water and blot it dry.

Figure 7.3

The endospore stain.

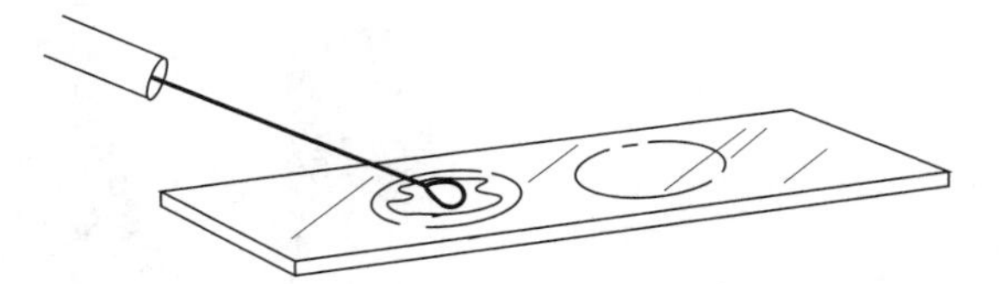

(a) Place one loopful of Congo red in each circle.

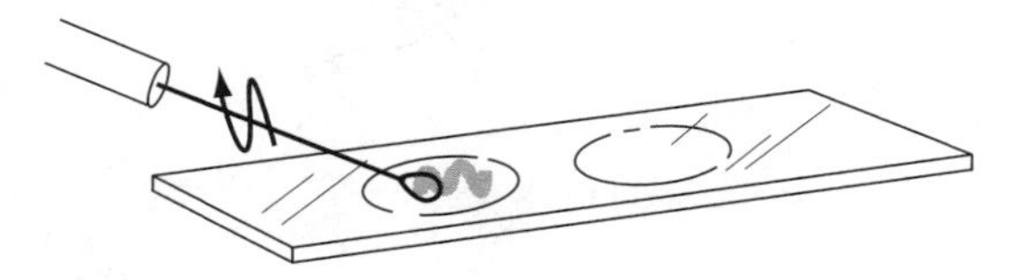

(b) Prepare a thick smear of bacteria in the Congo red.

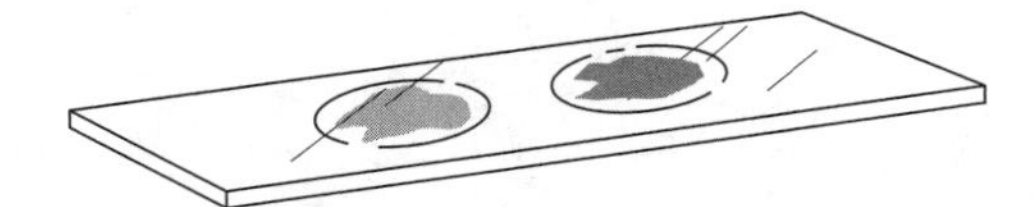

(c) Allow the smears to air-dry.

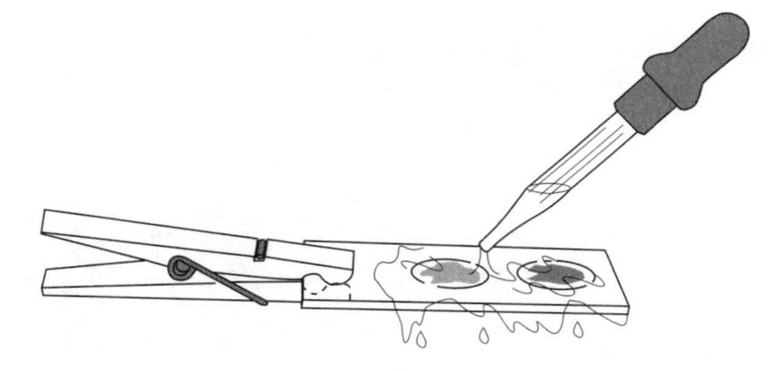

(d) Cover each smear with acid-alcohol for 15 seconds.

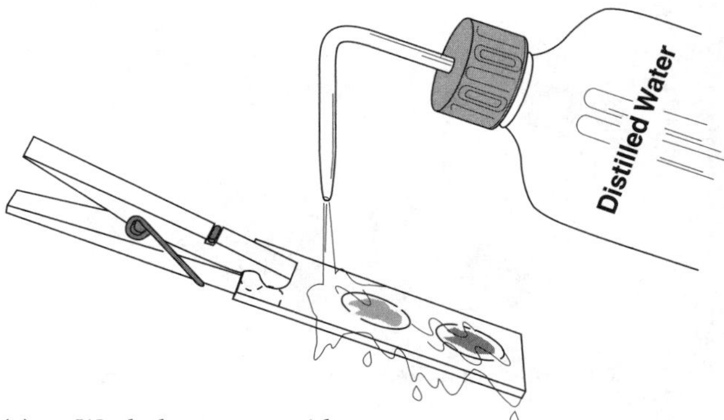

(e) Wash the smears with water.

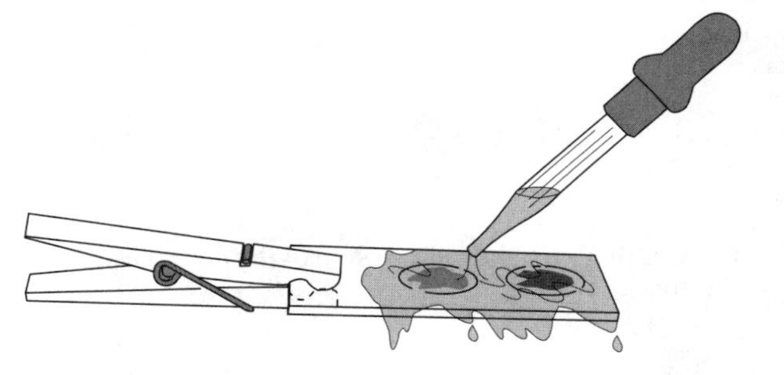

(f) Cover the smears with acid fuchsin for 1 minute.

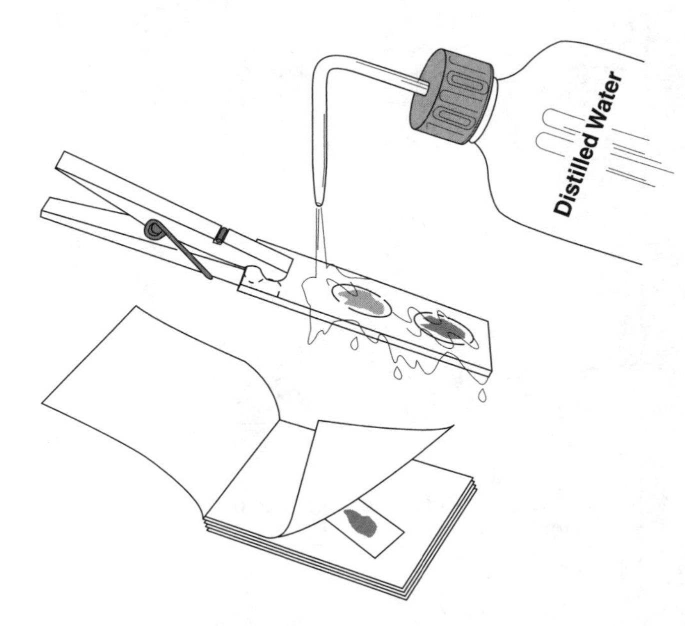

(g) Wash the smears with water and blot them dry.

Figure 7.4

The capsule stain.

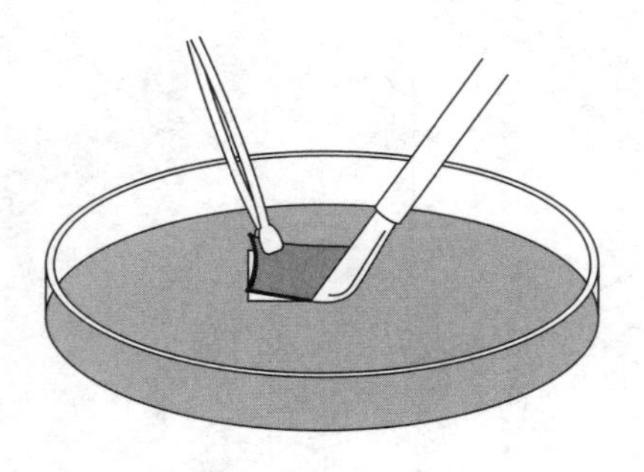

(a) Carefully cut a piece of the bacterial culture.

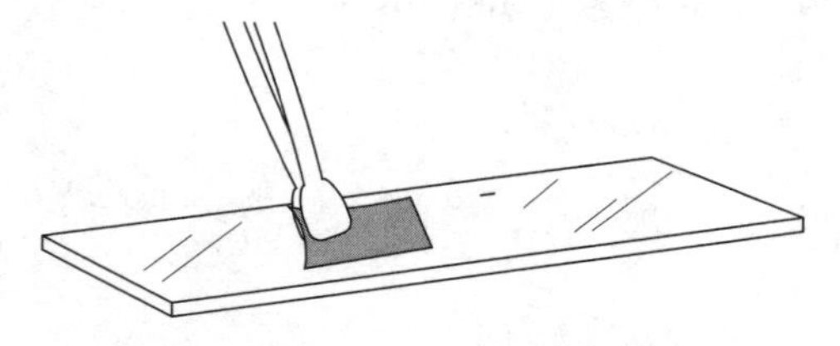

(b) Gently touch the culture side of the agar piece to a clean slide.

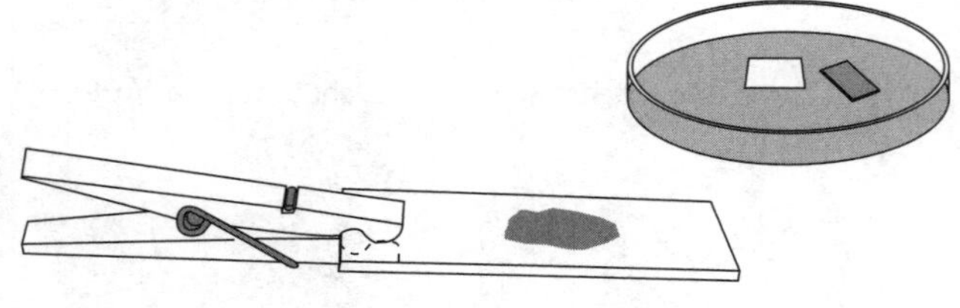

(c) Allow the imprint to air-dry.

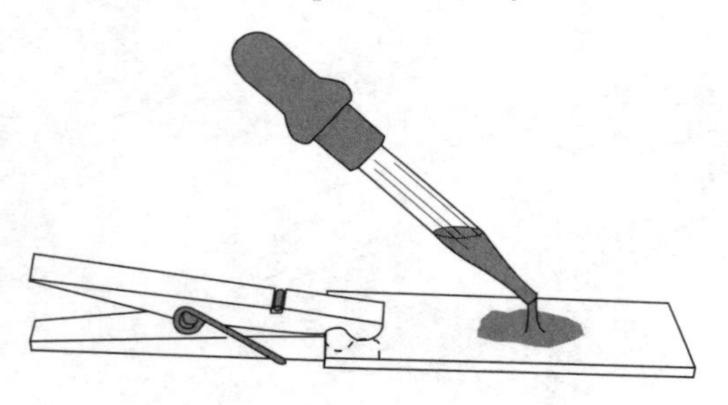

(d) Cover the slide with flagella mordant for 10 minutes.

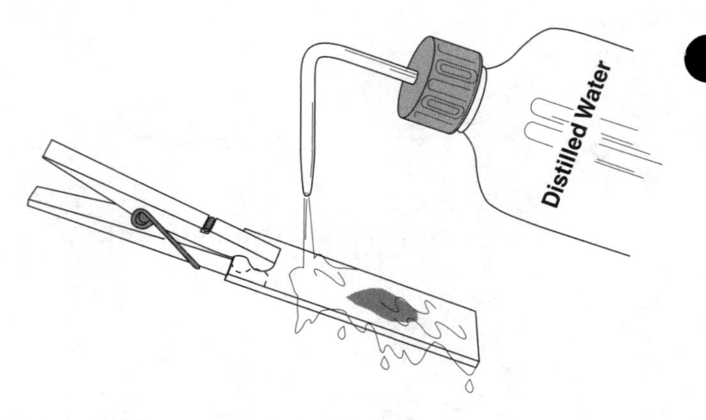

(e) Gently rinse the slide with water.

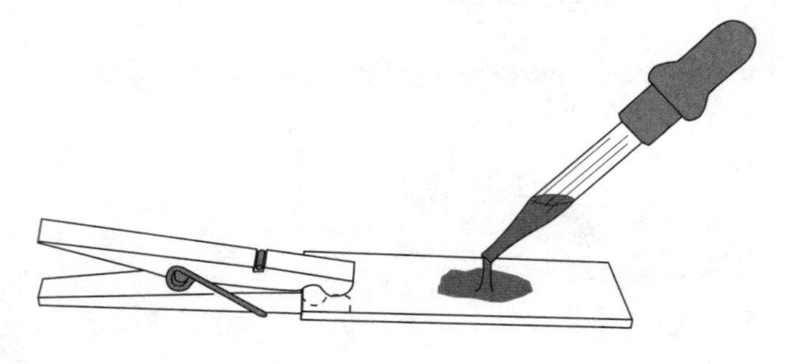

(f) Cover the slide with carbolfuchsin for 5 minutes.

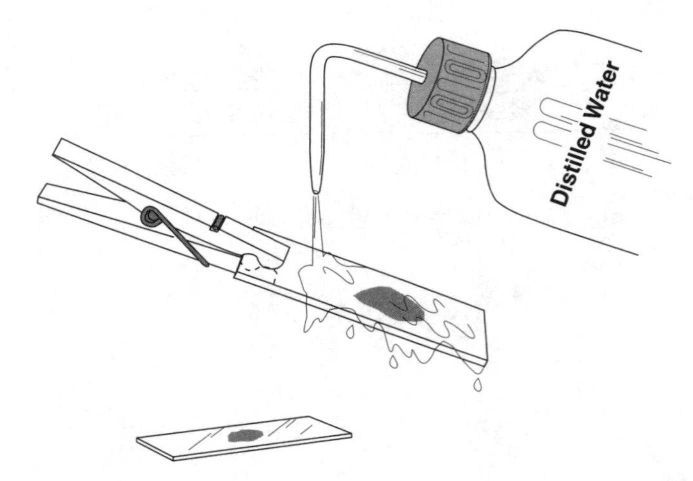

(g) Gently rinse the slide with water and let it air-dry.

Figure 7.5
The flagella stain.

Exercise 7

LABORATORY REPORT

Structural Stains (Endospore, Capsule, and Flagella)

NAME ______________________

DATE ______________________

LAB SECTION ______________________

Purpose ______________________

____

____

Data

Endospores

Sketch your results, and label the color in each diagram. Label the vegetative cells and endospores.

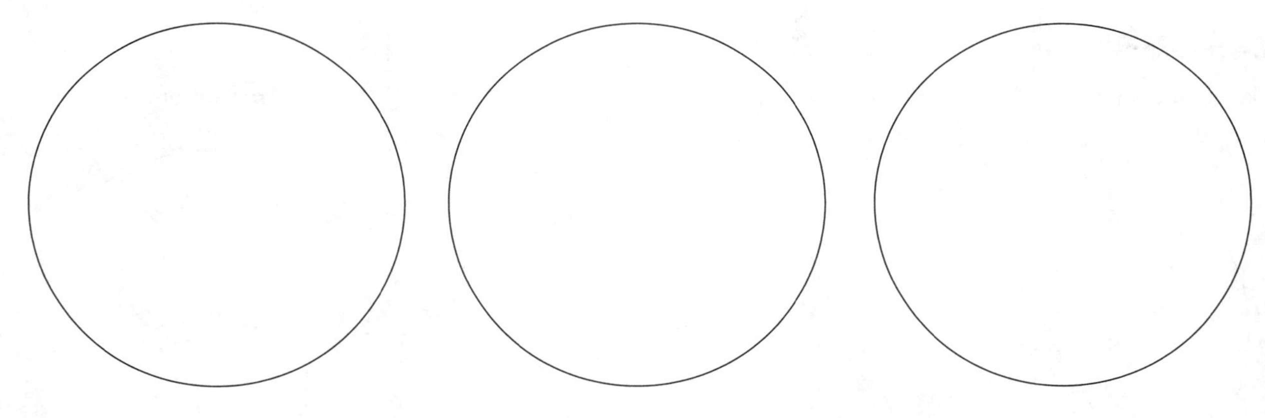

Bacteria:	*Bacillus megaterium*	*Bacillus subtilis* (24-hour)	*Bacillus subtilis* (72-hour)
Total magnification:	___ ×	___ ×	___ ×
Endospore position:	________	________	________

Demonstration Slides

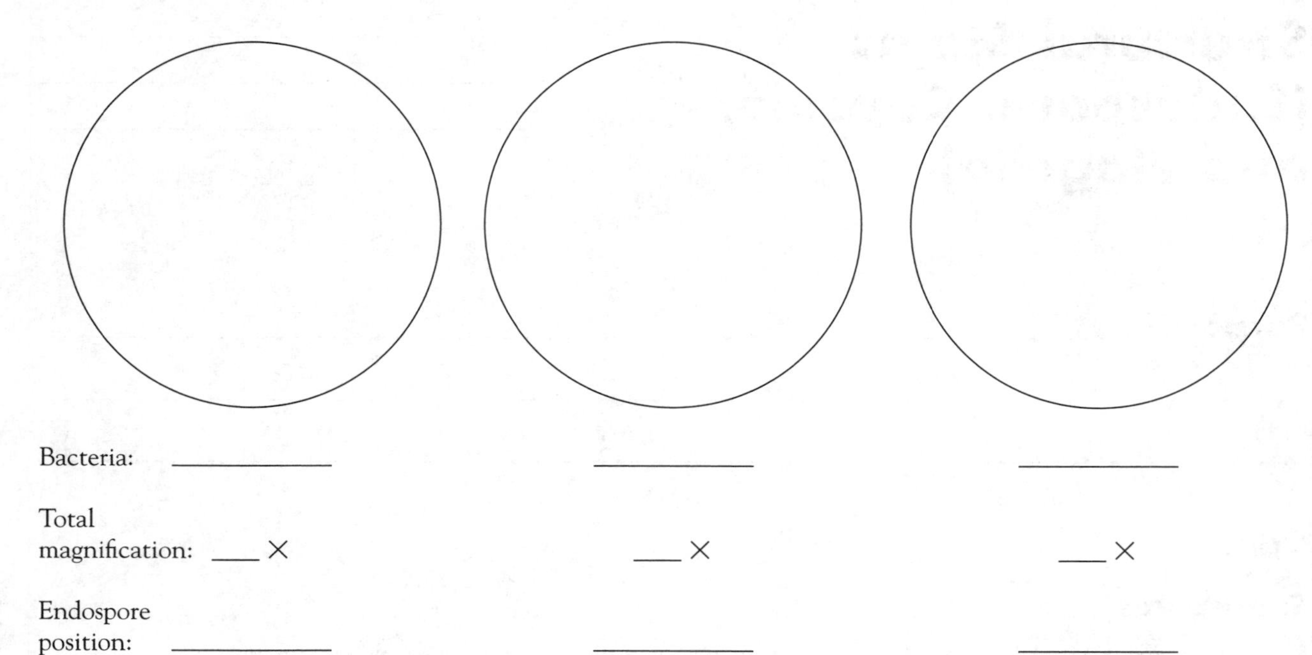

Bacteria: ____________ ____________ ____________

Total magnification: ___ × ___ × ___ ×

Endospore position: ____________ ____________ ____________

Capsules

Sketch and label the capsules and bacterial cells.

Demonstration Slide

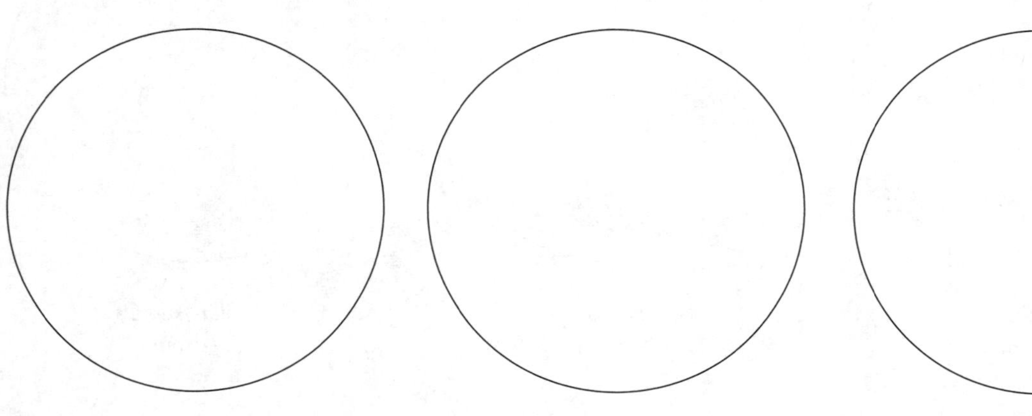

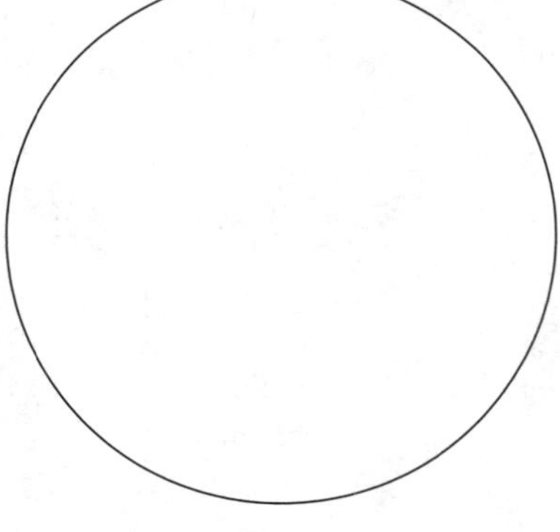

Bacteria: *Streptococcus salivarius* *Enterobacter aerogenes* ____________

Total magnification: ___ × ___ × ___ ×

Capsules present: ____________ ____________ ____________

Flagella

Sketch and label the flagella and bacteria. Demonstration Slides

Bacteria: *Proteus vulgaris* ____________ ____________

Total magnification: ___ × ___ × ___ ×

Flagella position: __________ __________ __________

Questions

1. What are the Gram reactions of *Clostridium* and *Bacillus*? ______________________________

____

2. How might a capsule contribute to pathogenicity? ______________________________

   How might flagella contribute to pathogenicity? ______________________________

3. Of what advantage to *Clostridium* is an endospore? ______________________________

____

____

4. Sketch each of the following flagellar arrangements:

   a. Monotrichous

   b. Lophotrichous

   c. Amphitrichous

   d. Peritrichous

5. How did the appearance of the 24-hour and 72-hour *Bacillus* cultures differ? How do you account for this difference? ___

6. Of what morphology are most bacteria possessing flagella? ___

   Which morphology usually does not have flagella? ___

7. What prevents the cell from appearing green in the finished endospore stain? ___

Critical Thinking

1. You can see endospores by simple staining. Why not use this technique?

2. How would an endospore stain of *Mycobacterium* appear?

3. What type of culture medium would increase the size of a bacterial capsule?

4. Describe the microscopic appearance of encapsulated *Streptococcus* if stained with safranin and nigrosin.

5. In the Dorner endospore stain, a smear covered with carbolfuchsin is steamed, then decolorized with acid-alcohol and counterstained with nigrosin. Describe the microscopic appearance after this procedure.

Exercise 8

EXERCISE 8

Morphologic Unknown

Dishonesty is knowing but ignoring the fact that the data are contradictory.
Stupidity is not recognizing the contradictions.

ANONYMOUS

Objective

After completing this exercise, you should be able to identify the morphology and staining characteristics of an unknown organism.

Background

Differential staining is usually the first step in the identification of bacteria. Morphology and structural characteristics obtained from microscopic examination are also useful for identification. *Bergey's Manual** is the most widely used reference for bacterial identification. In *Bergey's Manual*, bacteria are identified by differential staining, morphology, and several other characteristics. Additional testing is needed to identify bacterial species. You will learn these techniques later in this course.

You will be given an unknown culture of bacteria. Determine its morphologic and structural characteristics. The culture contains one species (rod or coccus) and is less than 24 hours old.

Quality control (QC) is an essential component of the microbiology laboratory. The accuracy of information from stains and other tests depends on a variety of factors, including the specimen, reagents, procedures, and the person doing the test. Generally during staining, known positive and negative bacteria are stained at the same time as unknown bacteria as a check on reagents and procedures. In this exercise, you will use your results from Exercises 5 through 7 as your QC.

Materials

Staining reagents

Culture

24-hour unknown slant culture of bacteria #_____

Techniques Required

Compound light microscopy, Exercise 1

Hanging-drop and wet-mount procedures, Exercise 2

Smear preparation, Exercise 3

Simple staining, Exercise 3

Negative staining, Exercise 4

Gram staining, Exercise 5

Acid-fast staining, Exercise 6

Endospore, capsule, and flagella staining, Exercise 7

Procedure

1. Record the number of your unknown in your Laboratory Report.
2. Determine the morphology, Gram reaction, and arrangement of your unknown. Perform a Gram stain and, if needed, an endospore stain, acid-fast stain, flagella stain, hanging-drop technique, and capsule stain. When are the latter needed? _____

   ____
3. Tabulate your results in the Laboratory Report.

**Bergey's Manual of Systematic Bacteriology*, 2nd ed., 5 vols. (2001), is the standard reference for classification of prokaryotic organisms. *Bergey's Manual of Determinative Bacteriology*, 9th ed. (1994), is the standard reference for identification of culturable bacteria and archaea.

Exercise 8 — LABORATORY REPORT

Morphologic Unknown

NAME ______________________

DATE ______________________

LAB SECTION ______________________

Results

Write *not necessary* by any category that does not apply.

Unknown # ________

Gram Stain

Sketch your unknown. Controls from Exercise 5

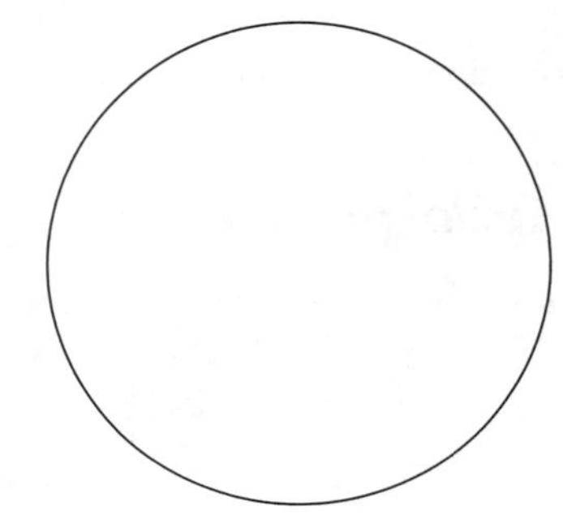

Specimen:	*Staphylococcus epidermidis*	*Escherichia coli*	Unknown
Total magnification:	___ ×	___ ×	___ ×
Gram reaction:	________	________	________
Morphology:	________	________	________
Predominant arrangement:	________	________	________

Acid-Fast Stain

Sketch your unknown. Controls from Exercise 6

Specimen:	*Mycobacterium phlei*	*Escherichia coli*	Unknown
Total magnification:	___ ×	___ ×	___ ×
Acid-fast reaction:	______	______	______

Endospore Stain

Sketch your unknown. Controls from Exercise 7

Specimen:	24-hr *Bacillus*	72-hr *Bacillus*	Unknown
Total magnification:	___ ×	___ ×	___ ×
Endospores present:	______	______	______

Capsule Stain

Sketch your unknown. Control from Exercise 7

Specimen:	*Streptococcus salivarius*	Unknown
Total magnification:	___ ×	___ ×
Capsules present:	____________	____________

Motility

Is your unknown motile? ________

How did you determine this? ____

Questions

1. Which two stains done in this experiment are differential stains? ____________________

____

2. Assume you have performed a Gram stain on a sample of pus from a patient's urethra. You see red, nucleated cells (>10 μm) and purple rods (2.5 μm). What can you conclude? ____________________

____

____

Critical Thinking

1. Using *Bergey's Manual* and your textbook, place the following genera in this flowchart: *Bacillus*, *Corynebacterium*, *Escherichia*, *Mycobacterium*, *Neisseria*, *Sporosarcina*, and *Staphylococcus*.

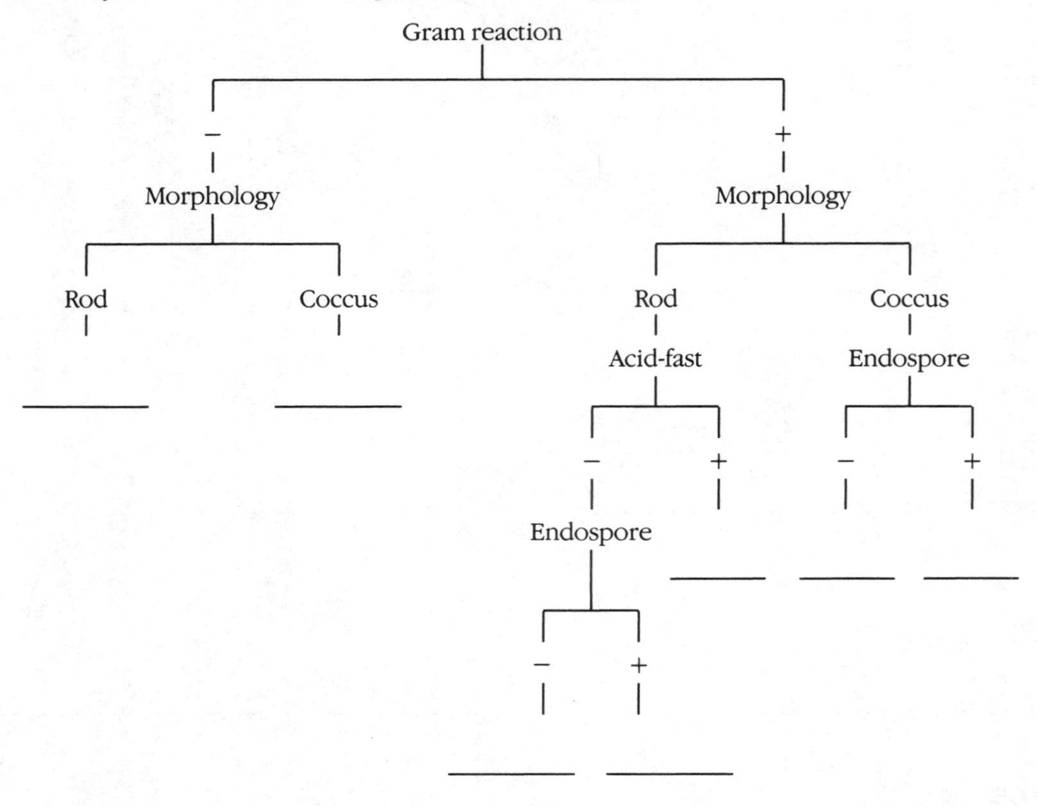

2. Which genera listed in the previous question should you test for capsules? For motility?

3. Using your textbook and this lab manual, fill in the Morphology column in the following table. Then use the information to construct a flowchart for these bacteria. Draw your flowchart in the space below the table.

	Morphology	Gram Reaction	Motile	Capsule	Arrangement	Endospore
Clostridium		+	+	–	Pairs, chains	+
Enterobacter		–	+	–	Singles, pairs	–
Klebsiella		–	–	+	Singles, pairs	–
Lactobacillus		+	Rarely	–	Chains	–
Staphylococcus		+	–	–	Pairs, clusters	–
Streptococcus		+	–	Some species	Pairs, chains	–

4. Provide an example from the previous question to show that microscopic examination alone cannot be used to identify bacteria.

Part Three

Cultivation of Bacteria

EXERCISES

Bacteria must be grown on culture media in order to characterize and identify them. This statement from the first edition of *Bergey's Manual* summarizes the need for cultivating bacteria:

> *The earlier writers classified the bacteria solely on their morphologic characters. A more detailed classification was not possible because the biologic characters of so few of the bacteria had been determined. With the accumulation of knowledge of the biologic characters of many bacteria it was realized that it is just as incorrect to group all rods under a single genus as to group all quadruped animals under one genus.**

Before attempting to culture bacteria, we must consider their nutritional requirements. Bacteria require sources of energy, carbon, nitrogen, minerals, and growth factors. Most bacteria are **chemoheterotrophs,** which require organic compounds for carbon and energy sources. Bacteria exhibit a wide range of nutritional requirements, and in many instances, enriching nutrients such as milk, serum, blood, or tomato juice must be added for fastidious organisms that require many growth factors. **Growth factors** are organic compounds such as vitamins or amino acids that are incorporated into a cell without alteration.

*Society of American Bacteriologists (D. H. Bergey, Chairperson). *Bergey's Manual of Determinative Bacteriology,* 1st ed. Baltimore: Williams & Wilkins, 1923, p. 1.

The first bacterial cultures were grown in *broth* (liquid) media such as infusions and blood. Robert Koch attempted to culture bacteria on potato slices when he realized the need for *solid media*. When some bacteria wouldn't grow on a potato, he added gelatin to the broth, but the gelatin liquefied under standard incubation conditions. Angelina Hess, a colleague of Koch, suggested the use of agar as a solidifying agent. Today, agar is the most commonly used solidifying agent in culture media (see the figure below).

The exercises in Part Three emphasize the use of **aseptic technique,** the prevention of unwanted microorganisms in laboratory and medical procedures.

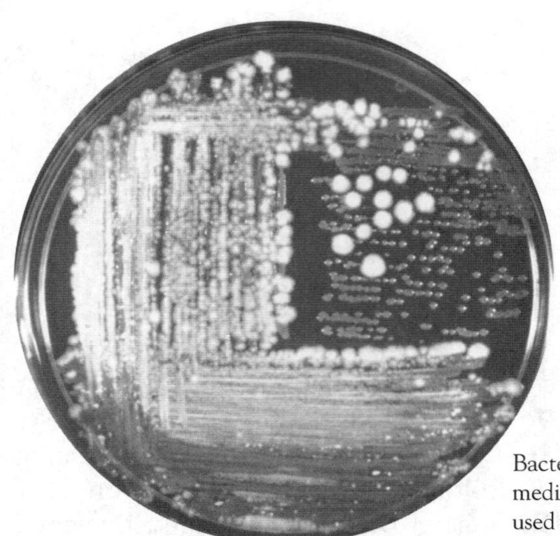

Bacteria grow into visible colonies on solid media. The appearance of colonies can be used to distinguish different bacteria.

Exercise 9

Microbes in the Environment

Whatever is worth doing at all is worth doing well.

PHILIP DORMER STANHOPE

Objectives

After completing this exercise, you should be able to:

1. Describe why agar is used in culture media.
2. Prepare nutrient broth and nutrient agar.
3. Compare bacterial growth on solid and liquid culture media.
4. Describe colony morphology using accepted descriptive terms.

Background

Microbes are everywhere; they are found in the water we drink, the air we breathe, and the earth on which we walk. They live in and on our bodies. Microbes occupy ecological niches on all forms of life and in most environments. In most situations, the ubiquitous microorganisms are harmless. However, in microbiology, work must be done carefully to avoid contaminating sterile media and materials with these microbes.

In this exercise, we will attempt to culture (grow) some microbes from the environment. When a medium is selected for culturing bacteria, macronutrients, an energy source, and any necessary growth factors must be provided. A medium whose exact chemical composition is known is called a **chemically defined medium** (Table 9.1).

Most chemoheterotrophic bacteria are routinely grown on **complex media**—that is, media for which the exact chemical composition varies slightly from batch to batch. Organic carbon, energy, and nitrogen sources are usually supplied by protein in the form of meat extracts and partially digested proteins called *peptones*. **Nutrient broth** is a commonly used liquid complex medium. When agar is added, it becomes a solid medium, called **nutrient agar** (Table 9.2).

Agar, an extract from marine red algae, has some unique properties that make it useful in culture media. Few microbes can degrade agar so it remains solid during microbial growth. It liquefies at 100°C and remains in a liquid state until cooled to 40°C. Once the agar has solidified, it can be incubated at temperatures of up to 100°C and remain solid.

Media must be sterilized after preparation. The most common method of sterilizing culture media that are heat stable is **steam sterilization,** or **autoclaving,** using steam under pressure. During this process, material to be sterilized is placed in the autoclave and heated to 121°C at 15 pounds of pressure (15 psi) for 15 minutes.

Culture media can be prepared in various forms, depending on the desired use. **Petri plates** containing solid media provide a large surface area for examination of colonies. The microbes will be **inoculated,** or intentionally introduced, onto nutrient agar and into nutrient broth. The bacteria that are inoculated into culture

Table 9.1

Glucose–Minimal Salts Broth

Ingredient	Amount/100 ml
Glucose	0.5 g
Sodium chloride (NaCl)	0.5 g
Ammonium dihydrogen phosphate ($NH_4H_2PO_4$)	0.1 g
Dipotassium phosphate (K_2HPO_4)	0.1 g
Magnesium sulfate ($MgSO_4$)	0.02 g
Distilled water	100 ml

Table 9.2

Nutrient Agar

Ingredient	Amount/100 ml
Peptone	0.5 g
Beef extract	0.3 g
Sodium chloride (NaCl)	0.8 g
Agar	1.5 g
Distilled water	100 ml

media increase in number during an **incubation period.** After suitable incubation, liquid media become **turbid,** or cloudy, due to bacterial growth. On solid media, colonies will be visible to the naked eye. A **colony** is a population of cells that arises from a single bacterial cell. A colony may arise from a group of the same microbes attached to one another, which is therefore called a **colony-forming unit.** Although many species of bacteria give rise to colonies that appear similar, each colony that appears different is usually a different species. (See Color Plate VI.1.)

Materials

250-ml Erlenmeyer flask with cap or plug

100-ml graduated cylinder

Distilled water

Nutrient broth powder

Agar

Glass stirring rod

5-ml pipette

Propipette or pipette bulb

Test tubes with caps (3)

Sterile Petri dishes (4)

Balance

Weighing paper or dish

Autoclave gloves

Hot plate

Tube containing sterile cotton swabs

Tube containing sterile water

Demonstration

Use of the autoclave

Techniques Required

Pipetting, Appendix A

Procedure

First Period

1. Preparing culture media
   a. Prepare 100 ml of nutrient broth in a 250-ml flask. Using the graduated cylinder, add 100 ml of distilled water to the flask. Read the preparation instructions on the nutrient broth bottle. Calculate the amount of nutrient broth powder needed for 100 ml. If the amount needed for 1000 ml is _____ grams, then _____ grams are needed for 100 ml. Weigh out the required amount and add it to the flask. Stir with a glass rod until the powder is dissolved.
   b. Attach a bulb or Propipette to the 5-ml pipette. Pipette 5 ml of the nutrient broth into each test tube and cap each tube. Label the tubes "nutrient broth." Place two in the To Be Autoclaved rack. Label the remaining tube "not sterilized," and incubate it at room temperature until the next period.
   c. Add agar (1.5% w/v) to the remaining 85 ml of nutrient broth. What quantity of agar will you need to add? _______________
   d. Bring the broth to a boil and continue boiling carefully until all the agar is dissolved. *Be careful:* Do not let the solution boil over. Stir often to prevent burning and boiling over.
   e. Stopper the flask, label it "nutrient agar," and place the flask and tube in the To Be Autoclaved basket.
   f. Listen to the instructor's demonstration of the use of the autoclave.
   g. After autoclaving, allow the flasks and tube to cool to room temperature, or proceed to part 2. What effect does the agar have on the culture medium? _______________
2. Pouring plates
   Transfer the melted sterile nutrient agar flasks to a 45°C water bath. Allow the flask of nutrient agar to cool to about 45°C (warm to the touch). If the agar has solidified, it will have to be reheated to liquefy it. To what temperature will it have to be heated? _______________
   The sterile nutrient agar must be poured into Petri dishes *aseptically*—that is, without letting microbes into the nutrient medium. *Read the following procedure before beginning* so that you can work quickly and efficiently.
   a. Set four sterile, unopened Petri dishes in front of you with the cover (larger half) on top. Have a lighted laboratory burner within reach on your workbench.

Keep the burner away from your hair and in the center of the bench.

   b. Holding the flask at an angle, remove the stopper with the fourth and fifth fingers of your other hand. Heat the mouth of the flask by passing it through the flame three times (Figure 9.1a). Why is it necessary to keep the flask at an angle through this procedure? _______________

c. Remove the cover from the first dish with the hand holding the plug. Quickly and neatly pour melted nutrient agar into the dish until the bottom is just covered to a depth of approximately 5 mm (Figure 9.1b). Keep the flask at an angle, and replace the dish cover; move on to the next plate until all the agar is poured.
d. When all the agar is poured, gently swirl the agar in each dish to cover any empty spaces; do not allow the agar to touch the dish covers.
e. To decrease condensation, leave the Petri plate covers slightly ajar for about 15 minutes until the agar solidifies.
f. Place the empty flask in the discard area.

3. Culturing microbes from the environment
   a. Design your own experiment. The purpose is to sample your environment and your body. Use your imagination. Here are some suggestions:
      1. You may use the lab, a washroom, or any place on campus for the environment.
      2. One nutrient agar plate might be left open to the air for 30 to 60 minutes.
      3. Inoculate a plate from an environmental surface such as the floor or workbench by wetting a cotton swab in sterile water, swabbing the environmental surface, and then swabbing the surface of the agar. Why is the swab first moistened in sterile water?

      ____

      ____

      After inoculation, the swab should be discarded in the container of disinfectant.
   b. Inoculate two plates from the environment. Inoculate one nutrient broth tube using a swab as described in step 3 of 3a. After swabbing the agar surface, place the swab in the nutrient broth and leave it there during incubation. You may need to break off part of the wooden handle to fit the swab into the nutrient broth.
   c. The plates and tube should be incubated at the approximate temperature of the environment sampled.
   d. Inoculate two plates from your body. You could:
      1. Place a hair on the agar.
      2. Obtain an inoculum by swabbing (see step 3 of 3a) part of your body with a wet swab.
      3. Touch the plate with your fingers.
   e. Incubate bacteria from your body at or close to your body temperature. What is human body temperature? _____°C
   f. Incubate all plates inverted, so water will condense in the lid instead of on the surface of the agar. Why is condensation on the agar undesirable? ______________________________

      ____

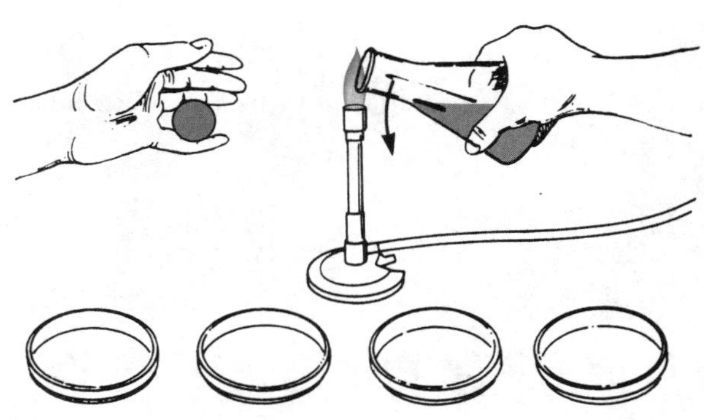

(a) Remove the stopper, and flame the mouth of the flask.

(b) Remove the cover from one dish, and pour nutrient agar into the dish bottom.

Figure 9.1

Petri plate pouring.

   g. Incubate all inoculated media until the next laboratory period.

Second Period

1. Observe and describe the resulting growth on the plates. Note each different-appearing colony, and describe the colony pigment (color) and morphology using the characteristics given in Figure 9.2. Determine the approximate number of each type of colony. When many colonies are present, record TNTC (too numerous to count) as the number of colonies.
2. Describe the appearance of the nutrient broth labeled "not sterile" and the broth you inoculated. Are they uniformly cloudy or turbid? __________ Look for clumps of microbial cells, called **flocculent.** Is there a membrane, or **pellicle,** across the surface of the broth? __________ See whether microbial cells have settled on the bottom of the tube, forming a **sediment.** (See Color Plate X.1.)
3. Save one turbid broth in the refrigerator for Exercise 11. Discard the plates and remaining tubes properly.

Place contaminated plates in the biohazard bag for autoclaving.

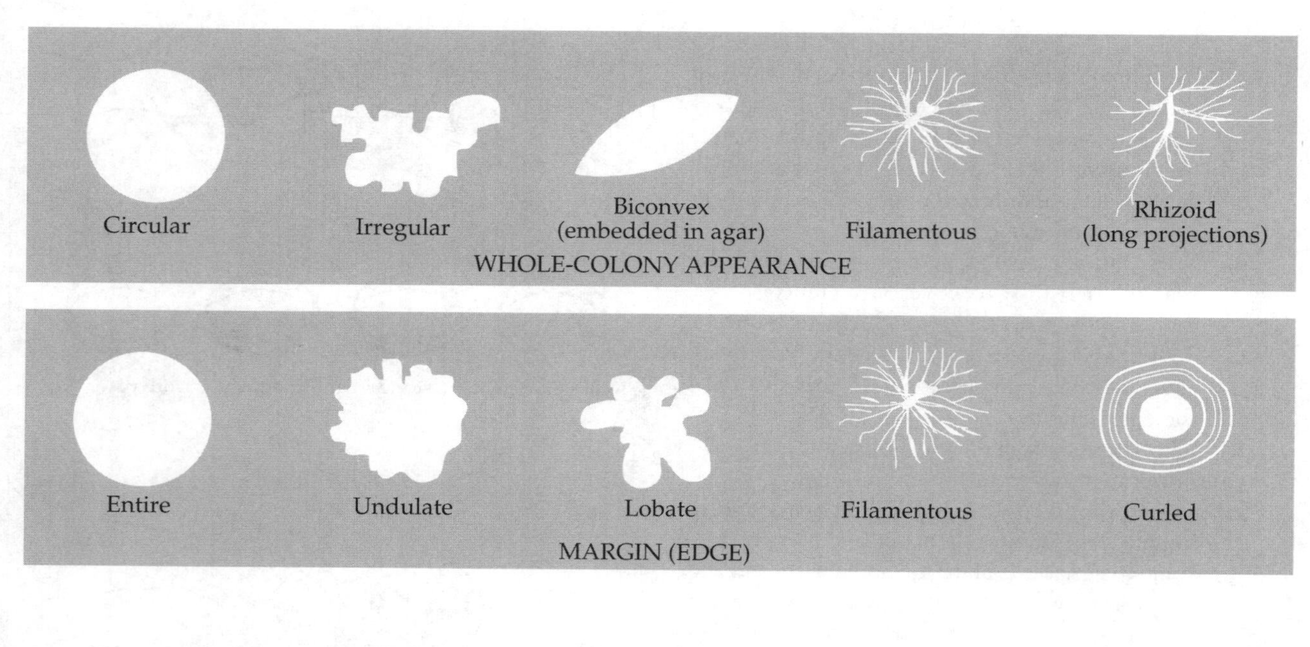

Figure 9.2
Colony descriptions.

Microbes in the Environment

NAME ______________________

DATE ______________________

LAB SECTION ______________________

Purpose

____

____

____

Data

Fill in the following table with descriptions of the bacterial colonies. Use a separate line for each different-appearing colony. Observe your classmates' plates if you didn't inoculate four plates.

	Colony Description					
	Diameter	Whole-Colony Appearance	Margin	Elevation	Pigment	Number of this Type
Area sampled:						
________						
Incubated at						
_____°C for						
_____ days						

Area sampled:						
________						
Incubated at						
_____°C for						
_____ days						

Area sampled:						
________						
Incubated at						
_____°C for						
_____ days						

	Colony Description					
	Diameter	Whole-Colony Appearance	Margin	Elevation	Pigment	Number of this Type
Area sampled:						
________						
Incubated at						
_____°C for						
_____ days						

Nutrient broths: Incubated at ____________°C for ____________ days

	Not-Sterilized Broth	Sterilized Broth Not Inoculated	Inoculated Broth Area Sampled: ________
Turbidity			
Flocculent present			
Sediment present			
Pellicle present			
Color			

Questions

1. How can you tell whether there is bacterial growth in the nutrient broth? ____________________

   ____

2. What is the minimum number of different bacteria present on one of your plates? ____________

   How do you know? ____

   ____

3. What is the value of Petri plates in microbiology? ____

   ____

4. What are bacteria using for nutrients in nutrient agar? ______________________________

____

What is the purpose of the agar? ______________________________

____

5. Which environment has the most total bacteria? ____________ The least? ____________ Provide a reason for the differences in total bacteria in these two places. ______________________________

____

____

6. Which environment has the most different bacteria? ____________ The least? ____________

Provide a reason for the differences in bacteria in these two places. ______________________________

____

____

Critical Thinking

1. Why is agar preferable to gelatin as a solidifying agent in culture media?

2. Did all the organisms living in or on the environments sampled grow on your nutrient agar? Briefly explain.

3. How could you determine whether the turbidity in your nutrient broth tube was from a mixture of different microbes or from the growth of only one kind of microbe?

Exercise 10

Transfer of Bacteria: Aseptic Technique

Study without thinking is worthless; thinking without study is dangerous.

CONFUCIUS

Objectives

After completing this exercise, you should be able to:

1. Provide the rationale for aseptic technique.
2. Differentiate among the following: broth culture, agar slant, and agar deep.
3. Aseptically transfer bacteria from one form of culture medium to another.

Background

In the laboratory, bacteria must be cultured in order to facilitate identification and to examine their growth and metabolism. Bacteria are **inoculated,** or introduced, into various forms of culture media in order to keep them alive and to study their growth. Inoculations must be done without introducing unwanted microbes, or **contaminants,** into the media. **Aseptic technique** is used in microbiology to exclude contaminants.

All culture media are **sterilized,** or rendered free of all life, prior to use. Sterilization is usually accomplished using an autoclave. Containers of culture media, such as test tubes or Petri plates, should not be opened until you are ready to work with them, and even then, they should not be left open.

Broth cultures provide large numbers of bacteria in a small space and are easily transported. **Agar slants** are test tubes containing solid culture media that were left at an angle while the agar solidified. Agar slants, like Petri plates, provide a solid growth surface, but slants are easier to store and transport than Petri plates. Agar is allowed to solidify in the bottom of a test tube to make an **agar deep.** Deeps are often used to grow bacteria that require less oxygen than is present on the surface of the medium. Semisolid agar deeps containing 0.5–0.7% agar instead of the usual 1.5% agar can be used to determine whether a bacterium is motile. Motile bacteria will move away from the point of inoculation, giving the appearance of an inverted Christmas tree.

Aseptic transfer and inoculation are usually performed with a sterile, heat-resistant, noncorroding Nichrome wire attached to an insulated handle. When the end of the wire is bent into a loop, it is called an **inoculating loop;** when straight, it is an **inoculating needle** (Figure 10.1). For special purposes, cultures may also be transferred with sterile cotton swabs, pipettes, glass rods, or syringes. These techniques will be introduced in later exercises.

Whether to use an inoculating loop or a needle depends on the form of the medium; after completing this exercise, you will be able to decide which instrument is to be used.

Materials

Tubes containing nutrient broth (3)

Tubes containing nutrient agar slants (3)

Tubes containing nutrient semisolid agar deeps (3)

Inoculating loop

Inoculating needle

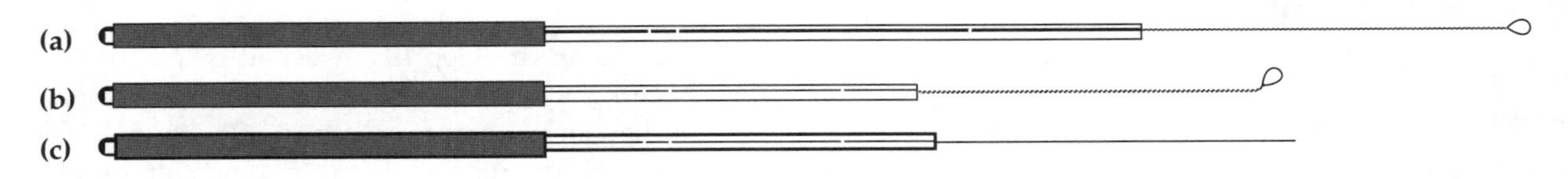

Figure 10.1

(a) An inoculating loop. **(b)** A variation of the inoculating loop in which the loop is bent at a 45° angle. **(c)** An inoculating needle.

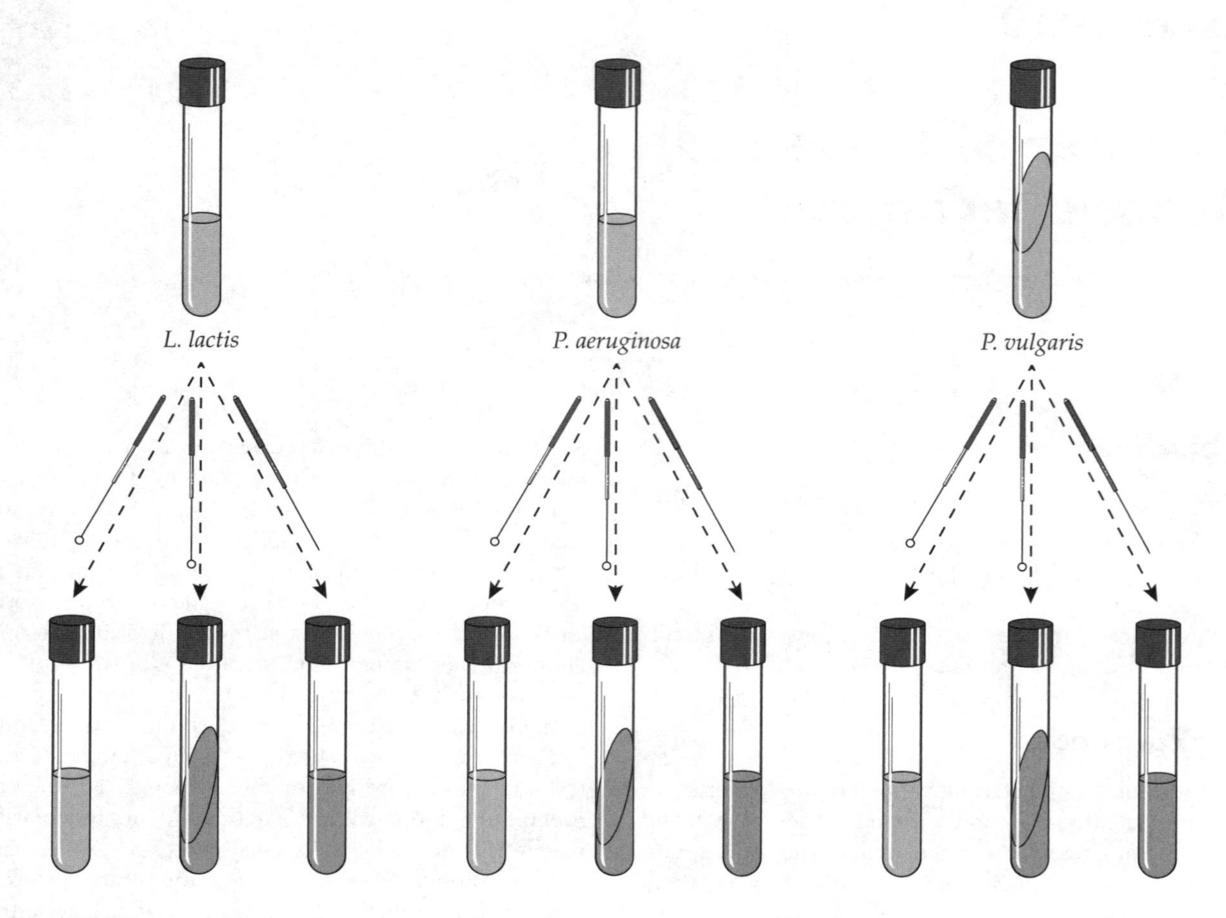

Figure 10.2
Inoculate one tube of each medium with each of the cultures using a loop or needle as indicated. Work with only one culture at a time to avoid contamination.

Test-tube rack

Gram-staining reagents

Cultures

Lactococcus lactis broth

Pseudomonas aeruginosa broth

Proteus vulgaris slant

Techniques Required

Compound light microscopy, Exercise 1

Smear preparation, Exercise 3

Gram staining, Exercise 5

Procedure (Figure 10.2)

1. Work with only one of the bacterial cultures at a time to prevent any mix-ups or cross-contamination. Label one tube of each medium with the name of the first culture, your name, the date, and your lab section. Inoculate each tube as described and then work with the next culture. Begin with one of the broth cultures, and gently tap the bottom of it to resuspend the sediment.
2. To inoculate nutrient broth, hold the inoculating loop in your dominant hand and one of the broth cultures of bacteria in the other hand.
   a. Sterilize the loop by holding the wire in a Bunsen burner flame (Figure 10.3a). Heat to redness. Why? ______________________
   b. Holding the loop like a pencil or paintbrush, curl the little finger of the same hand around the cap of the broth culture. Gently pull the cap off the tube while turning the culture tube (Figure 10.3b). If cotton stoppers are used, simply grasp the stopper with your finger. Do not set down the cap. Why not? ______________
   ____

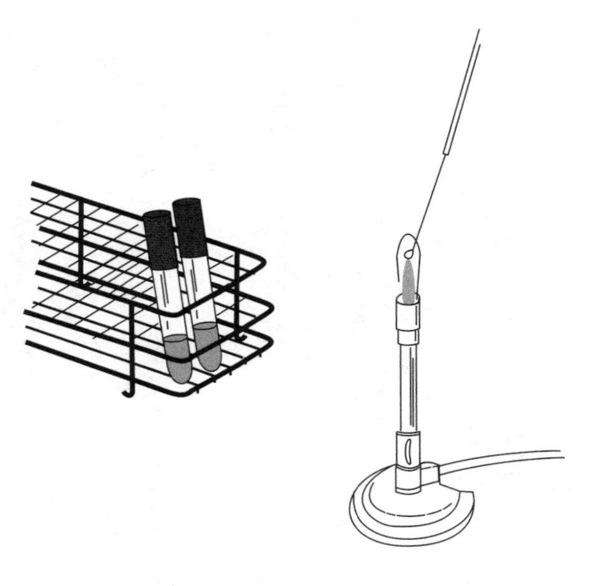

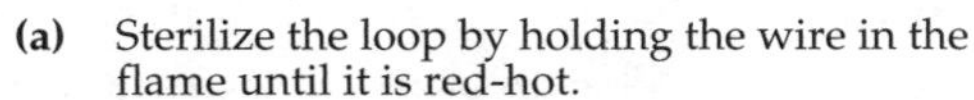

(a) Sterilize the loop by holding the wire in the flame until it is red-hot.

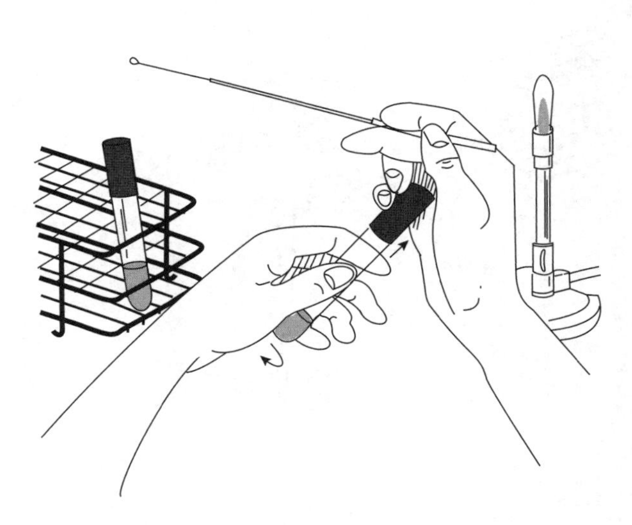

(b) While holding the sterile loop and the bacterial culture, remove the cap as shown.

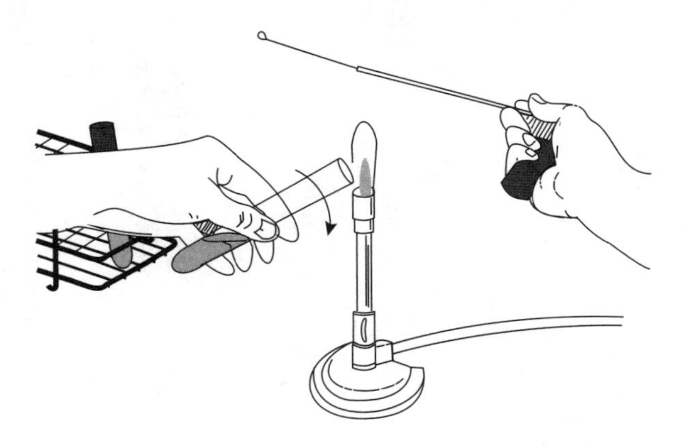

(c) Briefly pass the mouth of the tube through the flame three times before inserting the loop for an inoculum.

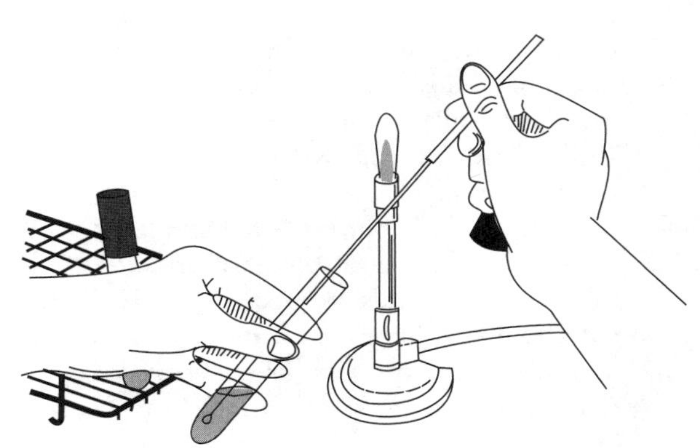

(d) Get a loopful of culture, heat the mouth of the tube, and replace the cap.

Figure 10.3

Inoculating procedures.

c. Holding the tube at an angle, pass the mouth of the tube through the flame three times (Figure 10.3c). What is the purpose of flaming the mouth of the tube? ____________
Always hold culture tubes and uninoculated tubes at about a 20° angle to minimize the amount of dust that could fall into them. Do not tip the tube too far, or the liquid will leak out around the loose-fitting cap. Do not let the top edge of the tube touch anything.

d. Immerse the sterilized, cooled loop into the broth culture to obtain a loopful of culture (Figure 10.3d). Why must the loop be cooled first?
____
Remove the loop, and while holding the loop, flame the mouth of the tube and recap it by turning the tube into the cap. Place the tube in your test-tube rack.

e. Remove the cap from a tube of sterile nutrient broth as previously described, and flame the

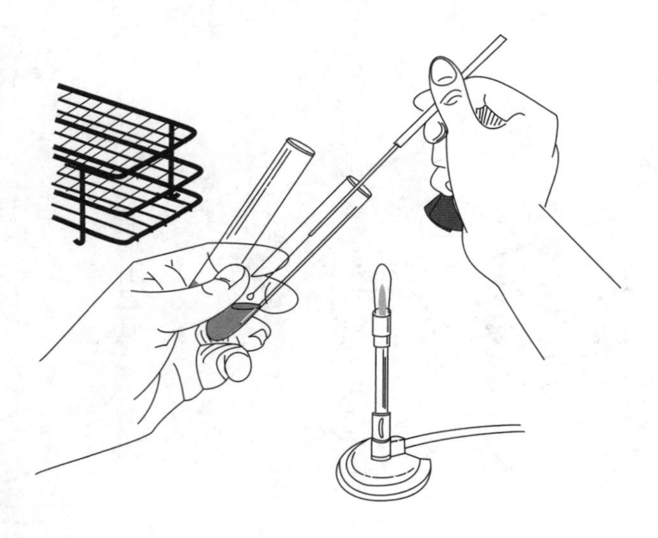

Figure 10.4

Experienced laboratory technicians can transfer cultures aseptically while holding multiple test tubes.

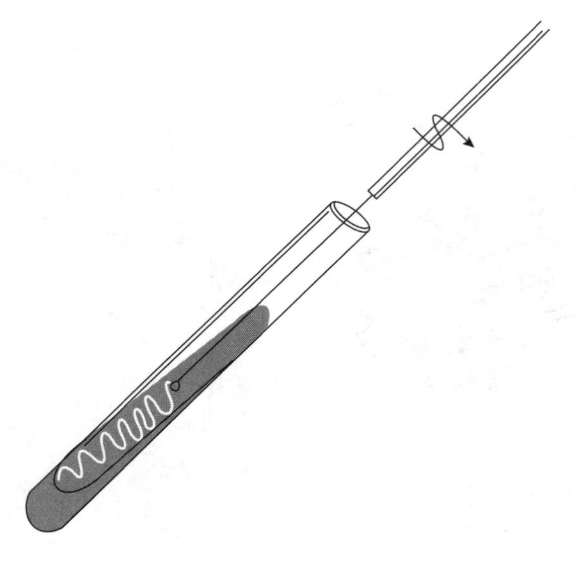

Figure 10.5

Inoculate a slant by streaking the loop back and forth across the surface of the agar.

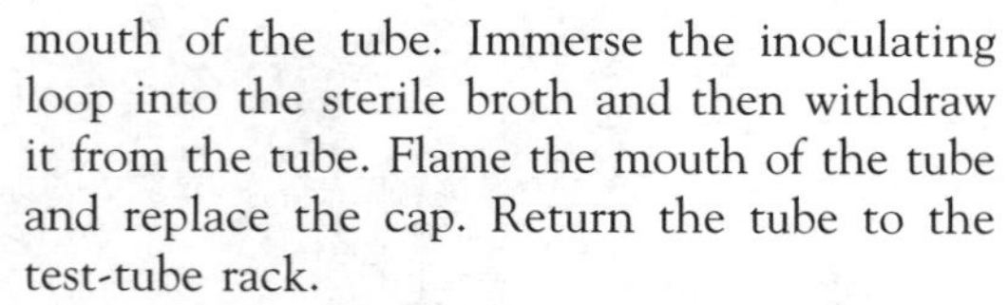

mouth of the tube. Immerse the inoculating loop into the sterile broth and then withdraw it from the tube. Flame the mouth of the tube and replace the cap. Return the tube to the test-tube rack.

f. Reflame the loop until it is red and let it cool. Some microbiologists prefer to hold several tubes in their hands at once (Figure 10.4). *Do not* attempt holding and transferring between multiple tubes until you have mastered aseptic transfer techniques.

3. Obtain a nutrient agar slant. Repeat steps 2a–2d, and inoculate the slant by moving the loop gently across the agar surface from the bottom of the slant to the top, being careful not to gouge the agar (Figure 10.5). Flame the mouth of the tube and replace the cap. Flame your loop and let it cool.
4. Obtain a nutrient semisolid agar deep, and using your inoculating *needle*, repeat steps 2a–2d. Inoculate the semisolid agar deep by plunging the needle straight down the middle of the deep and then pulling it out through the same stab, as shown in Figure 10.6. Flame the mouth of the tube and replace the cap. Flame your needle and let it cool.
5. Using the other broth culture, label one tube of each medium as described in step 1; inoculate a broth culture, agar slant, and semisolid agar deep, as described in steps 2, 3, and 4, using your inoculating loop and needle.
6. Label one tube of each medium with *Proteus vulgaris* as described in step 1. To transfer *Proteus vulgaris*, flame your loop and allow it to cool. Flame the mouth of the tube, and use your inoculating loop to carefully scrape a small amount of the culture off of the agar. Do not gouge the agar. Flame the mouth of the tube and replace the cap. Inoculate a broth and a slant as described in steps 2 and 3. Inoculate a semisolid agar deep with an inoculating needle. Carefully scrape a small amount of

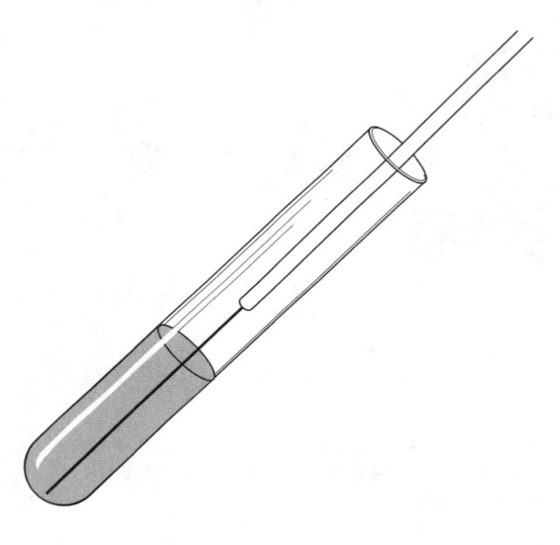

Figure 10.6

Inoculate an agar deep by stabbing into the agar with a needle.

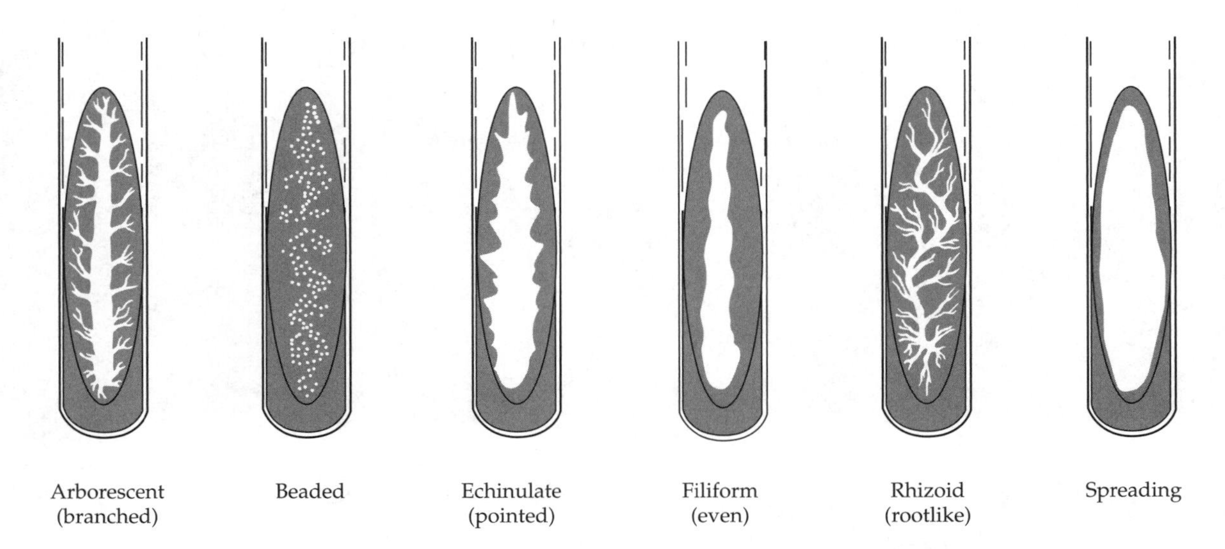

Figure 10.7
Patterns of growth on agar slants.

culture from the slant and inoculate the deep as described in step 4.

7. Incubate all tubes at 35°C until the next period.
8. Record the appearance of each culture, referring to Figure 10.7.
9. Make a smear of the *Lactococcus* broth culture and the *Lactococcus* slant culture. Perform a gram stain on both smears and compare them.

Exercise 10 **LABORATORY REPORT**

Transfer of Bacteria: Aseptic Technique

NAME ______________________

DATE ______________________

LAB SECTION ______________________

Purpose

____

Data

Nutrient Broth

Describe the nutrient broth cultures.

Bacterium	Is it turbid?	Is flocculent, pellicle, or sediment present?	Pigment
Lactococcus lactis			
Pseudomonas aeruginosa			
Proteus vulgaris			

Nutrient Agar Slant

Sketch the appearance of each culture. Note any pigmentation.

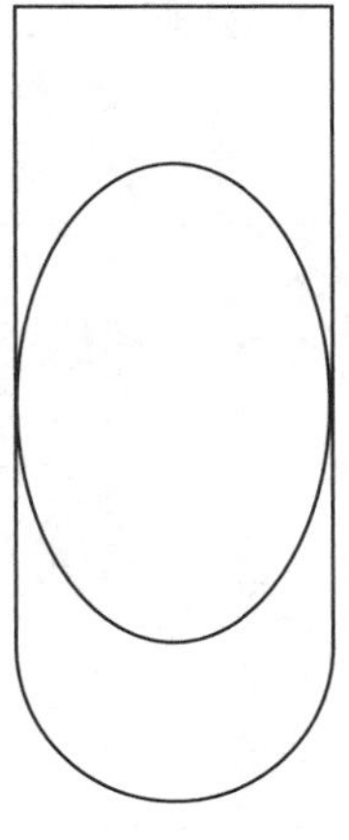

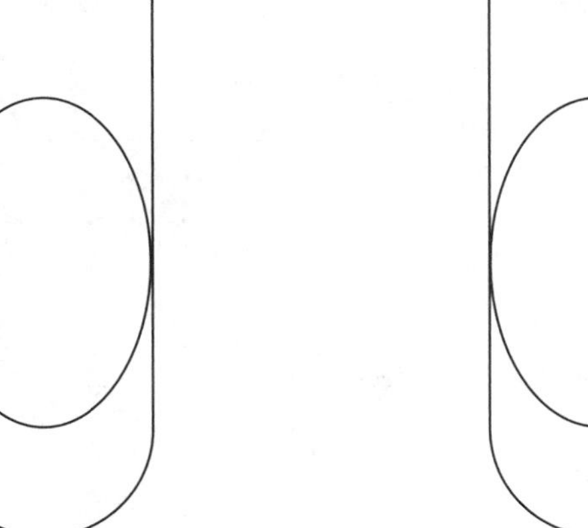

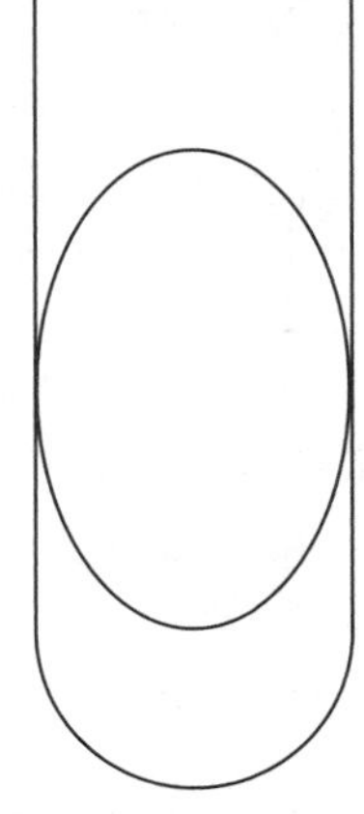

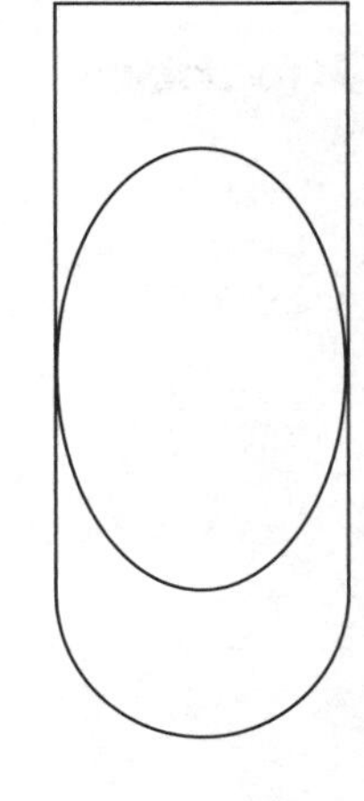

Bacteria:	*Lactococcus lactis*	*Pseudomonas aeruginosa*	*Proteus vulgaris*
Pattern of growth:	______	______	______

Nutrient Semisolid Agar Deep

Show the location of bacterial growth and note any pigment formation.

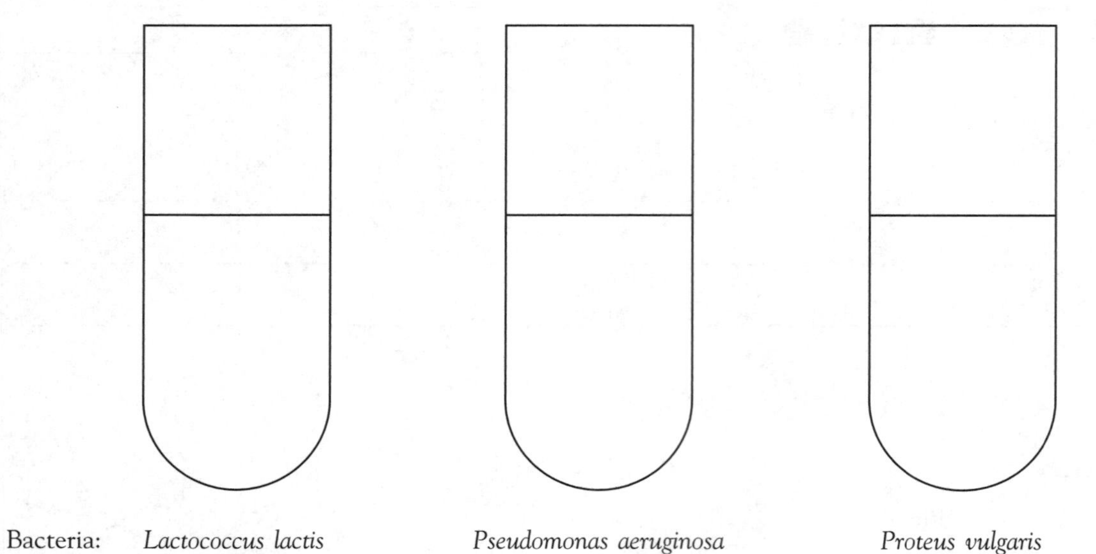

Bacteria: *Lactococcus lactis* *Pseudomonas aeruginosa* *Proteus vulgaris*

Comparison of Broth and Slant Cultures

	Lactococcus lactis	
	Broth Culture	Slant Culture
Gram stain		
Morphology		
Arrangement		

Questions

1. Did growth occur at different levels in the agar deep? ______________________________

____

____

2. Were any of the bacteria growing in the semisolid agar deeps motile? __________ Explain.

____

____

____

3. What other methods can be used to determine motility? ______________________________

____

4. What is the primary use of slants? ______________________________

Of deeps? ______________________________

Of broths? ______________________________

5. Can you determine whether a broth culture is pure (all one species) by visually inspecting it without a microscope? ______________ An agar deep culture? ______________

An agar slant culture? ______________

6. When is a loop preferable for transferring bacteria? Use an illustration from your results to answer. When is a needle preferable? ______________________________

____

____

7. What is the purpose of flaming the loop before use? After use? ______________________________

____

____

8. Why must the loop be cool before touching it to a culture? Should you set it down to let it cool? How do you determine when it is cool? ______________________________

____

____

9. Why is aseptic technique important? ______________________________

____

____

Critical Thinking

1. Why was the arrangement of *Lactococcus* from the broth culture different than that from the slant culture in the second period?

2. What evolutionary advantage would there be to the formation of a pellicle in a liquid medium by a bacterium?

3. How can you tell that the media provided for this exercise were sterile?

Exercise 11

Isolation of Bacteria by Dilution Techniques

Experience is the father of wisdom, and memory the mother.

THOMAS FULLER

Objectives

After completing this exercise, you should be able to:

1. Isolate bacteria by using the streak plate and pour plate techniques.
2. Prepare and maintain a pure culture.

Background

In nature most microbes are found growing in environments that contain many different organisms. Unfortunately, mixed cultures are of little use in studying microorganisms because of the difficulty they present in determining which organism is responsible for any observed activity. A **pure culture,** one containing a single kind of microbe, is required in order to study concepts such as growth characteristics, pathogenicity, metabolism, and antibiotic susceptibility. Because bacteria are too small to separate directly without sophisticated micromanipulation equipment, indirect methods of separation must be used.

In the 1870s, Joseph Lister attempted to obtain pure cultures by performing serial dilutions until each of his containers theoretically contained one bacterium. However, success was very limited, and **contamination,** the presence of unwanted microorganisms, was common. In 1880, Robert Koch prepared solid media, after which microbiologists could separate bacteria by dilution and trap them on the solid media. An isolated bacterium grows into a visible colony that consists of one kind of bacterium.

Currently there are three dilution methods commonly used for the isolation of bacteria: the streak plate, the spread plate, and the pour plate. In the **streak plate technique,** a loop is used to streak the mixed sample many times over the surface of a solid culture medium in a Petri plate. Theoretically, the process of streaking the loop repeatedly over the agar surface causes the bacteria to fall off the loop one by one and ultimately to be distributed over the agar surface, where each cell develops into a colony. The streak plate is the most common isolation technique in use today.

The spread plate and pour plate are quantitative techniques that allow determination of the number of bacteria in a sample. In the **spread plate technique,** a small amount of a previously diluted specimen is spread over the surface of a solid medium using a spreading rod.

In the **pour plate technique,** a small amount of diluted sample is mixed with melted agar and poured into empty, sterile Petri dishes. After incubation, bacterial growth is visible as colonies in and on the agar of a pour plate. To determine the number of bacteria in the original sample, a plate with between 25 and 250 colonies is selected. Fewer than 25 colonies is inaccurate because a single contaminant causes at least a 4% error. A plate with greater than 250 colonies is difficult to count. The number of bacteria in the original sample is calculated using the following equation:

$$\text{Colony-forming units per ml} = \frac{\text{Number of colonies}}{\text{Dilution}^* \times \text{Amount plated}}$$

Materials

Petri plates containing nutrient agar (2)

Tubes containing melted nutrient agar (3)

Sterile Petri dishes (3)

250-ml beaker

Sterile 1-ml pipettes (3)

Propipette or pipette bulb

Nutrient agar slant (second period)

*In this exercise, 1 ml of sample is put into each plate. Dilution refers to the dilution of the sample (Appendix B). For example, if 37 colonies were present on the 1:8000 plate, the calculation would be as follows:

$$\text{Colony-forming units per ml} = \frac{37}{1{:}8000 \times 1} = 37 \times 8000 = 296{,}000 = 2.96 \times 10^5$$

Cultures

Mixed broth culture of bacteria

Turbid nutrient broth from Exercise 9

Techniques Required

Compound light microscopy, Exercise 1

Aseptic technique, Exercise 10

Pipetting, Appendix A

Serial dilution techniques, Appendix B

Procedure

Streak Plate

1. Label the bottoms of two nutrient agar plates to correspond to the two broth cultures: mixed culture and turbid broth.
2. Flame the inoculating loop to redness, allow it to cool, and aseptically obtain a loopful of one broth culture.
3. The streaking procedure may be done with the Petri plate on the table (Figure 11.1a) or held in your hand (Figure 11.1b).
   a. To streak a plate (Figure 11.2), lift one edge of the Petri plate cover, and streak the first sector by making as many streaks as possible without overlapping previous streaks. Do not gouge the agar while streaking the plate. Hold the loop as you would hold a pencil or paintbrush, and gently touch the surface of the agar.
   b. Flame your loop and let it cool. Turn the plate so the next sector is on top. Streak through one area of the first sector, and then streak a few times away from the first sector.
   c. Flame your loop, turn the plate again, and streak through one area of the second sector. Then streak the third sector.
   d. Flame your loop, streak through one area of the third sector, and then streak the remaining area of the agar surface, being careful not to make additional contact with any streaks in the previous sections. Flame your loop before setting it down. Why? ____________________

      ____
4. Streak two plates: one of the mixed culture provided and one of the turbid broth from an environmental sample. Label each plate on the bottom with your name and lab section, the date, and the source of the inoculum.
5. Incubate the plates in an inverted position in the 35°C incubator (or at room temperature, depending on the inoculum) until discrete, isolated colonies develop (usually 24 to 48 hours). Why inverted?

   ____

   ____
6. After incubation, record your results. Use proper terms to describe the colonies. (Refer to Color Plates VI.2 and VI.3, and Figure 9.2.)
7. Prepare a subculture of one colony. Sterilize your needle by flaming it. Let it cool. Why use a needle instead of a loop? ____________________
   To subculture, touch the center of a small isolated colony located on a streak line, and then aseptically streak a sterile nutrient agar slant. How can you tell whether you touched only one colony and whether you have a pure culture? ____________________

   ____
8. Incubate the slant at 35°C until good growth is observed. Describe the growth pattern (Figure 10.7).

Pour Plate (Figure 11.3)

1. Label the bottoms of three empty, sterile Petri dishes with your name and lab section and the date. Label one dish "1:20," another "1:400," and the third one "1:8000." Place the labeled dishes on your workbench right-side up.
2. Fill a beaker with hot (45–50°C) water (about 3–6 cm deep), and place three tubes of melted nutrient agar in the beaker. Each tube contains 19 ml of nutrient agar.
3. Select a mixed broth culture.
4. Remove a pipette, attach a bulb, and aseptically transfer 1 ml of the broth to a tube of melted agar. Mix well, as shown in Figure 11.4. Using a different pipette, transfer 1 ml from this tube to a second tube. Work quickly so the agar does not solidify in the pipette. Aseptically pour the contents of the first tube into Petri dish 1:20. Discard the pipettes in the container of disinfectant.
5. Mix the second tube. With the third pipette, aseptically transfer 1 ml to the third tube. Pour the contents of the second tube into dish 1:400. Mix the third tube and pour its contents into the remaining dish, 1:8000.
6. Discard the tubes properly. Let the agar harden in the plates, and then incubate them at 35°C in an inverted position until growth is seen. *Suggestion:* When incubating multiple plates, use a rubber band to hold them together.
7. After incubation, count the number of colonies on the plates. Remember that more than 250 is too numerous to count, and less than 25 is too few to count.

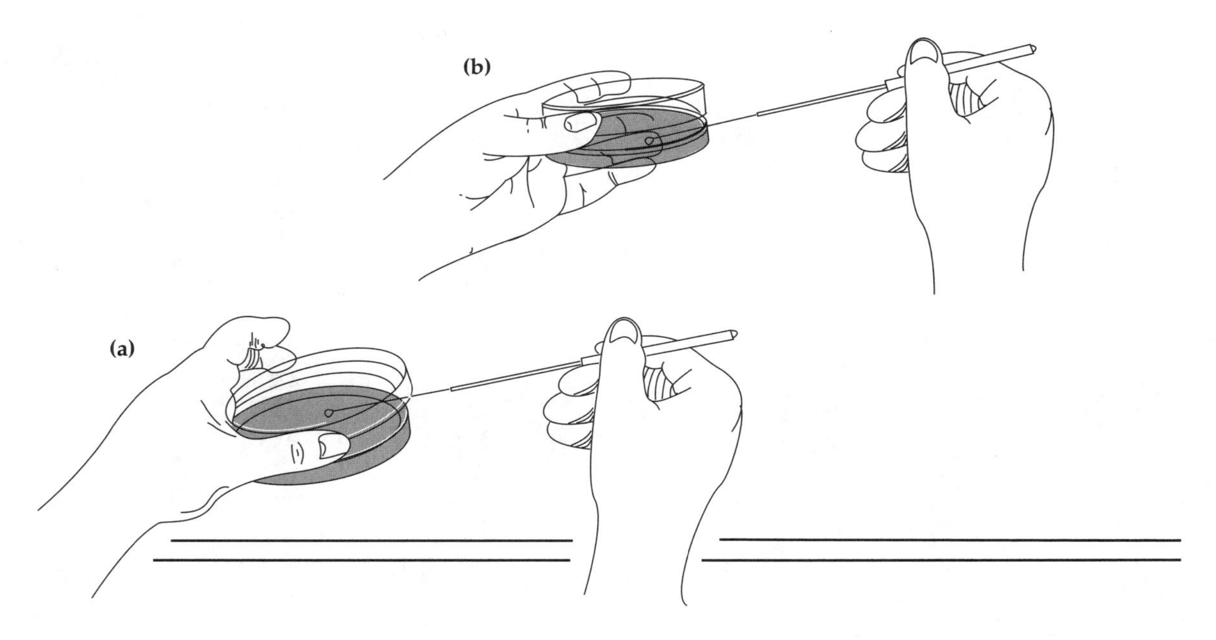

Figure 11.1

Inoculation of a solid medium in a Petri plate. Lift one edge of the cover while the plate **(a)** rests on the table or **(b)** is held.

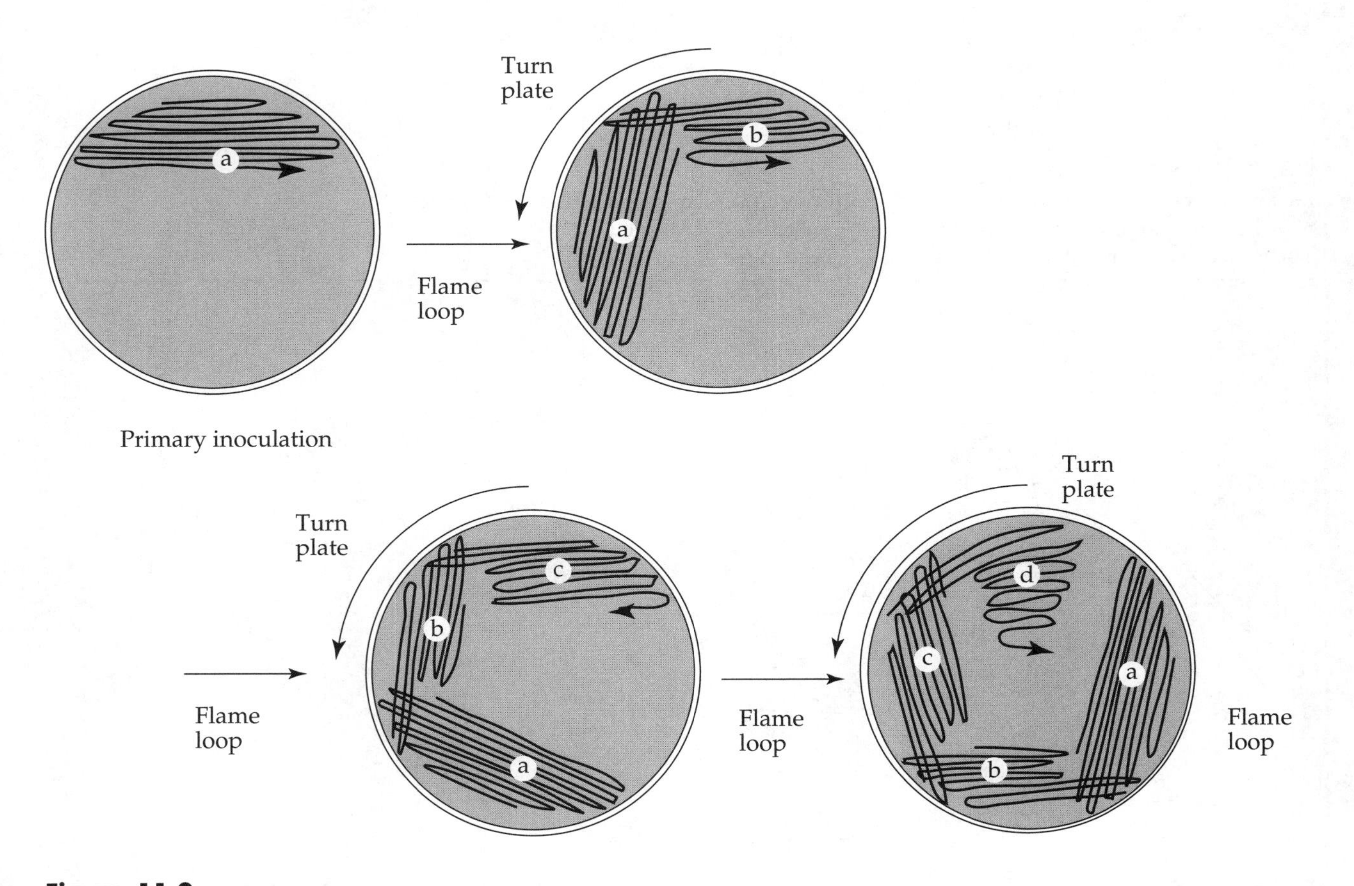

Figure 11.2

Streak plate technique for pure culture isolation of bacteria. The direction of streaking is indicated by the arrows. Between each section, sterilize the loop and reinoculate with a fraction of the bacteria by going back across the previous section.

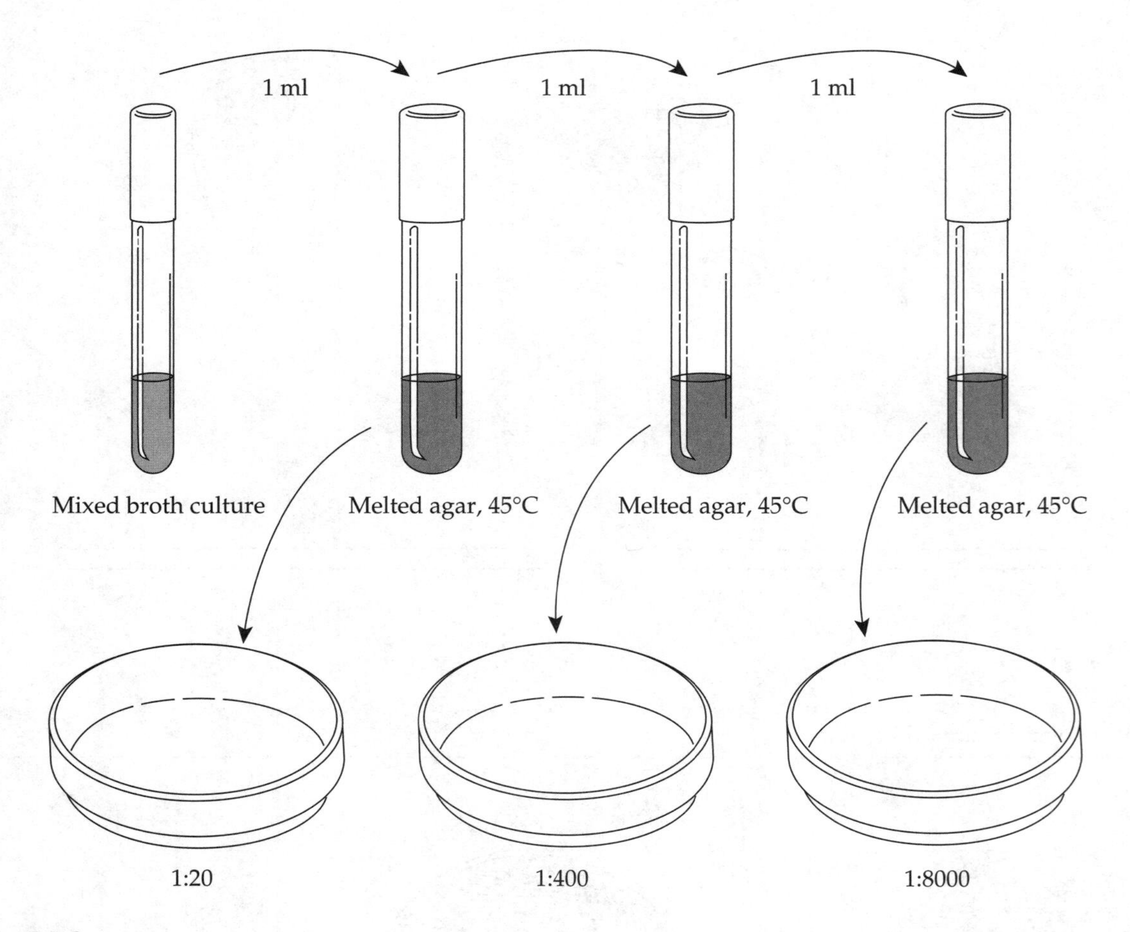

Figure 11.3

Pour plate technique. Bacteria are diluted through a series of tubes containing 19 ml of melted nutrient agar. The agar and bacteria are poured into sterile Petri dishes. The bacteria will form colonies where they are trapped in the agar.

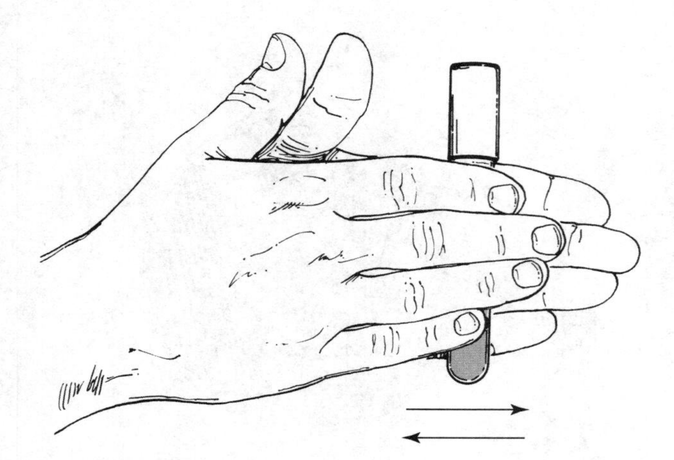

Figure 11.4

Mix the inoculum in a tube of melted agar by rolling the tube between your hands.

Exercise 11

LABORATORY REPORT

Isolation of Bacteria by Dilution Techniques

NAME ______________________

DATE ______________________

LAB SECTION ______________________

Purpose

____

____

____

Data

Streak Plate

Sketch the appearance of the streak plates.

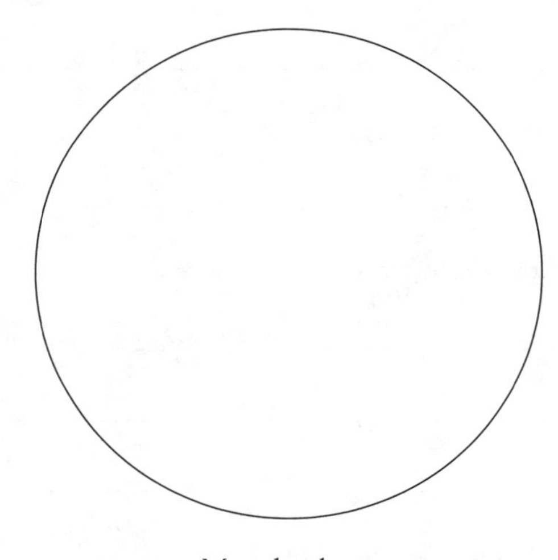

Mixed culture

Turbid broth

Fill in the following table using colonies from the most isolated streak areas.

Culture	Colony Description (Each different-appearing colony should be described.)				
	Diameter	Appearance	Margin	Elevation	Color
Mixed culture					
Turbid broth					

Pour Plate

Dilution	Number of Colonies
1:20	
1:400	
1:8000	

Calculate the number of colony-forming units per milliliter in the mixed culture. Which plate will you use for your calculations? ____________
Show your calculations.

________ Colony-forming units per ml

Subculture

Describe the growth on your slant. ____________

Do you appear to have a pure culture? ____________

Questions

1. How many different bacteria were in the mixed culture? ____________ How many in the turbid broth? ____________ How can you tell? ____________

   ____

2. How do the colonies on the surface of the pour plate differ from those suspended in the agar?

   ____

   ____

3. What is a contaminant? ____________

4. How would you determine whether a colony was a contaminant on a streak plate? ____________

   ____

   On a pour plate? ____________

   ____

5. What would happen if the plates were incubated a week longer? ____________

   A month? ____________

6. How could your streak plate technique be improved? ____________

   ____

   ____

Critical Thinking

1. Could some bacteria grow on the streak plate and not be seen using the pour plate technique? Explain.

2. What is a disadvantage of the streak plate technique? Of the pour plate technique?

3. Will the isolated colonies always be in the fourth sector on the streak plate?

Exercise 12

Special Media for Isolating Bacteria

Objectives

After completing this exercise, you should be able to:

1. Differentiate between selective and differential media.
2. Provide an application for enrichment and selective media.

Background

One of the major limitations of dilution techniques used to isolate bacteria is that organisms present in limited amounts may be diluted out on plates filled with dominant bacteria. For example, if the culture to be isolated has 1 million of bacterium A and only 1 of bacterium B, bacterium B will probably be limited to the first sector in a streak plate. To help isolate organisms found in the minority, various enrichment and selective culturing methods are available that either enhance the growth of some organisms or inhibit the growth of other organisms. **Selective media** contain chemicals that prevent the growth of unwanted bacteria without inhibiting the growth of the desired organism. **Enrichment media,** which are usually liquid media, contain chemicals that enhance the growth of desired bacteria. Other bacteria will grow, but the growth of the desired bacteria will be increased.

Another category of media useful in identifying bacteria is **differential media.** These media contain various nutrients that allow the investigator to distinguish one bacterium from another by how they metabolize or change the media with a waste product.

Because multiple methods and multiple media exist, you must be able to match the correct procedure to the desired microbe. For example, if bacterium B is salt-tolerant, a high concentration (>5%) of salt could be added to the culture medium. Physical conditions can also be used to select for a bacterium. If bacterium B is heat-resistant, the specimen could be heated before isolation. Dyes such as phenol red, eosin, or methylene blue are sometimes included in differential media. Products of bacterial metabolism can react with these dyes to produce a color change in the medium. You will study bacterial metabolism in the exercises in Part 4. The dyes (eosin and methylene blue) in eosin methylene blue (EMB) agar are also selective. These dyes inhibit the growth of some bacteria. Three culture media will be compared in this exercise (Table 12.1).

Table 12.1

Major Chemical Components of Media Used in This Exercise

	Nutrient Agar	Mannitol Salt Agar	EMB Agar
Peptone	0.5%	1.0%	1.0%
NaCl	0.8%	7.5%	
Agar	1.5%	1.5%	1.5%
Mannitol		1.0%	
Lactose, sucrose			0.5% each
Eosin			0.04%
Methylene blue			0.0065%
Phenol red		0.025%	

Materials

Petri plates containing nutrient agar (2)

Petri plates containing mannitol salt agar (2)

Petri plates containing EMB agar (2)

Gram-staining reagents

Cultures

Escherichia coli

Micrococcus luteus

Pseudomonas aeruginosa

Staphylococcus epidermidis

Unknown mixed culture

Techniques Required

Compound light microscopy, Exercise 1

Smear preparation, Exercise 3

Gram staining, Exercise 5

Aseptic technique, Exercise 10

Inoculating loop technique, Exercise 10

Streak plate procedure, Exercise 11

Procedure

1. Using a marker, divide one nutrient agar plate into four sections by labeling the bottom. Repeat to mark one mannitol salt plate and one EMB plate. Label one quadrant on each plate for each culture.

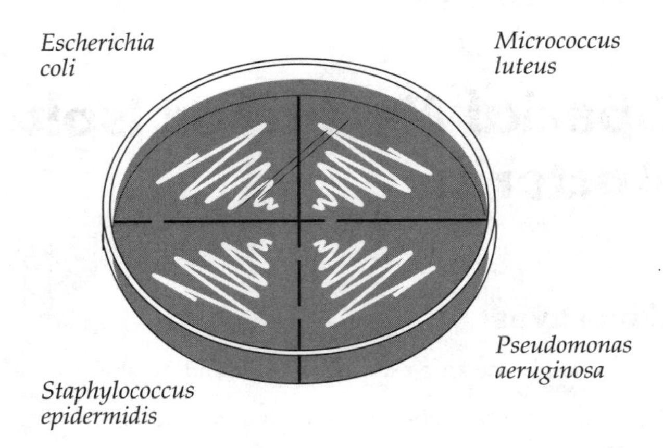

Figure 12.1

Divide a Petri plate into four sections by drawing lines on the bottom of the plate. Inoculate each section by streaking it with an inoculating loop.

2. Streak each culture on the agar, as shown in Figure 12.1.
3. Label the remaining nutrient agar, mannitol salt, and EMB plates with the number of your unknown. The unknown contains two different bacteria.
4. Streak your "unknown" onto the appropriate plates using the streaking technique shown in Figure 11.2.
5. Incubate the plates in an inverted position at 35°C. Record the results after 24–48 hours. (See Color Plates V.1, XII.1, XII.2, and XII.3.)
6. Perform a Gram stain on one colony of each different organism. Why don't you have to perform a Gram stain on each organism from every medium?

   ____

   ____

Exercise 12

Special Media for Isolating Bacteria

LABORATORY REPORT

NAME ______________________

DATE ______________________

LAB SECTION ______________________

Purpose

____

____

____

Data

Organism	Nutrient Agar		Mannitol Salt Agar		EMB Agar		Gram-Stain Results
	Growth: +/–	Appearance	Growth: +/–	Appearance	Growth: +/–	Appearance	
E. coli							
M. luteus							
P. aeruginosa							
S. epidermidis							
Unknown							

Questions

1. What two organisms are in your mixed culture? ______________________

   Could you identify them from the Gram stain? ______________________

   How did you identify them? ______________________

   ____

2. How did the results observed on the mannitol salt and EMB correlate to the Gram reaction of the bacteria?

   ____

   ____

3. Which medium is selective? ______________________

4. What is the purpose of peptone in the media? ______________________________

____

Of agar in the media? ______________________________

____

Critical Thinking

1. What ingredient makes mannitol salt selective? ______________________________

2. Fill in the blanks in this diagram to make a key to these bacteria: *Escherichia coli*, *Micrococcus luteus*, *Pseudomonas aeruginosa*, *Staphylococcus epidermidis*.

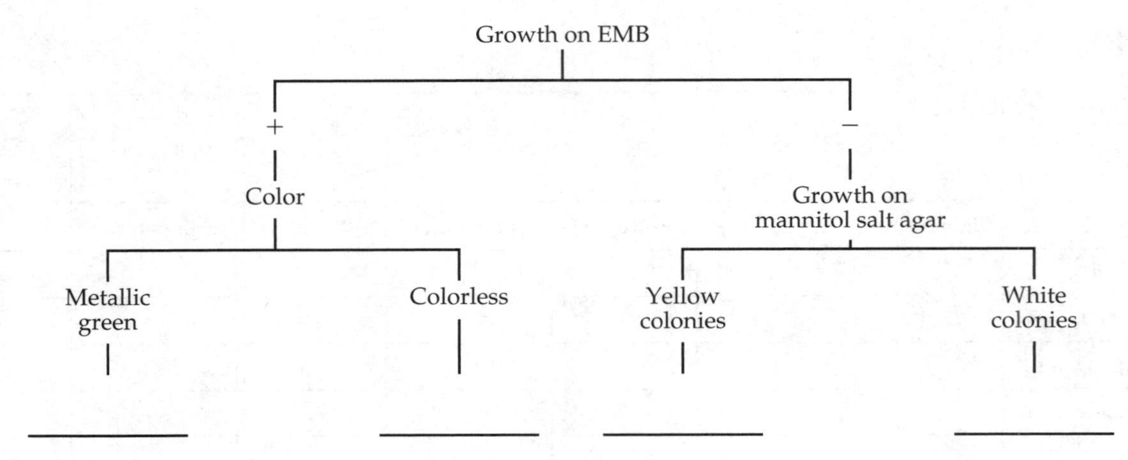

Circle or highlight the gram-positive bacteria in one color; use a different color to circle or highlight the gram-negative bacteria.

3. Design an enrichment medium to isolate a detergent-degrading bacterium that is found in soil.

Part Four Microbial Metabolism

EXERCISES

Van Leeuwenhoek saw microorganisms in wine, and Pasteur demonstrated that the microbes were living organisms. In 1872 Pasteur wrote:

> *It is impossible that the organic matter of the newly formed ferments [microorganisms] contain a single carbon atom which has not been derived from the fermented substance.**

The chemical reactions observed by Pasteur and the chemical reactions that occur within all living organisms are referred to as **metabolism.** Metabolic processes involve **enzymes,** which are proteins that catalyze biological reactions. The majority of enzymes function inside a cell—that is, they are **endoenzymes.** Many bacteria make some enzymes, called **exoenzymes,** that are released from the cell to catalyze reactions outside of the cell (see the illustration on the next page).

*Quoted in H. A. Lechevalier and M. Solotorovsky. *Three Centuries of Microbiology*. New York: Dover Publications, 1974, p. 26.

Some bacteria use particular metabolic pathways in the presence of oxygen (**aerobic**) and other pathways in the absence of oxygen (**anaerobic**). Pasteur continued:

> *I have demonstrated that since this ferment [microorganism] survived in the presence of some free oxygen, it lost its fermentative abilities in proportion to the concentration of this gas.*

Because many bacteria share the same colony and cell morphology, additional factors, such as metabolism, are used to identify them. Moreover, studying the metabolic activities of microbes helps us understand their role in ecology.

The first five exercises in Part Four introduce concepts in microbial metabolism and laboratory tests used to detect various metabolic activities. In Exercise 18, unknown bacteria will be identified on the basis of metabolic characteristics by using commercial rapid identification methods.

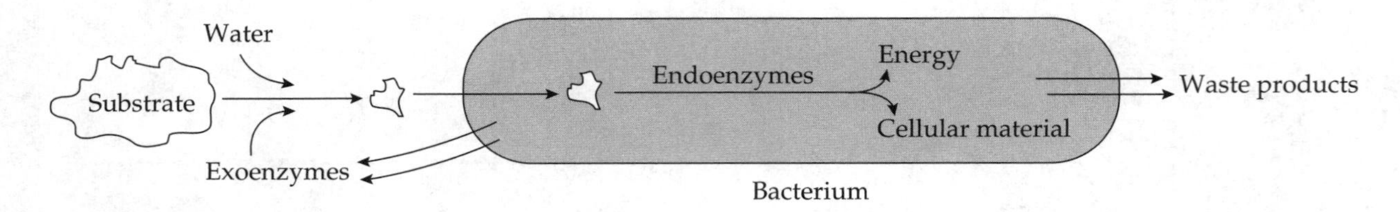

Large molecules are broken down outside of a cell by exoenzymes. Smaller molecules released by this reaction are taken into the cell and further degraded by endoenzymes.

Exercise 13

Carbohydrate Catabolism

The men who try to do something and fail are infinitely better than those who try nothing and succeed.

LLOYD JONES

Objectives

After completing this exercise, you should be able to:

1. Define the following terms: carbohydrate, catabolism, and hydrolytic enzymes.
2. Differentiate between oxidative and fermentative catabolism.
3. Perform and interpret microbial starch hydrolysis and OF tests.

Background

Chemical reactions that release energy from the decomposition of complex organic molecules are referred to as **catabolism.** Most bacteria catabolize carbohydrates for carbon and energy. **Carbohydrates** are organic molecules that contain carbon, hydrogen, and oxygen in the ratio $(CH_2O)_n$. Carbohydrates can be classified based on size: monosaccharides, oligosaccharides, and polysaccharides. **Monosaccharides** are simple sugars containing from three to seven carbon atoms. **Oligosaccharides** are composed of two to about 20 monosaccharide molecules; *disaccharides* are the most common oligosaccharides. **Polysaccharides** consist of eight or more monosaccharide molecules.

Exoenzymes are mainly **hydrolytic enzymes** that leave the cell and break down, by the addition of water, large substrates into smaller components, which can then be transported into the cell. Amylase hydrolyzes the polysaccharide starch into smaller carbohydrates. Glucose, a monosaccharide, can be released by hydrolysis (Figure 13.1). In the laboratory, the presence of an exoenzyme is determined by looking for a change in the substrate outside of a bacterial colony.

Glucose can enter a cell and be catabolized; some bacteria, using endoenzymes, catabolize glucose oxidatively, producing carbon dioxide and water. **Oxidative catabolism** requires the presence of molecular oxygen (O_2). Most bacteria, however, can ferment glucose without using oxygen. **Fermentative catabolism** does not require oxygen but may occur in its presence. The metabolic end-products of fermentation are small organic molecules, usually organic acids. Some bacteria produce gases from the fermentation of carbohydrates.

Whether an organism is oxidative or fermentative can be determined by using Rudolph Hugh and Einar Leifson's OF basal media with the desired carbohydrate added. **OF medium** is a nutrient semisolid agar deep containing a *high* concentration of carbohydrate and a *low* concentration of peptone. The peptone will support the growth bacteria that don't use the carbohydrate. Two tubes are used: one open to the air and one sealed to keep air out. OF medium contains the indicator bromthymol blue, which turns yellow in the presence of acids, indicating catabolism of the carbohydrate. Alkaline conditions, due to the use of peptone and not the carbohydrate, are indicated by a dark blue color due to ammonia production. If the carbohydrate is metabolized in both tubes, fermentation has occurred. An organism that can only use the carbohydrate under aerobic conditions will produce acid in the open tube only. Acids are produced as intermediates in respiration, and the indicator will turn yellow in the top of the open tube. This organism is called "oxidative" in an OF test.

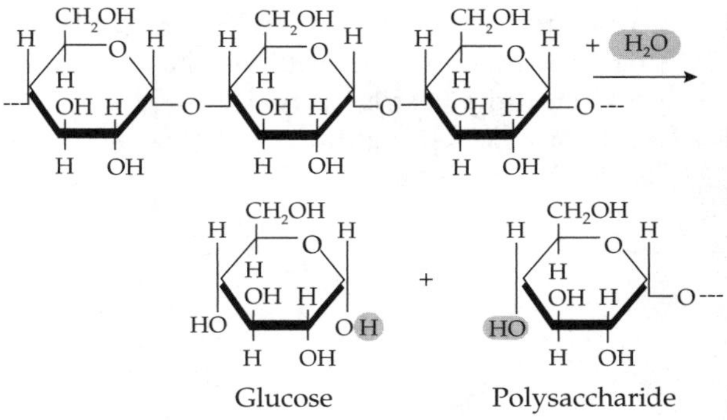

CH_2OH H O H H OH H O CH_2OH H O H H OH H O CH_2OH H O H H OH H O --- + H_2O →
H OH H OH H OH

CH_2OH H O H H OH H HO OH H OH + CH_2OH H O H H OH H O--- HO H OH

Glucose Polysaccharide

Figure 13.1

Starch hydrolysis. A molecule of water is used when starch is hydrolyzed.

Carbohydrate catabolism will be demonstrated in this exercise. These differential tests will be important in identifying bacteria in later exercises.

Materials

Petri plate containing nutrient starch agar

OF-glucose deeps (2)

Mineral oil

Second Period

Gram's iodine

OF-glucose deep

Cultures (as assigned)

Bacillus subtilis

Escherichia coli

Pseudomonas aeruginosa

Alcaligenes faecalis

Techniques Required

Inoculating loop and needle technique, Exercise 10

Aseptic technique, Exercise 10

Procedure

Starch Hydrolysis

1. With a marker, divide the starch agar into three sectors by labeling the bottom of the plate.
2. Using a loop, streak a short single line of *Bacillus*, *Escherichia*, and *Pseudomonas*.
3. Incubate the plate, inverted, at 35°C for 24 hours. After growth occurs, the plate may be refrigerated at 5°C until the next lab period.
4. Record any bacterial growth, and then flood the plate with Gram's iodine (Figure 13.2). Areas of starch hydrolysis will appear clear, while unchanged starch will stain dark blue. (See Color Plate V.2.) Record your results.

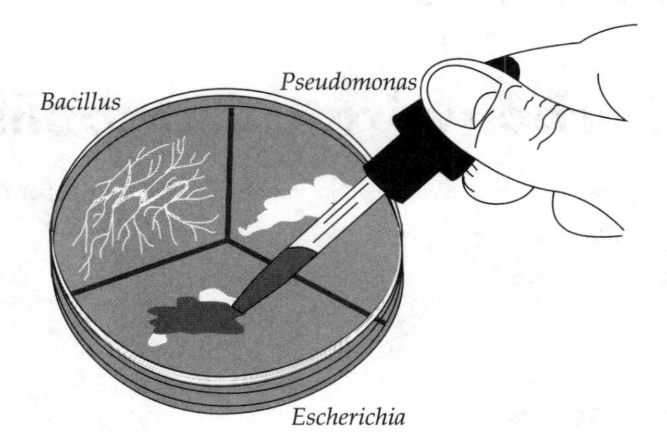

Figure 13.2

Starch hydrolysis test. After incubation, add iodine to the plate to detect the presence of starch.

OF-Glucose

1. Using an inoculating needle, inoculate two tubes of OF-glucose media with the assigned bacterial culture (*Escherichia*, *Pseudomonas*, or *Alcaligenes*).
2. Pour about 5 mm of mineral oil over the medium in one of the tubes. Replace the cap.
3. Incubate both tubes at 35°C until the next lab period.
4. Compare the inoculated tubes and an uninoculated OF-glucose tube. Record the following: the presence of growth, whether glucose was used, and the type of metabolism. Motility can also be ascertained from the OF tubes. How? ____________________

    ____
5. Observe and record the results from the microorganisms you did not culture. (See Color Plate III.1.)

Exercise 13

LABORATORY REPORT

Carbohydrate Catabolism

NAME ______________________

DATE ______________________

LAB SECTION ______________________

Purpose ______________________

____

____

Data

Record your results in the following tables.

Starch Hydrolysis

Organism	Growth	Color of Medium Around Colonies After Addition of Iodine	Starch Hydrolysis: (+) = yes (−) = no
Bacillus subtilis			
Pseudomonas aeruginosa			
Escherichia coli			

OF-Glucose

Color of uninoculated medium: ____________

Organism	Growth		Color		Fermenter (F), Oxidative (O), Neither (−)	Motile
	Aerobic Tube	Anaerobic Tube	Aerobic Tube	Anaerobic Tube		
Pseudomonas aeruginosa						
Alcaligenes faecalis						
Escherichia coli						

Questions

1. Which organism(s) gave a positive test for starch hydrolysis? ______________________

____

How can you tell? ______________________

2. What would be found in the clear area that would not be found in the blue area of a starch agar plate after the addition of iodine? ______________________

____

3. How can you tell amylase is an exoenzyme and not an endoenzyme? ______________________

____

4. How can you tell from OF-glucose medium whether an organism uses glucose aerobically? ______________________

____

Ferments glucose? ______________________

Doesn't use glucose? ______________________

5. If an organism grows in the OF-glucose medium that is exposed to air, is the organism oxidative or fermentative? Explain. ______________________

____

6. Why is it important to first determine whether growth occurred in a differential medium, such as starch agar, before examining the plate for starch hydrolysis? ______________________

____

____

Critical Thinking

1. Aerobic organisms degrade glucose, producing carbon dioxide and water. What acid turns the indicator yellow?

2. How can organisms that don't use starch grow on a starch agar plate?

3. If iodine were not available, how would you determine whether starch hydrolysis had occurred?

4. Locate the H^+ and OH^- ions from the water molecule that was split (hydrolyzed) in this reaction, showing esculin hydrolysis.

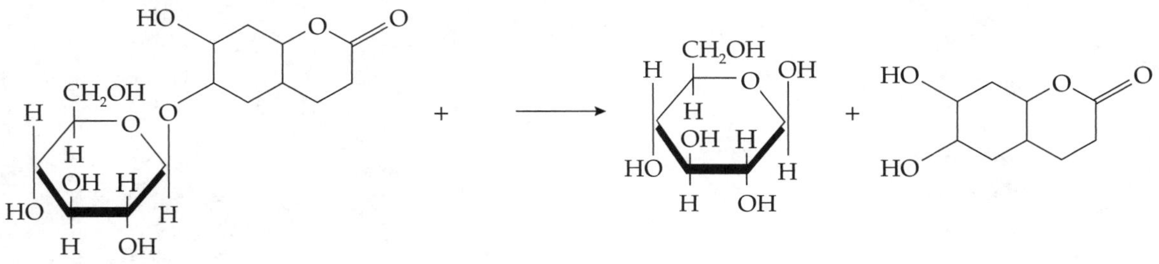

Esculin $\longrightarrow$ Glucose + Esculetin

Exercise 14

Fermentation

Objectives

After completing this exercise, you should be able to:

1. Define fermentation.
2. Perform and interpret carbohydrate fermentation tests.
3. Perform and interpret the MR and V–P tests.
4. Perform and interpret the citrate test.

Background

Once a bacterium has been determined to be fermentative by the OF test, further tests can determine which carbohydrates, in addition to glucose, are fermented; in some instances, the end-products can also be determined. Many carbohydrates—including monosaccharides such as glucose, disaccharides like sucrose, and polysaccharides such as starch—can be fermented. Many bacteria produce organic acids (for example, lactic acid) and hydrogen and carbon dioxide gases from carbohydrate fermentation (Figure 14.1). A **fermentation tube** is used to detect acid and gas production from carbohydrates. The fermentation medium contains peptone, an acid–base indicator (phenol red), an inverted tube to trap gas, and 0.5–1.0% of the desired carbohydrate. In Figure 14.2, the phenol red indicator is red (neutral) in an uninoculated fermentation tube; fermentation that results in acid production will turn the indicator yellow (pH of 6.8 or below). When gas is produced during fermentation, some will be trapped in the inverted, or Durham, tube. Fermentation occurs with or without oxygen present; however, during prolonged incubation periods (greater than 24 hours), many bacteria will begin growing oxidatively on the peptone after exhausting the carbohydrate supplied, causing

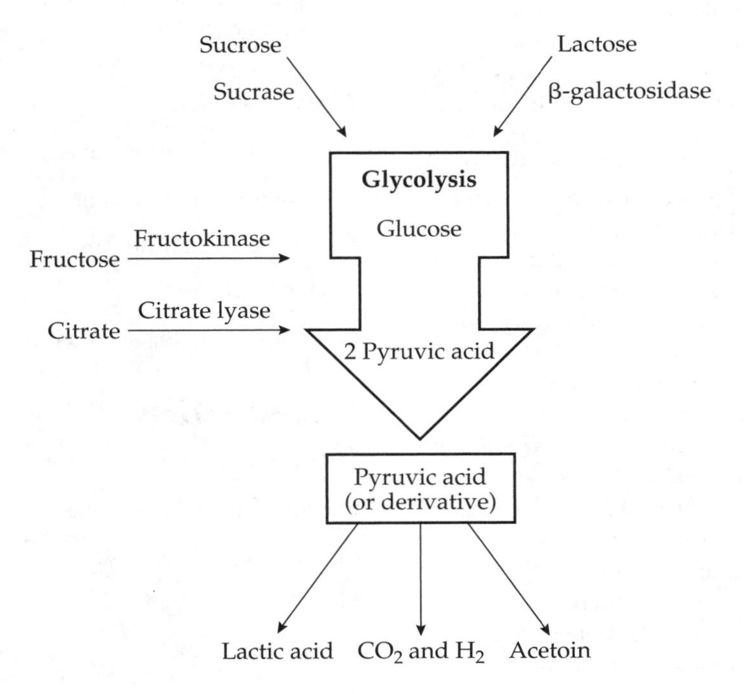

Figure 14.1

Fermentation. Bacteria are often identified by their enzymes. These enzymes can be detected by observing a bacterium's ability to grow on specific compounds. For example, *E. coli* and *Salmonella* are distinguished because *E. coli* can ferment lactose, and typical *Salmonella* cannot.

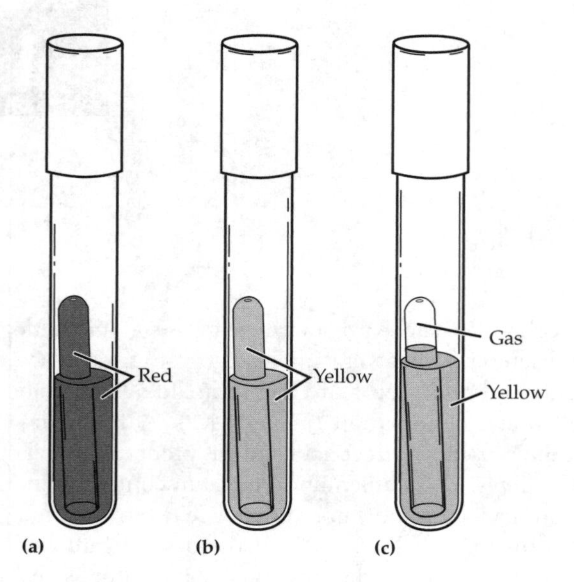

Figure 14.2

Carbohydrate fermentation tube. **(a)** The phenol red indicator is red in a neutral or alkaline solution. **(b)** Phenol red turns yellow in the presence of acids. **(c)** Gases are trapped in the inverted tube while the indicator shows the production of acid.

neutralization of the indicator and turning it red due to ammonia production.

Fermentation processes can produce a variety of end-products, depending on the substrate, the incubation, and the organism. In some instances, large amounts of acid may be produced, while in others, a majority of neutral products may result (Figure 14.3a). The **MRVP test** is used to distinguish between organisms that produce large amounts of acid from glucose and those that produce the neutral product *acetoin*. MRVP medium is a glucose-supplemented nutrient broth used for the **methyl red (MR) test** and the **Voges–Proskauer (V–P) test.** If an organism produces a large amount of organic acid from glucose, the medium will remain red when methyl red is added in a positive MR test, indicating that the pH is below 4.4. If neutral products are produced, methyl red will turn yellow, indicating a pH above 6.0. The production of acetoin is detected by the addition of potassium hydroxide and α-naphthol. If acetoin is present, the upper part of the medium will turn red; a negative V–P test will turn the medium light brown. The chemical process is shown in Figure 14.3b. The production of acetoin is dependent on the length of incubation, as shown in Figure 14.4.

The ability of some bacteria to ferment citrate can be useful for identifying bacteria. When citric acid or sodium citrate is in solution, it loses a proton or Na^+ to form a citrate ion. Bacteria with the enzyme citrate lyase can break down citrate to form pyruvate, which can be reduced in fermentation. Simmons citrate agar (Table 14.1) contains citrate as the only carbon source and ammonium (NH_4^+) as the only nitrogen source. When bacteria use citrate and ammonium, the medium is alkalized because of ammonia (NH_3) produced from NH_4^+. The indicator bromthymol blue changes to blue when the medium is alkalized, indicating a positive citrate utilization test.

Table 14.1

Simmons Citrate Agar

Ingredient	Amount
Sodium citrate	0.2%
Sodium chloride	0.5%
Monoammonium phosphate	0.1%
Dipotassium phosphate	0.1%
Magnesium sulfate	0.02%
Agar	1.5%
Bromthymol blue	0.0008%

Materials

Glucose fermentation tube

Lactose fermentation tube

Sucrose fermentation tube

MRVP broths (4)

Simmons citrate agar slants (2)

Second Period

MRVP broth

Glucose fermentation tube

Simmons citrate agar slant

Parafilm squares

Empty test tube

Methyl red

V–P reagent I, α-naphthol solution

V–P reagent II, potassium hydroxide (40%)

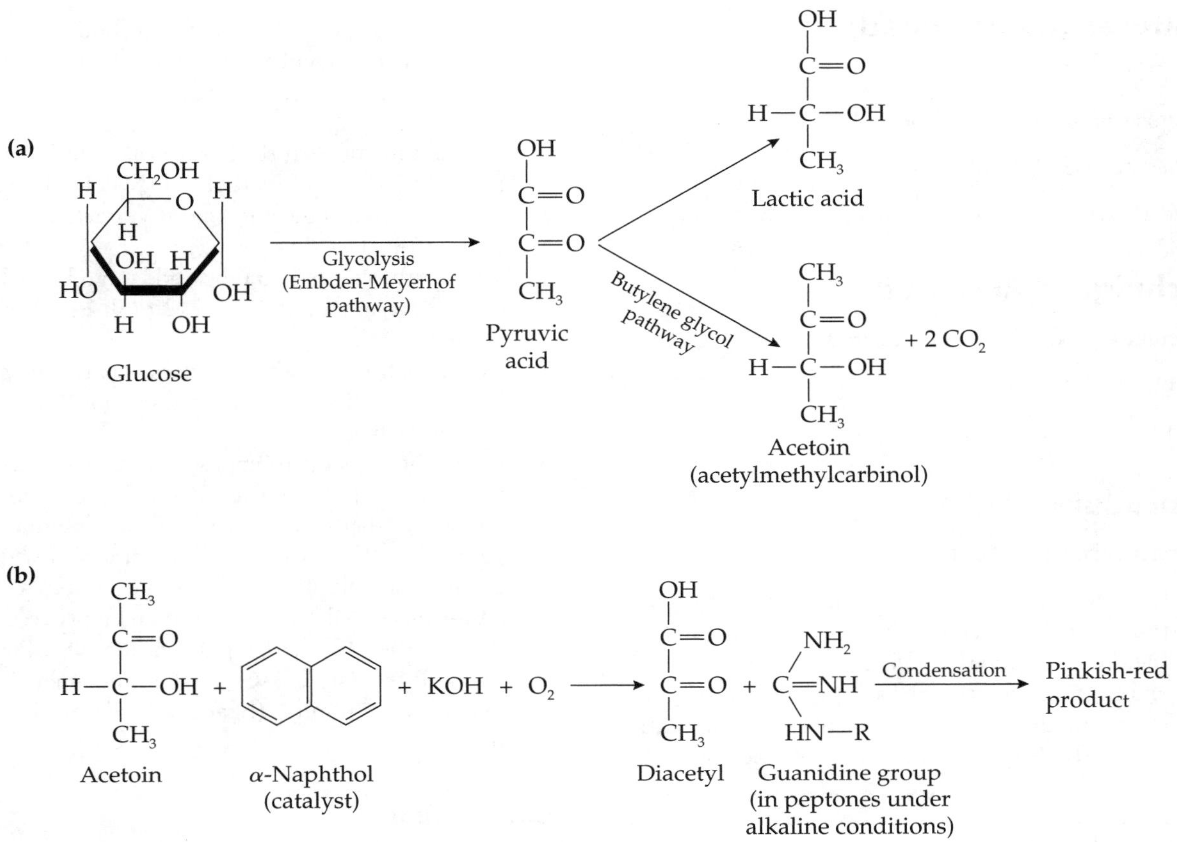

(a)

Glucose → Glycolysis (Embden-Meyerhof pathway) → Pyruvic acid → Lactic acid

Pyruvic acid → Butylene glycol pathway → Acetoin (acetylmethylcarbinol) + 2 CO_2

(b)

Acetoin + α-Naphthol (catalyst) + KOH + O_2 → Diacetyl + Guanidine group (in peptones under alkaline conditions) → Condensation → Pinkish-red product

Figure 14.3

MRVP test. **(a)** Organic acids, such as lactic acid, or neutral products, such as acetoin, may result from fermentation. **(b)** Potassium hydroxide (KOH) and α-naphthol are used to detect acetoin.

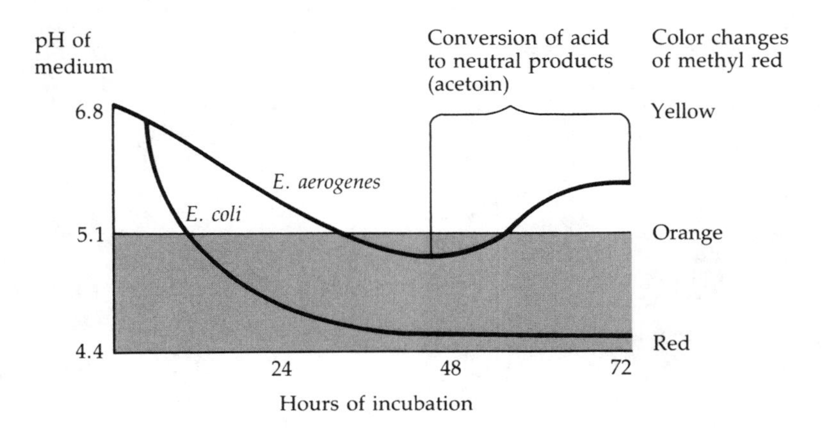

Figure 14.4

The production of acetoin is dependent on incubation time and pH.

Cultures (as assigned)

Escherichia coli

Enterobacter aerogenes

Alcaligenes faecalis

Proteus vulgaris

Techniques Required

Inoculating loop technique, Exercise 10

Aseptic technique, Exercise 10

OF test, Exercise 13

Procedure

Fermentation Tubes

1. Use a loop to inoculate the fermentation tubes with the assigned bacterial culture.
2. Incubate the tubes at 35°C. Examine them at 24 and 48 hours for growth, acid, and gas production. Compare them to an uninoculated fermentation tube. Why is it important to record the presence of growth? ______________________________

   ____
3. Record your results with this culture, as well as your results for the other species tested. (See Color Plate III.2.)

MRVP Tests

1. Using a loop, inoculate two MRVP tubes with *Escherichia* and two with *Enterobacter*.
2. Incubate the tubes at 35°C for 48 hours or longer. Why is time of incubation important (Figure 14.4)?

   ____

   ____
3. To one tube of each set (*Escherichia* and *Enterobacter*), add 5 drops of methyl red. Record the resulting color. Red indicates a positive methyl red test (Figure 14.3a).
4. To the other set of two tubes, add 0.6 ml (12 drops) of V–P reagent I and 0.2 ml (2 or 3 drops) of V–P reagent II.
5. Cover each tube with a Parafilm square, and shake the tubes carefully. Discard the Parafilm in the disinfectant jar.
6. Leave the caps off to expose the media to oxygen in order to oxidize the acetoin (Figure 14.3b). Allow the tubes to stand for 15 to 30 minutes. A positive V–P test will develop a pinkish-red color.
7. Create controls. Pour half of the contents of an uninoculated MRVP broth into an empty test tube. Perform the MR test (step 3) on one tube. Perform the V–P test (steps 4–6) on the other tube.
8. Record your results. (See Color Plates III.3 and III.4.)

Citrate Test

1. Using a loop, inoculate one citrate slant with *Escherichia coli* and the other slant with *Enterobacter aerogenes*.
2. Incubate the tubes at 35°C until the next lab period.
3. Compare the tubes to an uninoculated citrate slant and record your results. (See Color Plate III.5.)

Exercise 14

Fermentation

LABORATORY REPORT

NAME ______________________

DATE ______________________

LAB SECTION ______________________

Purpose

____

____

____

Data

Fermentation Tubes

Color of uninoculated medium: ______________

Organism		Carbohydrate											
		Glucose				Lactose				Sucrose			
		Growth	Color	Acid	Gas	Growth	Color	Acid	Gas	Growth	Color	Acid	Gas
Escherichia coli	24 hr												
	48 hr												
Enterobacter aerogenes	24 hr												
	48 hr												
Alcaligenes faecalis	24 hr												
	48 hr												
Proteus vulgaris	24 hr												
	48 hr												

MRVP Tests

Organism	Growth	MR		V–P	
		Color	+ or −	Color	+ or −
Escherichia coli					
Enterobacter aerogenes					
Controls					

Citrate Test

Organism	Growth	Color	+ or −
Escherichia coli			
Enterobacter aerogenes			
Controls			

Questions

1. Why are fermentation tubes evaluated at 24 and 48 hours? ______________________________

   ____

   What would happen if an organism used up all the carbohydrate in a fermentation tube? __________

   ____

   What would the organism use for energy? ________________ What color would the indicator be then?

   ____

2. If an organism metabolizes glucose aerobically, what result will occur in the fermentation tubes?

   ____

   ____

3. Were these media differential or selective? ______________________________

   ____

Critical Thinking

1. Could an organism be both MR and V–P positive? Explain.

2. Could an organism be a fermenter and also be both MR and V–P negative? Explain.

3. How could you determine whether a bacterium fermented the following carbohydrates: mannitol, sorbitol, adonitol, or arabinose?

4. If a bacterium cannot ferment glucose, why not test its ability to ferment other carbohydrates?

EXERCISE 15

Exercise 15

Protein Catabolism, Part 1

Objectives

After completing this exercise, you should be able to:

1. Determine a bacterium's ability to hydrolyze gelatin.
2. Test for the presence of urease.

Background

Proteins are large organic molecules that include cellular enzymes and many structures. The subunits that make up a protein are called **amino acids** (Figure 15.1). Amino acids consist of carbon, hydrogen, oxygen, nitrogen, and sometimes, sulfur. Amino acids bond together by **peptide bonds** (Figure 15.2), forming a small chain (a **peptide**) or a larger molecule (a **polypeptide**).

Bacteria can hydrolyze the peptides or polypeptides to release amino acids (Figure 15.2). They use the amino acids as carbon and energy sources when carbohydrates are not available. However, amino acids are primarily used in anabolic reactions.

Large protein molecules, such as gelatin, are hydrolyzed by exoenzymes, and the smaller products of hydrolysis are transported into the cell. Hydrolysis of gelatin can be demonstrated by growing bacteria in nutrient gelatin. Nutrient gelatin dissolves in warm water (50°C), solidifies (**gels**) when cooled below 25°C, and liquefies (**sols**) when heated to about 25°C. When an exoenzyme hydrolyzes gelatin, it liquefies and does not solidify even when cooled below 20°C.

Urea is a waste product of protein digestion in most vertebrates and is excreted in the urine. Presence of the enzyme **urease,** which liberates ammonia from urea (Figure 15.3), is a useful diagnostic test for identifying bacteria. **Urea agar** contains peptone, glucose, urea, and phenol red. The pH of the prepared medium is 6.8 (phenol red turns yellow). During incubation, bacteria possessing urease will produce ammonia, which raises the pH of the medium, turning the indicator fuchsia (hot pink) at pH 8.4.

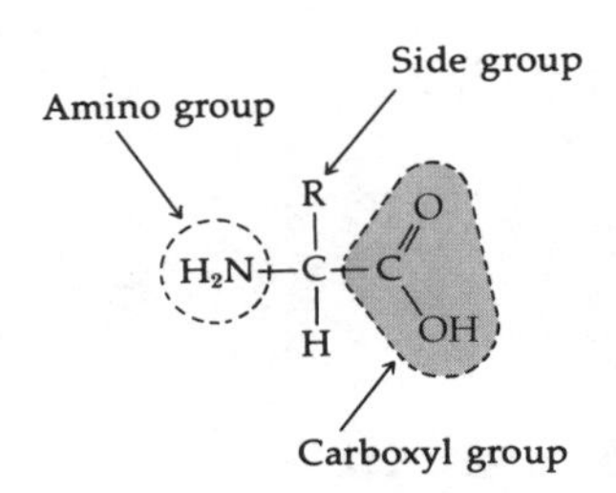

Figure 15.1

General structural formula for an amino acid. The letter R stands for any of a number of groups of atoms. Different amino acids have different R groups.

Peptide bond

$H_3C-O-C(=O)-CH(CH_2C_6H_5)-NH-C(=O)-CH(NH_2)-CH_2-COOH + H_2O \longrightarrow H_3C-O-C(=O)-CH(CH_2C_6H_5)-N(H)-H + HO-C(=O)-CH(NH_2)-CH_2-C(=O)-OH$

Aspartame (NutraSweet), a dipeptide — Methylated phenylalanine — Aspartic acid

Figure 15.2

Hydrolysis of peptides. A molecule of water is used when the peptide is hydrolyzed.

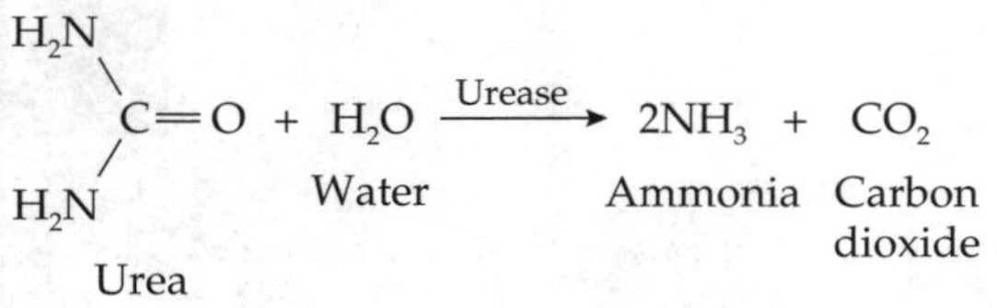

$$\begin{array}{c} H_2N \\ \quad \backslash \\ \quad C{=}O \\ \quad / \\ H_2N \end{array} + H_2O \xrightarrow{\text{Urease}} 2NH_3 + CO_2$$

Urea — Water — Ammonia — Carbon dioxide

Figure 15.3
Urea hydrolysis.

We will investigate bacterial action on nutrient gelatin and urea agar in this exercise and amino acid metabolism in Protein Catabolism, Part 2.

Materials

Tubes containing nutrient gelatin (2)

Tubes containing urea agar (2)

Cultures

Pseudomonas aeruginosa

Proteus vulgaris

Techniques Required

Inoculating loop technique, Exercise 10

Aseptic technique, Exercise 10

Procedure

1. Label one tube of each medium "*Pseudomonas*" and the other tube "*Proteus*."
2. Perform the gelatin hydrolysis. Examine the nutrient gelatin: Is it solid or liquid? ______________
   What is the temperature of the laboratory? _____
   If the gelatin is solid, what would you need to do to liquefy it? ______________________
   To resolidify it? ______________________
   a. Inoculate one tube with *Pseudomonas* and one with *Proteus*, using your inoculating needle.
   b. Incubate the tubes at room temperature, and record your observations at 2 to 4 days and again at 4 to 7 days. Do not agitate the tube when the gelatin is liquid. Why? __________
   ____
   c. If the gelatin has liquefied, place the tube in a beaker of crushed ice for a few minutes. Is the gelatin still liquefied? ______________
   Record your results. Indicate liquefaction or hydrolysis by (+). (See Color Plate III.6.)
3. Conduct a urease test.
   a. Inoculate one urea agar slant with *Pseudomonas* and one with *Proteus*.
   b. Incubate the tubes for 24 to 48 hours at 35°C. (See Color Plate III.7.) Record your results: (+) for the presence of urease (red) and (−) for no urease. What color is phenol red at pH 6.8 or below? ______________
   At pH 8.4 or above? ______________

Exercise 15

LABORATORY REPORT

Protein Catabolism, Part 1

NAME ______________________

DATE ______________________

LAB SECTION ______________________

Purpose

____

____

____

Data

Fill in this table and the one on the next page.

Controls	Appearance of Uninoculated Tube
Gelatin	
Urea agar	

Diagram the appearance of the gelatin.

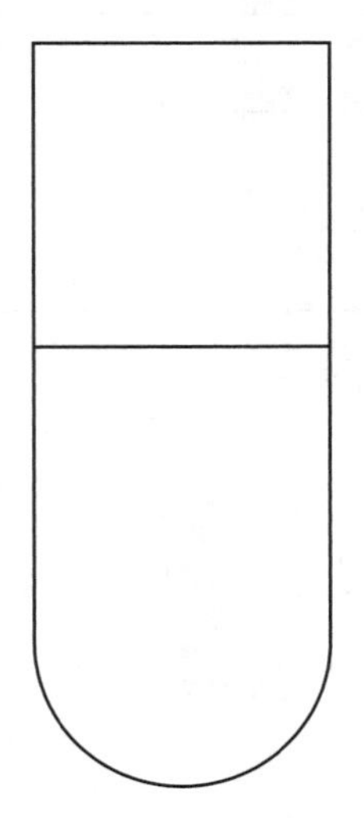

Pseudomonas aeruginosa

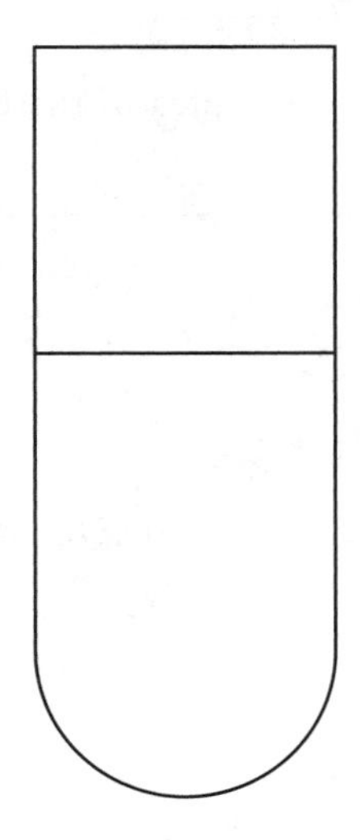

Proteus vulgaris

Test	Results	
	Pseudomonas aeruginosa	*Proteus vulgaris*
Gelatin hydrolysis at ______ days Growth		
Hydrolysis		
Urea agar Growth		
Hydrolysis		

Questions

1. Nutrient gelatin can be incubated at 35°C. What would have to be done to determine hydrolysis after incubation at 35°C? ______________________________

____

____

2. What is the source of urea in the body? ______________________________

____

____

Critical Thinking

1. Why is agar used as a solidifying agent in culture media instead of gelatin?

2. What would you find in the liquid of hydrolyzed gelatin?

3. When changing a baby's wet diapers, one smells ammonia. Why?

4. *Helicobacter pylori* bacteria grow in the human stomach. These bacteria produce a large amount of urease. Of what value is this urease to *Helicobacter*?

Exercise 16

Protein Catabolism, Part 2

Objectives

After completing this exercise, you should be able to:

1. Define the following terms: deamination and decarboxylation.
2. Explain the derivation of H_2S in decomposition.
3. Perform and interpret an indole test.
4. Interpret results from MIO medium.

Background

Once amino acids are taken into a bacterial cell, various metabolic processes can occur using endoenzymes. Before an amino acid can be used as a carbon and energy source, the amino group must be removed. The removal of an amino group is called **deamination.** The amino group is converted to ammonia, which can be excreted from the cell. Deamination results in the formation of an organic acid. Deamination of the amino acid phenylalanine can be detected by forming a colored ferric ion complex with the resulting acid (Figure 16.1). Deamination can also be ascertained by testing for the presence of ammonia using *Nessler's reagent,* which turns deep yellow in the presence of ammonia.

Various amino acids may be decarboxylated, yielding products that can be used for synthesis of other cellular components. **Decarboxylation** is the removal of carbon dioxide from an amino acid. The presence of a specific decarboxylase enzyme results in the breakdown of the amino acid with the formation of the corresponding amine, liberation of carbon dioxide, and a shift in pH to alkaline. Media for decarboxylase reactions consist of glucose, nutrient broth, a pH indicator, and the desired amino acid. In Figure 16.2, bromcresol purple is used as a pH indicator, and a positive decarboxylase test yielding excess amines is indicated by purple. (Bromcresol purple is yellow in acidic conditions.) The names given to some of the amines, such as putrescine, indicate how foul smelling they are. Cadaverine was the name given to the foul-smelling diamine derived from decarboxylation of lysine in decomposing bodies on a battlefield.

Some bacteria liberate **hydrogen sulfide (H_2S)** from the sulfur-containing amino acids: cystine, cysteine, and methionine. H_2S can also be produced from the reduction of inorganic compounds, such as thiosulfate ($S_2O_3^{2-}$). H_2S is commonly called rotten-egg gas because of the copious amounts liberated when eggs decompose. To detect H_2S production, a heavy-metal salt containing ferrous ion (Fe^{2+}) is added to a nutrient culture medium. When H_2S is produced, the sulfide (S^{2-}) reacts with the metal salt to produce a visible black precipitate. The production of hydrogen sulfide from cysteine is shown in Figure 16.3.

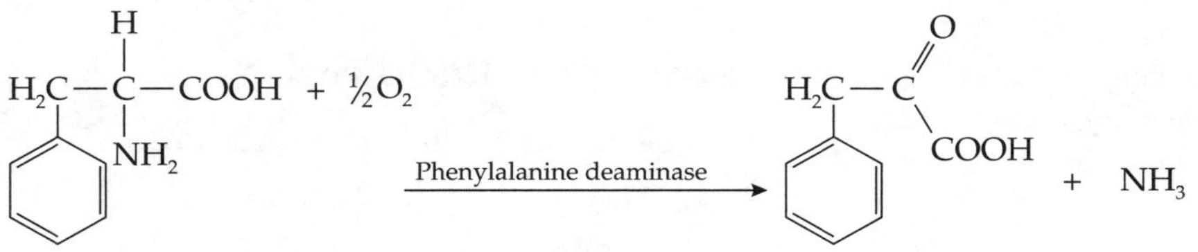

$$\text{L-Phenylalanine} + \tfrac{1}{2}O_2 \xrightarrow{\text{Phenylalanine deaminase}} \text{Phenylpyruvic acid} + \text{Ammonia } (NH_3)$$

$$\text{Phenylpyruvic acid} + \text{Ferric ion } (Fe^{3+}) \longrightarrow \text{Green complex}$$

Figure 16.1

Phenylpyruvic acid resulting from the deamination of phenylalanine is detected by the addition of ferric ion (Fe^{3+}).

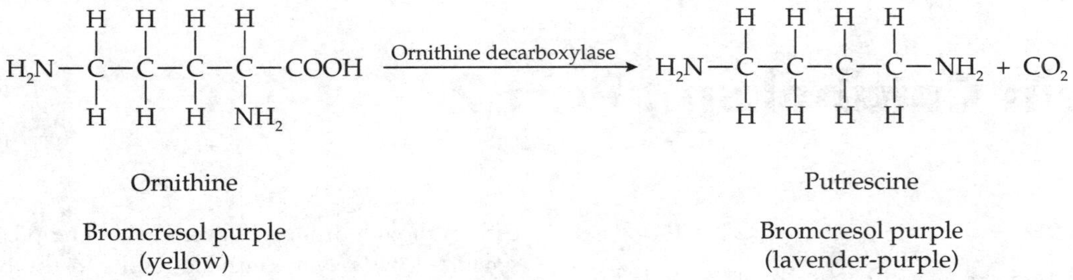

$$H_2N{-}CH_2{-}CH_2{-}CH_2{-}CH(NH_2){-}COOH \xrightarrow{\text{Ornithine decarboxylase}} H_2N{-}CH_2{-}CH_2{-}CH_2{-}CH_2{-}NH_2 + CO_2$$

Ornithine — Bromcresol purple (yellow)

Putrescine — Bromcresol purple (lavender-purple)

Figure 16.2

Decarboxylation of the amino acid ornithine causes a change in the bromcresol purple indicator.

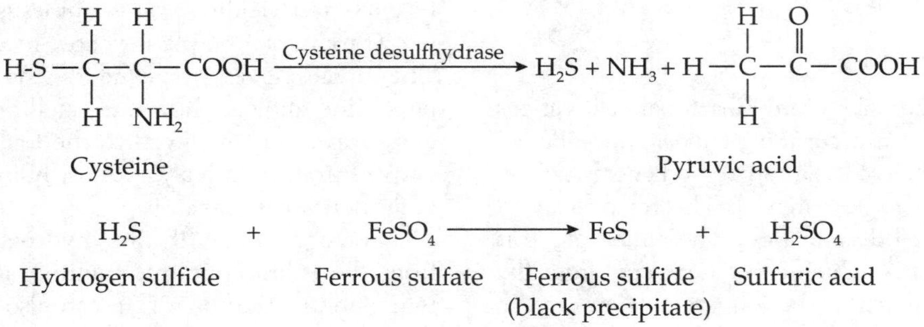

$$HS{-}CH_2{-}CH(NH_2){-}COOH \xrightarrow{\text{Cysteine desulfhydrase}} H_2S + NH_3 + CH_3{-}CO{-}COOH$$

Cysteine — Pyruvic acid

$$H_2S + FeSO_4 \longrightarrow FeS + H_2SO_4$$

Hydrogen sulfide + Ferrous sulfate → Ferrous sulfide (black precipitate) + Sulfuric acid

Figure 16.3

The release of H_2S from the amino acid cysteine is detected by the formation of ferrous sulfide.

The ability of some bacteria to convert the amino acid tryptophan to indole or a blue compound called indigo is a useful diagnostic tool (Figure 16.4). The **indole test** is performed by inoculating a bacterium into tryptone broth and detecting indole by the addition of dimethylaminobenzaldehyde (**Kovacs reagent**). In this exercise, a differential screening medium, motility indole ornithine (MIO) agar, will be used. **MIO** is a single culture medium in which motility, indole production, and ornithine decarboxylase activity can be determined.

Materials

Phenylalanine slants (2)

Peptone iron deeps (2)

MIO deeps (2)

Second Period

Ferric chloride reagent

Kovacs reagent

Cultures (as needed)

Escherichia coli

Proteus vulgaris

Enterobacter aerogenes

Techniques Required

Inoculating loop technique, Exercise 10

Aseptic technique, Exercise 10

Procedure

Phenylalanine Deamination

1. Streak one phenylalanine slant heavily with *Proteus vulgaris* and the other with one of the remaining bacterial cultures.
2. Incubate the tubes for 1 to 2 days at 35°C. Observe for the presence of growth.
3. Add 4 or 5 drops of ferric chloride reagent to the top of the slant, allowing the reagent to run

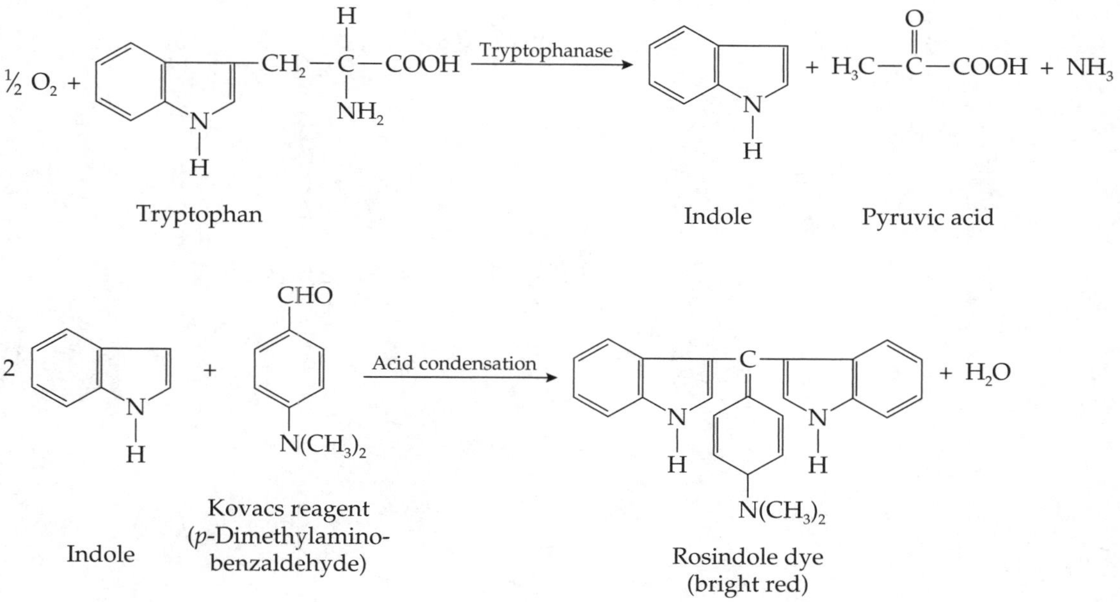

Figure 16.4
Many bacteria produce indole from the amino acid tryptophan.

through the growth on the slant. A positive test gives a dark green color. (See Color Plate III.10.)

Hydrogen Sulfide Production

1. Stab one peptone iron deep with *Escherichia coli* and the other with *Proteus vulgaris*.
2. Incubate the tubes at 35°C for up to 7 days. Observe initially at 24 or 48 hours.
3. Observe for the presence of growth. Blackening in the butt of the tube indicates a positive test. (See Color Plate III.9.)

MIO

1. Stab one MIO deep with *Enterobacter aerogenes* and the other with *Proteus vulgaris*.
2. Incubate the tubes for 24 hours at 35°C.
3. Compare the inoculated tube with an uninoculated tube. Observe for the presence of growth. Motility is demonstrated by growth diffusing out from the stab inoculation line or by clouding of the medium.
4. The ornithine decarboxylation reaction is indicated by a purple color; a negative is yellow. Why?

   ____

   ____
5. Add 4 to 5 drops of Kovacs reagent. Mix the tube gently. A cherry red color indicates a positive indole test. (See Color Plate III.8.)

Results

Record your results for each test in the Laboratory Report.

Exercise 16

Protein Catabolism, Part 2

LABORATORY REPORT

NAME ______________________

DATE ______________________

LAB SECTION ______________________

Purpose

____

____

____

Data

Fill in the following table.

	Test						MIO					
	Phenylalanine Deaminase			Hydrogen Sulfide			Motility		Indole		Ornithine Decarboxylase	
	Original color: ______			Original color: ______			Original color: ______					
Organism	Growth	Color	Reaction (+ or −)	Growth	Color	Reaction (+ or −)	Growth	Motility (+ or −)	Color	Reaction (+ or −)	Color	Reaction (+ or −)
Escherichia coli							Not tested		Not tested		Not tested	
Proteus vulgaris												
Enterobacter aerogenes				Not tested								

Questions

1. When spoilage of canned foods occurs, what causes the blackening of the cans? ______________________

____

____

____

2. Show how lysine could be decarboxylated to give the end-products indicated.

$$H_2N-CH_2-CH_2-CH_2-CH_2-CH(NH_2)-COOH \longrightarrow \quad + \quad$$

Lysine — Cadaverine — ____________

3. Show how alanine could be deaminated to give the end-products indicated.

$$H_2N-CH(CH_3)-COOH \longrightarrow \quad + \quad$$

Alanine — Pyruvic acid — ____________

Of what value is deamination to a microbe?

____

Critical Thinking

1. Why look for black precipitate (FeS) in the butt instead of on the surface of an H_2S test?

2. Decarboxylation of an amino acid results in the evolution of carbon dioxide. Would a gas trap, such as that used in a fermentation test, be an accurate measure of decarboxylation?

3. In blue diaper syndrome, a baby's urine turns blue in its diapers following the administration of oral tryptophan. The baby's serum and urine levels of tryptophan are very low. What is causing the blue pigment indigo to appear?

Exercise 17

Respiration

Objectives

After completing this exercise, you should be able to:

1. Define reduction.
2. Compare and contrast the following terms: aerobic respiration, anaerobic respiration, and fermentation.
3. Perform nitrate reduction, catalase, and oxidase tests.

Background

Molecules that combine with electrons liberated during metabolic processes are called **electron acceptors.** Electron acceptors become **reduced** when they gain electrons. Electrons are formed from the ionization of a hydrogen atom, as shown here:

$$\underset{\text{Hydrogen atom}}{H} \longrightarrow \underset{\text{Hydrogen ion}}{H^+} + \underset{\text{Electron}}{e^-}$$

When an electron acceptor picks up an electron, it becomes negatively charged and combines with the positively charged hydrogen ion. **Reduction,** then, is a gain of electrons or hydrogen atoms.

Organic molecules act as electron acceptors in fermentative metabolism. Inorganic molecules serve as electron acceptors in oxidative metabolism or **respiration.** Molecular oxygen (O_2) is the final electron acceptor in **aerobic respiration.**

In aerobic bacteria, cytochromes carry electrons to O_2. Four general classes of bacterial cytochromes have been identified, and the **oxidase test** is used to determine the presence of one of these, cytochrome *c*. The oxidase test is useful in identifying bacteria.

During aerobic respiration, hydrogen atoms may combine with oxygen, forming hydrogen peroxide (H_2O_2), which is lethal to the cell. Most aerobic organisms produce the enzyme **catalase,** which breaks down hydrogen peroxide to water and oxygen, as shown here:

$$\underset{\text{Hydrogen peroxide}}{2H_2O_2} \xrightarrow{\text{Catalase}} \underset{\text{Water}}{2H_2O} + \underset{\text{Oxygen}}{O_2\uparrow}$$

In the process of **anaerobic respiration,** inorganic compounds other than O_2 act as final electron acceptors. (A few bacteria use organic electron acceptors in anaerobic respiration.) During anaerobic respiration, some bacteria reduce nitrates to nitrites; others further reduce nitrites to nitrous oxide, and some bacteria reduce nitrous oxide to nitrogen gas.

Nitrate broth (nutrient broth plus 0.1% potassium nitrate) is used to determine a bacterium's ability to reduce nitrates (Figure 17.1). Nitrites are detected by the addition of dimethyl-α-naphthylamine and sulfanilic acid to nitrate broth. A red color, indicating that nitrite is present, is a positive test for nitrate reduction. A negative test (no nitrites) is further checked for the presence of nitrate in the broth by the addition of zinc. If nitrates are present, reduction has not taken place. Zinc will reduce nitrate to nitrite, and a red color will appear. If neither nitrate nor nitrite is present, the nitrogen has been reduced to nitrous oxide (N_2O) or nitrogen gas (N_2).

The reduction of some inorganic compounds in respiration is shown in Figure 17.1. In this exercise, we will examine the reduction of nitrate, the catalase test, and the oxidase test.

Materials

Tubes containing nitrate broth (3)

Petri plates containing trypticase soy agar (2)

Second Period

Tube containing nitrate broth

Nitrate reagent A (dimethyl-α-naphthylamine)

Nitrate reagent B (sulfanilic acid)

Hydrogen peroxide, 3%

Sterile toothpick

Oxidase reagent, oxidase strip, or oxidase DrySlide

Zinc dust

Cultures

Bacillus subtilis

Escherichia coli

Pseudomonas aeruginosa

Lactococcus lactis

Techniques Required

Inoculating loop technique, Exercise 10

Aseptic technique, Exercise 10

Aerobic respiration:

$$\frac{1}{2}O_2 + 2H^+ + 2e^- \longrightarrow H_2O$$

Oxygen — Water

Anaerobic respiration:

$$SO_4^{2-} + 10H^+ + 10e^- \longrightarrow H_2S + 4H_2O$$

Sulfate ion — Hydrogen sulfide

$$CO_3^{2-} + 10H^+ + 10e^- \longrightarrow CH_4 + 3H_2O$$

Carbonate ion — Methane

$$NO_3^- + 2H^+ + 2e^- \longrightarrow NO_2^- + H_2O \xrightarrow{6H^+ + 6e^-} N_2O \xrightarrow{2H^+ + 2e^-} N_2$$

Nitrate ion — Nitrite ion — Nitrous oxide — Nitrogen gas

Figure 17.1

Chemical equations showing the reduction of some electron acceptors in bacterial respiration.

Plate streaking, Exercise 11

Carbohydrate catabolism, Exercise 13

Procedure

Nitrate Reduction Test

1. Label three tubes of nitrate broth, and inoculate one with *Lactococcus*, one with *Pseudomonas*, and one with *Bacillus*.
2. Incubate the tubes at 35°C for 2 days. Examine the tubes for gas production. (See Color Plate III.12.)
3. Add 5 drops of nitrate A and 5 drops of nitrate B to each tube and to an uninoculated nitrate broth tube. Shake the tubes gently.
4. A red color within 30 seconds is a positive test. If the broth turns red, what compound is present?

   ____
5. If it does not turn red, add a small pinch of zinc dust; if it turns red after 20–30 seconds, the test is negative. If not, it is positive for nitrate reduction. Why? ______________________

   ____
6. Record your results.

Oxidase Test

1. Divide one plate in half. Label one half "*Escherichia*" and the other "*Pseudomonas*." Streak each organism on the appropriate half.
2. Incubate the plate, inverted, at 35°C for 24 to 48 hours.
3. To test for cytochrome oxidase, do one of the following:
   a. Drop oxidase reagent on the colonies and observe them for a color change to pink within 1 minute and then blue to black. Oxidase-negative colonies will not change color.
   b. Place an oxidase test strip in a Petri dish and moisten an area of the strip with water. Smear three or four well-isolated colonies onto the moistened area with a loop or a toothpick. Oxidase-positive bacteria will turn blue or purple within 10 to 20 seconds. (See Color Plate III.11.)
   c. Using a loop or toothpick, smear a little of one colony onto the oxidase DrySlide. Oxidase-positive bacteria will turn a dark-purple color within 2 minutes.

Catalase Test

1. Divide the other plate in half, and inoculate one half with *Lactococcus* and the other half with one short streak or spot of *Bacillus*.
2. Incubate the plate, inverted, at 35°C for 24 hours. After growth occurs, the plate may be kept at room temperature until the next lab period.
3. To test for catalase, do one of the following:
   a. Drop hydrogen peroxide on the colonies and observe them for bubbles.
   b. With a sterile toothpick, touch the center of a colony and transfer it to a clean glass slide. Discard the toothpick in the To Be Autoclaved area. Drop hydrogen peroxide on the organisms and observe them for bubbles (catalase-positive). (See Color Plate XIV.3.) What gas is in the bubbles?

      ____

Exercise 17

Respiration

LABORATORY REPORT

NAME ____________

DATE ____________

LAB SECTION ____________

Purpose

____

Data

Nitrate Reduction Test

Organism	Color After Adding Nitrate Reagents A and B	Color After Adding Zinc	Gas	NO_3^- Reduction
Uninoculated tube				
Lactococcus lactis				
Pseudomonas aeruginosa				
Bacillus subtilis				

Oxidase Test

Organism	Color After Adding Reagent	Oxidase Reaction
Escherichia coli		
Pseudomonas aeruginosa		

Catalase Test

Organism	Appearance After Adding H_2O_2	Catalase Reaction
Lactococcus lactis		
Bacillus subtilis		

Questions

1. Define reduction. ______________________________

2. Why does hydrogen peroxide bubble when it is poured on a skin cut? ______________________________

3. Differentiate between aerobic respiration and anaerobic respiration. ______________________________

4. Differentiate between fermentation and anaerobic respiration. ______________________________

Critical Thinking

1. Would nitrate reduction occur more often in the presence or absence of molecular oxygen? Explain.

2. In the nitrate reduction test, what does the presence of gas indicate?

3. Is nitrate reduction beneficial or harmful to farmers?

Exercise 18

EXERCISE 18

Rapid Identification Methods

Objectives

After completing this exercise, you should be able to:

1. Evaluate three methods of identifying enterics.
2. Name three advantages of the "systems approach" over conventional tube methods.

Background

The microbiology laboratory must identify bacteria quickly and accurately. Accuracy is improved by using a series of standardized tests. The IMViC tests were developed as a means of separating members of the Enterobacteriaceae (enterics)*, particularly the coliforms, to determine whether water was contaminated with sewage. The IMViC uses a standard combination of four tests, with each capital letter in **IMViC** representing a test; the *i* is added for easier pronunciation. The tests are as follows:

I for indole production from tryptophan

M for methyl red test for acid production from glucose

V for the Voges–Proskauer test for production of acetoin from glucose

C for the use of citrate as the sole carbon source

Although variation among strains does exist, IMViC reactions for selected species of the Enterobacteriaceae are given in Table 18.1.

Table 18.1

IMViC Reactions for Selected Species of the Enterobacteriaceae

Species	Indole	Methyl Red	Voges-Proskauer	Citrate
Escherichia coli	+(v)	+	−	−
Citrobacter freundii	−	+	−	+
Enterobacter aerogenes	−	−	+	+
Enterobacter cloacae	−	−	+	+
Serratia marcescens	−	+ or −*	+	+
Proteus vulgaris	+	+	−	−(v)
Proteus mirabilis	−	+	− or +**	+(v)

v = variable
*Majority of strains give + results
**Majority of strains give − results

Rapid identification methods have been developed that provide a large number of results from one inoculation. Examples are Enterotube II** and API 20E*** for identifying oxidase-negative, gram-negative bacteria belonging to the family Enterobacteriaceae. Enterotube II (see Color Plate III.14) is divided into 12 compartments, each containing a different substrate in agar. A similar tube called Oxi/Ferm is used to identify oxidase-positive, gram-negative rods. API 20E (see Color Plate III.13) consists of 20 microtubes containing dehydrated substrates. The substrates are rehydrated by adding a bacterial suspension. No culturing beyond the initial isolation is necessary with these systems. Comparisons between these rapid identification methods and conventional culture methods show that they are as accurate as conventional test-tube methods.

Computerized analysis of test results increases accuracy because each test is given a point value. Tests that are more important than others get more points. IMViC uses only four tests of equal value.

As commercial identification systems are developed, they can provide greater standardization in identification because they overcome the limitations of hunting through a key, preparing media, and evaluating tests within a laboratory or between different laboratories. They also save time, money, and labor.

*Enterobacteriaceae are aerobic or facultatively anaerobic, gram-negative, nonendospore-forming, rod-shaped bacteria. Coliforms are Enterobacteriaceae that ferment lactose with acid and gas formation within 48 hours at 35°C.

**BD Bioscience, San Jose, CA 95131.

***bioMérieux, Inc., Durham, NC 27712.

Materials

Petri plate containing nutrient agar

Oxidase reagent

IMViC tests and reagents

Enterotube II and reagents

API 20E tray and reagents

Tube containing:

- 5.0 ml sterile saline
- 5-ml pipette
- Mineral oil
- Sterile Pasteur pipette

Culture

Unknown enteric #_____

Techniques Required

Inoculating loop and needle techniques, Exercise 10

Aseptic technique, Exercise 10

Plate streaking, Exercise 11

MRVP tests, Exercise 14

Fermentation tests, Exercise 14

Protein catabolism, Parts 1 and 2, Exercises 15 and 16

Respiration, Exercise 17

Procedure

Isolation

1. Streak the nutrient agar plate with your unknown for isolation and to determine purity of the culture. Incubate the plate, inverted, at 35°C for 24 to 48 hours. Record the appearance of the colonies.
2. Determine the oxidase reaction of one of the colonies on the plate. Why?________________

   ____

   How will you determine the oxidase reaction? ____

   ____

   ____

IMViC Tests

1. Inoculate tubes of tryptone broth (indole test), MRVP broths, and Simmons citrate agar with your unknown.
2. Incubate the tubes at 35°C for 48 hours or longer; perform the appropriate tests. Add 4 to 5 drops of

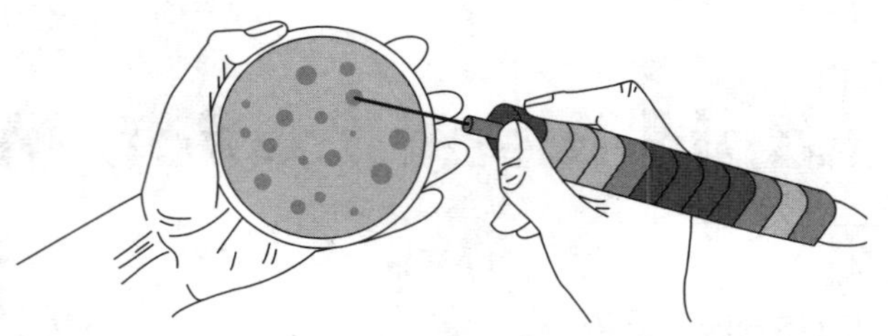

(a) Pick a well-isolated colony with the inoculating end of the wire.

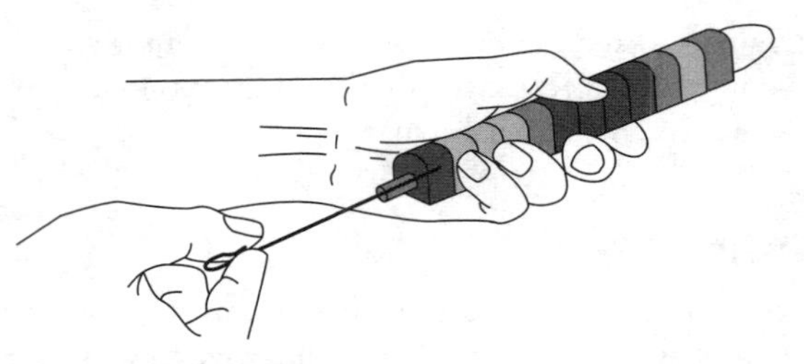

(b) Hold the bent end of the wire, and withdraw the needle through all 12 compartments with a turning motion.

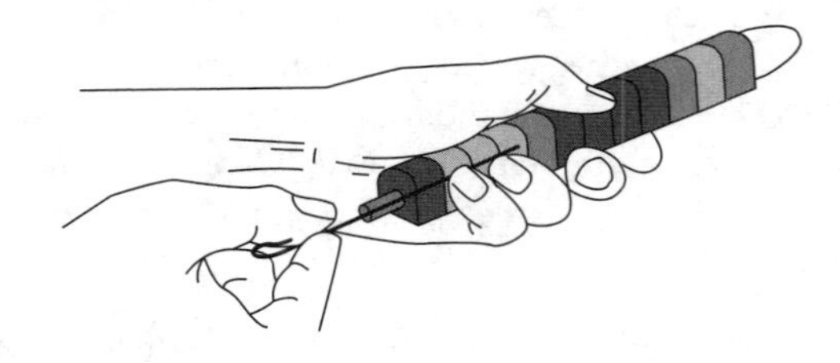

(c) Reinsert the wire through all 12 compartments. Then withdraw it to the notch on the wire. Break the wire at the notch.

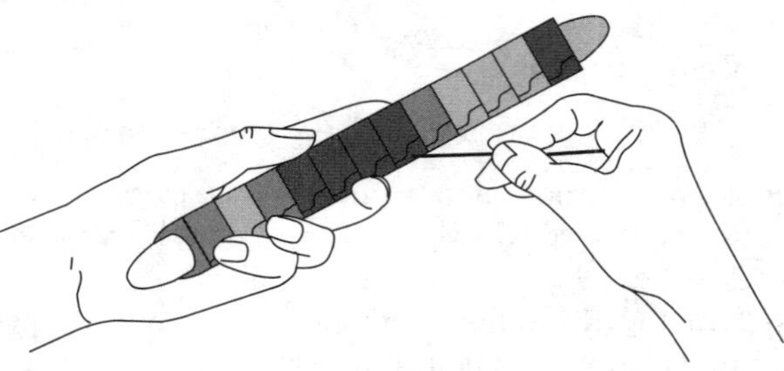

(d) Using the broken wire, punch holes through the foil covering the air inlets in the last eight compartments. Replace the caps loosely.

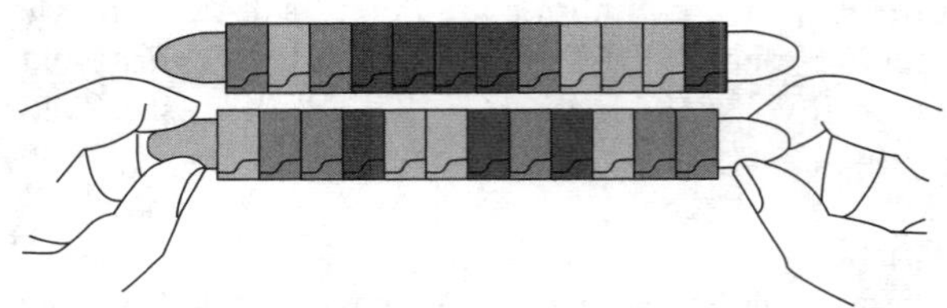

(e) After incubation, compare the tube to an uninoculated one to record results.

Figure 18.1

Inoculating an Enterotube II.

Table 18.2

Enterotube II Biochemical Reactions

Test	Comments	Indicator Changed From	Indicator Changed To
GLU	Acid from glucose	Red	Yellow
GAS	Gas produced from fermentation of glucose trapped in this compartment, causing separation of the wax		
LYS	Lysine decarboxylase	Yellow	Purple
ORN	Ornithine decarboxylase	Yellow	Purple
H_2S	Ferrous ion reacts with sulfide ions, forming a black precipitate		
IND	Kovacs reagent is added to the H_2S/IND compartment to detect indole	Beige	Red
ADON	Adonitol fermentation	Red	Yellow
LAC	Lactose fermentation	Red	Yellow
ARAB	Arabinose fermentation	Red	Yellow
SORB	Sorbitol fermentation	Red	Yellow
V–P	Voges–Proskauer reagents detect acetoin	Beige	Red
DUL	Dulcitol fermentation	Green	Yellow
PA	Phenylpyruvic acid released from phenylalanine after its deamination combines with iron salts to form a black precipitate		
UREA	Ammonia changes the pH of the medium	Yellow	Pink
CIT	Citric acid used as a carbon source	Green	Blue

Source: BD Bioscience, San Jose, CA 95131.

Kovacs reagent to the tryptone broth to test for indole. Mix the tube gently. A cherry red color indicates a positive test. (See Color Plate III.8.) Record your results.

Enterotube II (Figure 18.1)

1. Remove both caps from the Enterotube II. The straight end of the wire is used to pick up the inoculum; the bent end is the handle. Holding the Enterotube II, pick a well-isolated colony with the inoculating end of the wire (Figure 18.1a). Avoid touching the agar with the needle.
2. Inoculate the Enterotube II by holding the bent end of the wire and twisting; the tip of the wire should be visible in the citrate compartment. Withdraw the needle through all 12 compartments using a turning motion (Figure 18.1b).
3. Reinsert the needle into the Enterotube II, using a turning motion, through all 12 compartments until the notch on the wire is aligned with the opening of the tube. Break the wire at the notch by bending it (Figure 18.1c). The portion of the needle remaining in the tube maintains anaerobic conditions necessary for fermentation, production of gas, and decarboxylation.
4. Punch holes with the broken-off wire through the foil covering the air inlets of the last eight compartments (adonitol through citrate) to provide aerobic conditions (Figure 18.1d). Replace the caps on both ends of the tube.

Discard the wire in disinfectant.

5. Incubate the tube lying on its flat surface at 35°C for 24 hours.
6. Interpret and record all reactions (see Table 18.2) in the Laboratory Report. Read all the other tests *before* the indole test, which follows.
   a. Indole test. Place the Enterotube II horizontally and melt a small hole in the plastic film covering the H_2S/indole compartment by using a warm inoculating loop. Add 1 to 2 drops of

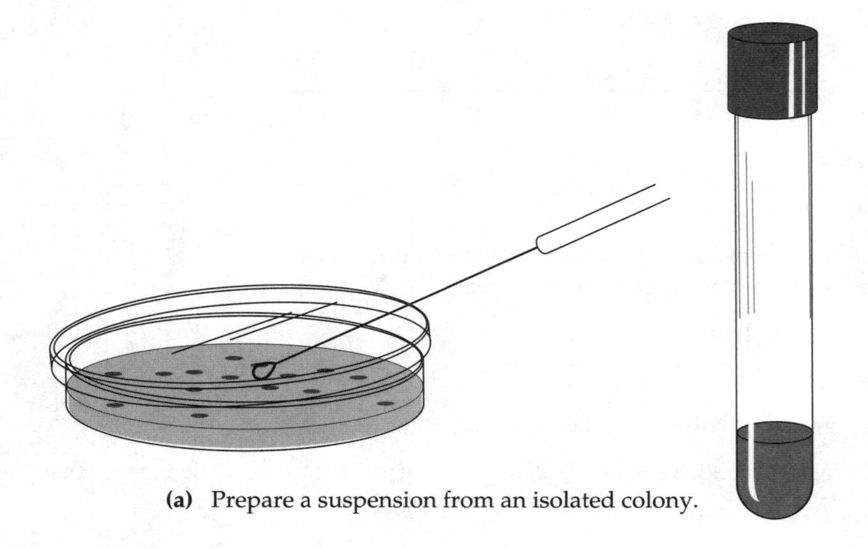

(a) Prepare a suspension from an isolated colony.

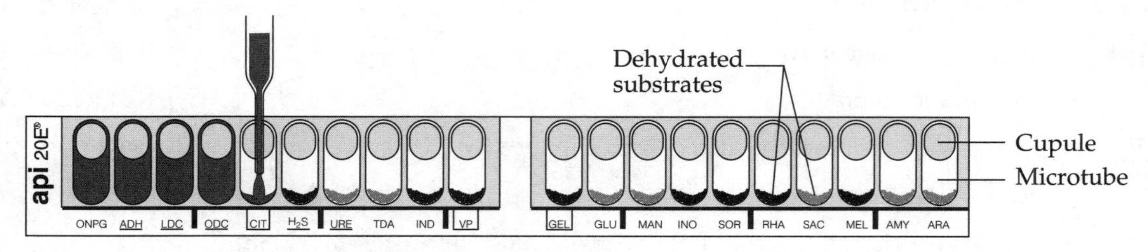

(b) Aseptically inoculate the microtubes with the bacterial suspension.
Carefully fill the tube and cupule sections of the CIT, VP, and GEL tubes (boxed tests).

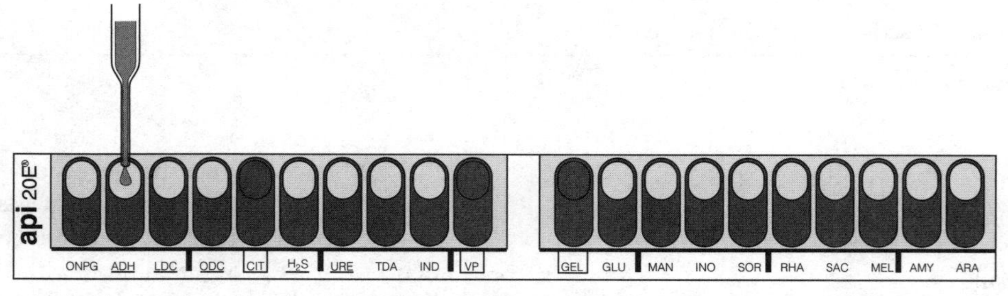

(c) Add mineral oil to the ADH, LDC, ODC, H_2S, and URE cupules (underlined tests).

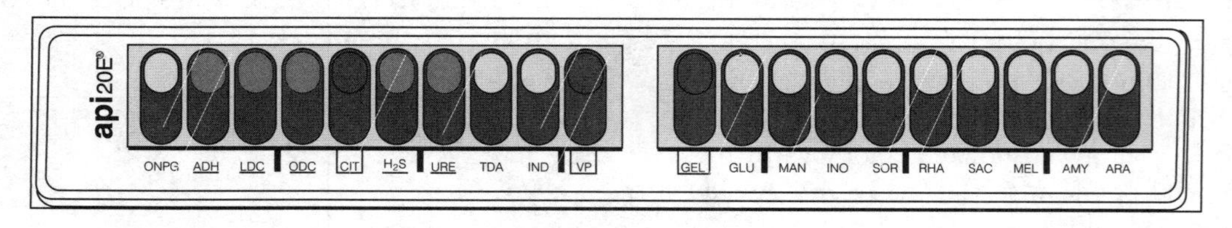

(d) Incubate the strip in its plastic tray.

Figure 18.2
Inoculating the API 20E system.

Kovacs reagent, and allow the reagent to contact the agar surface. A positive test is indicated by a red color within 10 seconds.

b. V–P test if needed. Add 2 drops of 20% KOH containing 5% α-naphthol to the V–P compartment. A positive test is indicated by development of a red color within 20 minutes.

7. Indicate each positive reaction by circling the number appearing below the appropriate compartment of the Enterotube II diagram in the Laboratory Report. Add the circled numbers only within each bracketed section, and enter this sum in the space provided below the arrow. The V–P test may be needed as a confirming test. Read the five numbers in these spaces across as a five-digit number in the *Computer Coding and Identification System*.*
8. Dispose of the Enterotube II by placing it in the autoclave bag.

API 20E (Figure 18.2)

1. Prepare a bacterial suspension by touching the center of a well-isolated colony with a sterile loop and thoroughly mixing the inoculum in 5 ml of sterile saline (Figure 18.2a).
2. Place 5 ml of tap water into the corrugated incubation tray to provide a humid atmosphere during incubation.
3. Using a sterile Pasteur pipette, tilt the API 20E tray and fill the tube section of the microtubes with the bacterial suspension. Fill the tube *and* cupule sections of the CIT, VP, and GEL tubes (Figure 18.2b). Place the tip of the pipette against the side wall of the cupule and carefully inoculate without introducing bubbles into the cupule.
4. After inoculation, completely fill the cupule section of the ADH, LDC, ODC, H_2S, and URE tubes with mineral oil to create anaerobic conditions (Figure 18.2c).
5. Place the plastic lid on the tray and incubate the strip for 24 hours at 35°C (Figure 18.2d). If the strip cannot be read after 24 hours, remove the strip from the incubator and refrigerate it.
6. Interpret and record all reactions (see Table 18.3) in the Laboratory Report. Read all the other tests before the TDA, VP, and IND tests, which follow.
   a. TDA test. Add 1 drop of 10% ferric chloride. A positive test is brownish-red. Indole-positive organisms may produce an orange color; this is a negative TDA reaction.
   b. V–P test. Add 1 drop of V–P reagent II (KOH), and then 1 drop of V–P reagent I (α-napthol). A positive reaction produces a red color (not pale pink) after 10 minutes.
   c. Indole test. Add 1 drop of Kovacs reagent. A red ring after 2 minutes indicates a positive reaction.
   d. Nitrate reduction. Before adding reagents, look for bubbles in the GLU tube. Bubbles indicate reduction of nitrate to N_2. Add 2 drops of nitrate reagent A (dimethyl-α-napthylamine) and 2 drops of nitrate reagent B (sulfanilic acid). A positive reaction (red) may take 2 to 3 minutes to develop. A negative test can be confirmed by adding zinc dust.
7. Indicate each positive reaction with a (+) in the appropriate compartment of the Laboratory Report. Add the points for each positive reaction within each bold-outlined section. Read the seven numbers across as a seven-digit number in the *API 20E Analytical Profile Index*.** Nitrate reduction is a confirming test and not part of the seven-digit code.
8. Dispose of the API strip, tray, and lid by placing them in the autoclave bag.

*BD Bioscience, www.bd.com.

**bioMérieux, www.biomerieux-usa.com

Table 18.3
API 20E Biochemical Reactions

Test	Comments	Indicator	
		Positive	Negative
ONPG	O-nitrophenyl-β-D-galactopyranoside is hydrolyzed by the enzyme that hydrolyzes lactose	Yellow	Colorless
ADH	Arginine dihydrolase transforms arginine into ornithine, NH_3, and CO_2	Red	Yellow
LDC	Decarboxylation of lysine liberates cadaverine	Red	Yellow
ODC	Decarboxylation of ornithine produces putrescine	Red	Yellow
CIT	Citric acid used as sole carbon source	Dark blue	Light green
H_2S	Blackening indicates reduction of thiosulfate to H_2S	Black	No blackening
URE	Urea is hydrolyzed by the enzyme urease to NH_3 and CO_2	Red	Yellow
TDA	Deamination of tryptophan produces indole and pyruvic acid	Brown	Yellow
IND	Kovacs reagent is added to detect indole	Red ring	Yellow
VP	Addition of KOH and α-naphthol detects the presence of acetoin	Red	Colorless
GEL	Gelatin hydrolysis	Diffusion of pigment	No diffusion
GLU MAN INO SOR RHA SAC MEL AMY ARA	Fermentation of glucose Fermentation of mannitol Fermentation of inositol Fermentation of sorbitol Fermentation of rhamnose Fermentation of sucrose Fermentation of melibiose Fermentation of amygdalin Fermentation of arabinose	Yellow or yellow-green	Blue or green
NO_2, N_2 gas, N_2O	Nitrate reduction	Red Bubbles Yellow after addition of zinc	Yellow No bubbles Orange after reagents and zinc

Source: bioMérieux, Inc., Durham, NC 27712.

Exercise 18

Rapid Identification Methods

LABORATORY REPORT

NAME ______________________

DATE ______________________

LAB SECTION ______________________

Purpose ______________________

____

____

Data

Unknown # ______________________

Appearance on nutrient agar: ______________________

Oxidase reaction: ______________________

IMViC

Indicate positive (+) and negative (−) results for each test.

Indole: ______________________

Methyl red: ______________________

V–P: ______________________

Citrate: ______________________

Enterotube II

Circle the number corresponding to each positive reaction below the appropriate compartment. Then determine the five-digit code.

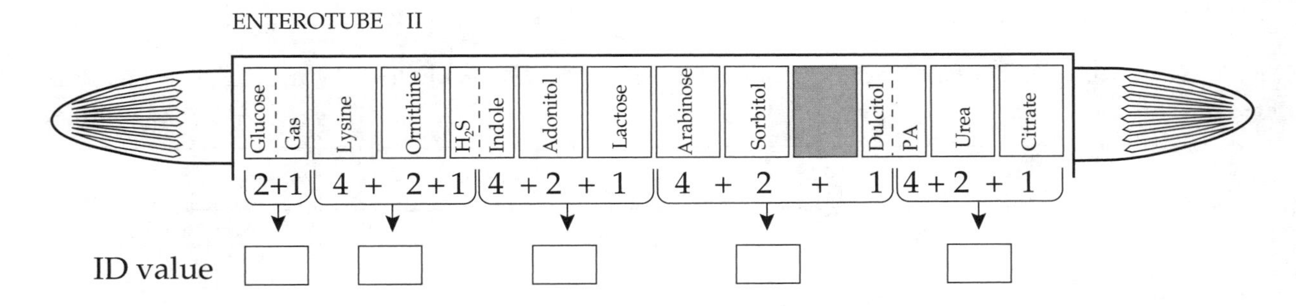

V–P: ______________________

API 20E

Indicate positive (+) and negative (−) results in the Results line. Then determine the seven-digit code.

	ONPG 1	ADH 2	LDC 4	ODC 1	CIT 2	H_2S 4	URE 1	TDA 2	IND 4	VP 1	GEL 2	GLU 4
Results												
Profile number												

	MAN 1	INO 2	SOR 4	RHA 1	SAC 2	MEL 4	AMY 1	ARA 2	Oxidase 4	NO_2	N_2 gas
Results											
Profile number											

Questions

1. What species was identified in unknown # ________ by the IMViC tests? ________

   By the Enterotube II? ________

   By the API 20E? ________

2. Did all three methods agree? ________ If not, explain any discrepancies. ________

3. Which method did you prefer? ________ Why? ________

4. Why are systems developed to identify Enterobacteriaceae? ________

5. Why is an oxidase test performed on a culture before using API 20E and Enterotube II to identify the culture?

Critical Thinking

1. Why should the first digit in the five-digit Enterotube II ID value be equal to or greater than 2, and the fourth digit in the API 20E profile number be equal to or greater than 4?

2. Why can one species have two or more numbers in a rapid identification system? (For example, *E. coli* is Enterotube II numbers 34540 and 34560; *Citrobacter braakii* is API 20E numbers 3504552 and 3504553.)

3. Use Table 18.1 to give an example of a limitation of the IMViC tests.

Part Five Microbial Growth

EXERCISES

In previous exercises, we studied bacterial growth by observing colonies or turbidity, making visual determinations on an all-or-none basis with the naked eye. In Exercise 20, we will use instrumentation to detect the slight changes in bacterial numbers that occur over time—changes that are not visible to the naked eye—to measure lag and log phases of growth (see the graph on the next page).

Bacteria have nutritional, physical, chemical, and environmental requirements that must be met in order for growth to occur. A knowledge of the conditions necessary for microbial growth can facilitate culturing microorganisms and controlling unwanted organisms.

In 1861, Jean Baptiste Dumas presented a paper to the Academy of Sciences on behalf of Pasteur. In this paper, he stated:

> *The existence of infusoria having the characteristics of ferments is already a fact which seems to deserve attention, but a characteristic which is even more interesting is that these infusoria-animalcules live and multiply indefinitely in absence of the smallest amount of air or free oxygen. . . . This is, I believe, the first known example of animal ferments and also of animals living without free oxygen gas.**

When we speak of *air* as a growth requirement, we are referring to the oxygen in air. Furthermore, when we refer to oxygen, we usually mean the molecular

*In H. A. Lechevalier and M. Solotorovsky. *Three Centuries of Microbiology*. New York: Dover Publications, 1974, p. 45.

oxygen (O_2) that acts as an electron acceptor in aerobic respiration. The significance of the presence or absence of oxygen is investigated in Exercise 19.

Most bacteria in aqueous environments such as tap water, oceans, or body fluids are not growing in colonies as they do on laboratory media. Bacteria are not usually swimming around as they do on a microscope slide. Most bacteria grow in mixed assemblages of species called biofilms (Exercise 21).

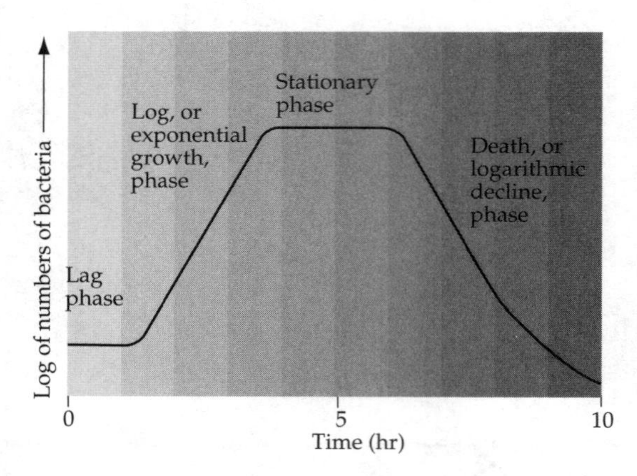

Bacterial growth curve, showing the four typical phases of growth.

Exercise 19

Oxygen and the Growth of Bacteria

What is education but a process by which a person begins to learn how to learn?

PETER USTINOV

Objectives

After completing this exercise, you should be able to:

1. Identify the incubation conditions for each of the following types of organisms: aerobes, obligate anaerobes, aerotolerant anaerobes, microaerophiles, and facultative anaerobes.
2. Describe three methods of culturing anaerobes.
3. Cultivate anaerobic bacteria.

Background

The presence or absence of molecular oxygen (O_2) can be very important to the growth of bacteria (Figure 19.1). Some bacteria, called obligate **aerobic bacteria,** require oxygen, while others, called **anaerobic bacteria,** do not use oxygen. One reason **obligate anaerobes** cannot tolerate the presence of oxygen is that they lack catalase, and the resultant accumulation of hydrogen peroxide is lethal. **Aerotolerant anaerobes** cannot use oxygen but tolerate it fairly well, although their growth may be enhanced by microaerophilic conditions. Most of these bacteria use fermentative metabolism.

Some bacteria, the **microaerophiles,** grow best in an atmosphere with increased carbon dioxide (5% to 10%) and lower concentrations of oxygen. Microaerophiles will grow in a solid nutrient medium at a depth to which small amounts of oxygen have diffused into the medium (Figure 19.1e). In order to culture microaerophiles on Petri plates and nonreducing media, a **CO_2 jar** is used. Inoculated plates and tubes are placed in a large jar with a lighted candle or CO_2-generating packet. (See Figure 48.1 on page 342.)

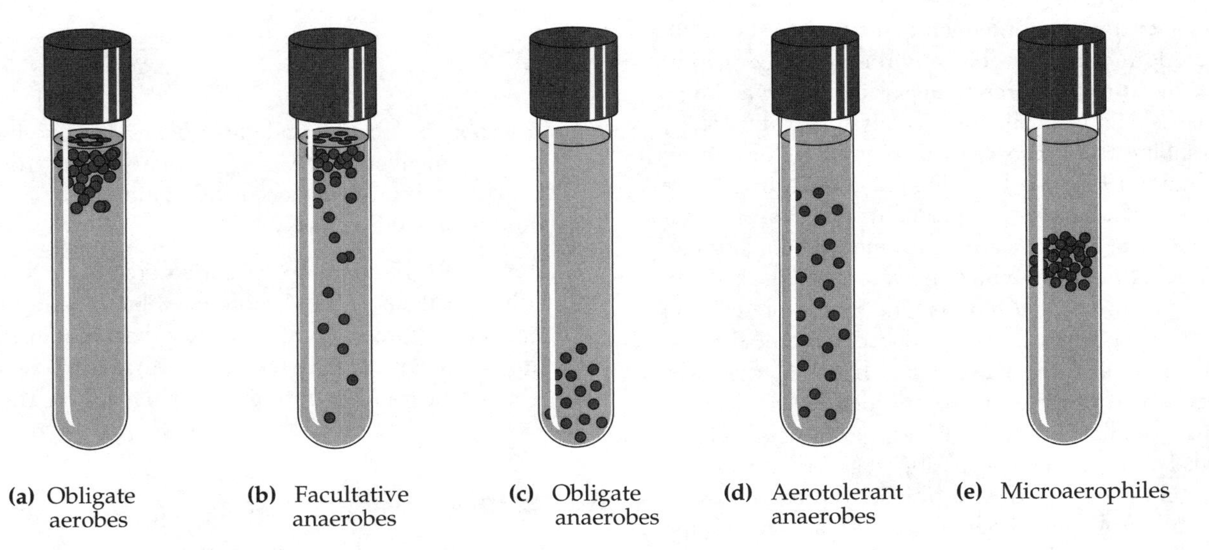

Figure 19.1

Effect of oxygen concentration on the growth of various types of bacteria in a tube of solid medium. Oxygen diffuses only a limited distance from the atmosphere into the solid medium.

The majority of bacteria are capable of living with or without oxygen; these bacteria are called **facultative anaerobes** (Figure 19.1b).

Five genera of bacteria lacking catalase are *Streptococcus, Enterococcus, Leuconostoc, Lactobacillus,* and *Clostridium*. Species of *Clostridium* are obligate anaerobes, but members of the other four genera are aerotolerant anaerobes. The four genera of aerotolerant anaerobes lack the cytochrome system to produce hydrogen peroxide and therefore do not need catalase. Determining the presence or absence of catalase can help in identifying bacteria. When a few drops of 3% hydrogen peroxide are added to a microbial colony and catalase is present, molecular oxygen is released as bubbles.

$$2H_2O_2 \longrightarrow 2H_2O + O_2\uparrow$$

In the laboratory, we can culture anaerobes either by excluding free oxygen from the environment or by using reducing media. Many anaerobic culture methods involve both processes. Some of the anaerobic culturing methods are as follows:

1. **Reducing media** contain reagents that chemically combine with free oxygen, reducing the concentration of oxygen. In thioglycolate broth, **sodium thioglycolate** ($HSCH_2COONa$) will combine with oxygen. A small amount of agar is added to increase the viscosity, which reduces the diffusion of air into the medium. Usually dye is added to indicate where oxygen is present in the medium. Resazurin, which is pink in the presence of excess oxygen and colorless when reduced, or methylene blue (see number 2) are commonly used indicators.
2. Conventional, nonreducing media can be incubated in an anaerobic environment. Oxygen is excluded from a **Brewer anaerobic jar** by adding a GasPak and a palladium catalyst to catalytically combine the hydrogen with oxygen to form water (Figure 19.2). Carbon dioxide and hydrogen are given off when water is added to the GasPak envelope of sodium bicarbonate and sodium borohydride. A methylene blue indicator strip is placed in the jar; methylene blue is blue in the presence of oxygen and white when reduced. One of the disadvantages of the Brewer jar is that the jar must be opened to observe or use one plate. An inexpensive modification of the Brewer jar has been developed using disposable plastic bags for one or two culture plates. In this technique, the bag is coated with antifogging chemicals, and a wet sodium bicarbonate tablet is added to generate carbon dioxide. Iron (Fe^0, e.g., steel wool) activated with water removes O_2 as iron oxide (Fe_2O_3) is produced. One plate can be observed without opening the bag.

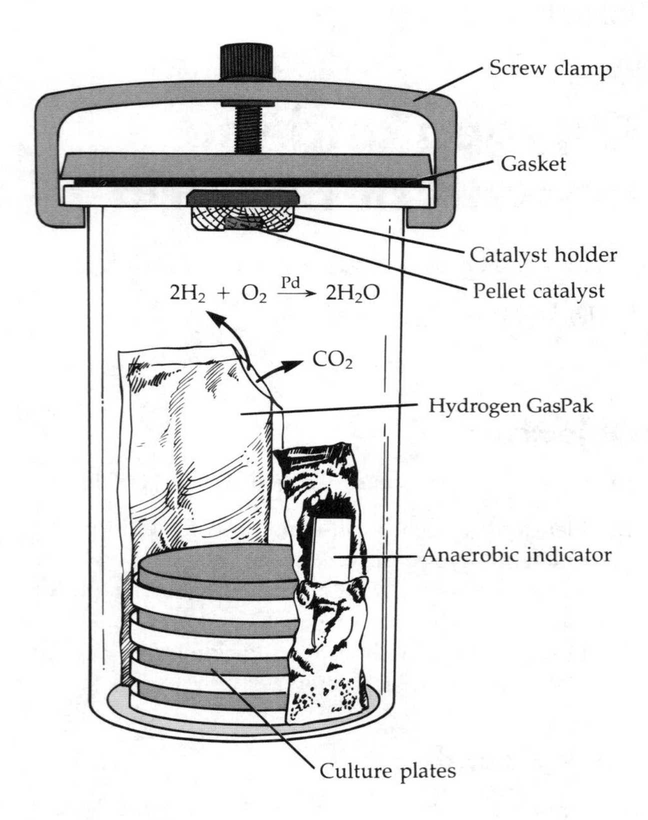

Figure 19.2

Brewer anaerobic jar. Carbon dioxide and hydrogen are generated by the addition of water to a chemical packet. Hydrogen gas combines with atmospheric oxygen in the presence of a palladium catalyst (in the lid or in the GasPakPlus packet) to form H_2O. The anaerobic indicator contains methylene blue, which is blue when oxidized and turns colorless when the oxygen is removed from the container.

3. **Anaerobic incubators** and **glove boxes** can also be used for incubation. Air is evacuated from the chamber and can be replaced with a mixture of carbon dioxide and nitrogen.

All methods of anaerobic culturing are only effective if the specimen or culture of anaerobic organisms is collected and transferred in a manner that minimizes exposure to oxygen. In this exercise, we will try two methods of anaerobic culturing and will perform the catalase test.

Materials

Petri plates containing nutrient agar (4)

Tubes containing thioglycolate broth + indicator (4)

Brewer anaerobic jars (2 per lab section)

3% hydrogen peroxide (H_2O_2) (second period)

Cultures

Alcaligenes faecalis

Clostridium sporogenes

Enterococcus faecalis

Escherichia coli

Unknown bacterium #______

Techniques Required

Inoculating loop technique, Exercise 10

Aseptic technique, Exercise 10

Plate streaking, Exercise 11

Selective media, Exercise 12

Catalase test, Exercise 17

Procedure

1. Don't shake the thioglycolate. Why? __________ What should you do to salvage it if you do shake it? ____

____
2. Label four tubes of thioglycolate broth, and aseptically inoculate one with a loopful of *Alcaligenes*, one with *Clostridium*, one with *Enterococcus*, and one with *Escherichia*.
3. Incubate the tubes at 35°C until the next lab period.
4. Record the appearance of growth in each tube.
5. With a marker, divide two nutrient agar plates into four sectors on the bottom of the plates. Label one plate "Aerobic" and the other "Anaerobic."
6. Streak a single line of *Alcaligenes*, *Clostridium*, *Enterococcus*, and *Escherichia* (Figure 19.3).
7. Label the remaining nutrient agar plates "Aerobic" and "Anaerobic." Then inoculate each plate with your unknown organism by streaking for isolation (Figure 11.2).
8. Incubate the "Aerobic" plates, inverted, at 35°C until the next lab period. Place the "Anaerobic" plates, inverted, in the Brewer jar.

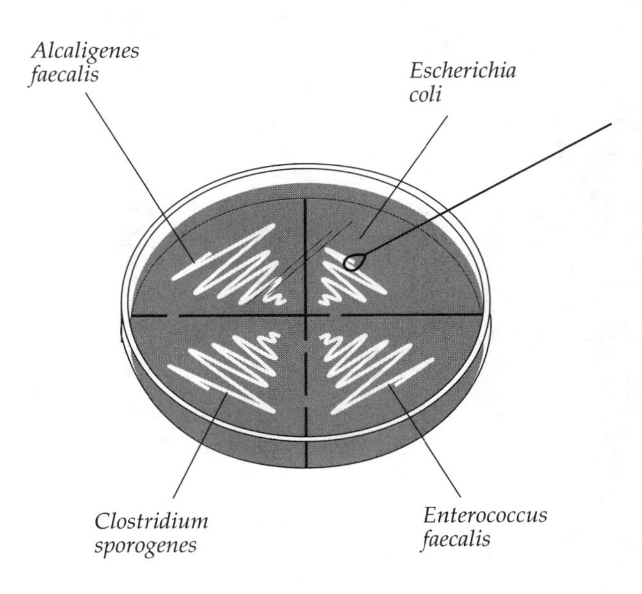

Figure 19.3

Divide a Petri plate into quadrants by drawing lines on the bottom of the plate. Inoculate each quadrant by streaking it with an inoculating loop.

9. Your instructor will demonstrate how the jar is rendered anaerobic once it is filled with plates. Ten ml of water are added to a GasPak generator and placed in the jar with a methylene blue indicator. What causes the condensation that forms on the sides of the jar? ______________________________

____
10. Incubate the jar at 35°C until the next lab period.
11. Record the growth on each plate.
12. Perform the catalase test by adding a few drops of 3% H_2O_2 to the different colonies on the nutrient agar. A positive catalase test produces a bubbling white froth. A dissecting microscope (Appendix E) can be used if more magnification is required to detect bubbling. The catalase test may also be done by transferring bacteria to a slide and adding the H_2O_2 to it. Bacterial growth on blood agar must be tested this way. Why? ______________________________

____

Exercise 19

Laboratory Report

Oxygen and the Growth of Bacteria

Name ______________________

Date ______________________

Lab Section ______________________

Purpose

____

____

____

Expected Results

Before the next lab period, indicate the amount of growth you *expect* to see on each plate and the catalase reaction.

Bacteria	Aerobic		Anaerobic	
	Growth	Catalase Reaction	Growth	Catalase Reaction
Alcaligenes				
Clostridium				
Enterococcus				
Escherichia				

Data

Thioglycolate

Sketch the location of bacterial growth and note where the indicator has turned blue or pink.

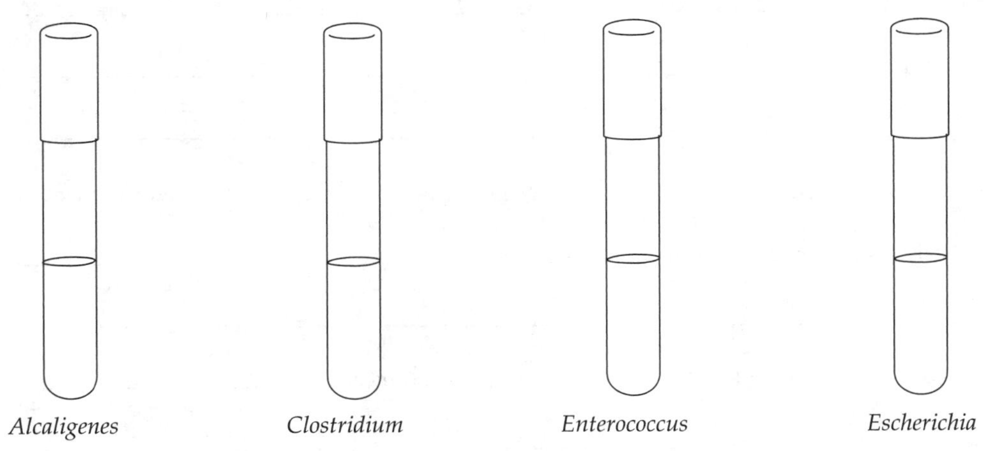

Nutrient Agar Plates

Bacteria	Aerobic		Anaerobic	
	Growth	Catalase Reaction	Growth	Catalase Reaction
Alcaligenes				
Clostridium				
Enterococcus				
Escherichia				
Unknown				

Does the anaerobic plate have an odor? ______________________________

Conclusions

Categorize the five organisms used in this exercise in terms of their oxygen requirements:

Alcaligenes: ______________________

Clostridium: ______________________

Enterococcus: ______________________

Escherichia: ______________________

Unknown #: ______________________

Questions

1. Did your results differ from your expected results? ______________________

   Briefly explain why or why not. ______________________

   ____

2. Can aerobic bacteria grow in the absence of O_2? ________ What would you need to do to determine whether bacteria growing on a Petri plate from the Brewer jar are anaerobes? ______________________

   ____

3. What does the appearance of a blue or pink color in a thioglycolate tube mean? ______________________

   ____

   ____

4. Why will obligate anaerobes grow in thioglycolate? ____

____

____

5. To what do you attribute the odors of anaerobic decomposition? ____________________________

____

____

Critical Thinking

1. *Bergey's Manual* describes *Streptococcus* and *Escherichia* as facultative anaerobes. How do the oxygen requirements of these organisms differ? Which one could correctly be called an aerotolerant anaerobe?

2. How can the aerobe *Pseudomonas aeruginosa* grow in the absence of oxygen?

3. The catalase test is often used clinically to distinguish between two genera of gram-positive cocci, ____________ and ______________________________. It is also used to distinguish between two genera of gram-positive rods, ______________________________ and ______________________________.

4. The following genera of anaerobes are commonly found in fecal samples. Using *Bergey's Manual* and your textbook, place these genera in this flowchart to differentiate them: *Bacteroides*, *Fusobacterium*, *Clostridium*, *Enterococcus*, *Veillonella*, and *Lactobacillus*.

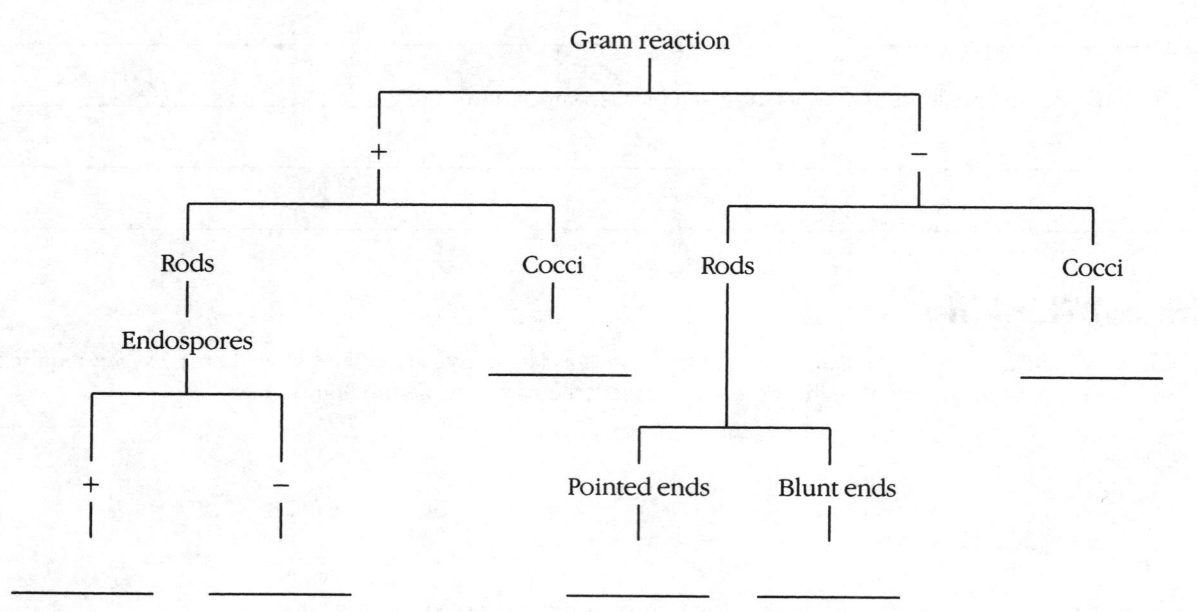

Exercise 20

Determination of a Bacterial Growth Curve: The Role of Temperature

Objectives

After completing this exercise, you should be able to:

1. Identify the four phases of a typical bacterial growth curve.
2. Measure bacterial growth turbidimetrically.
3. Interpret growth data plotted on a graph.
4. Determine the effect of temperature on bacterial growth.

Background

Most bacteria grow within a particular temperature range (Figure 20.1). The *minimum growth temperature* is the lowest temperature at which a species will grow. A species grows fastest at its *optimum growth temperature*. And the highest temperature at which a species can grow is its *maximum growth temperature*. Some heat is necessary for growth. Heat probably increases the rate of collisions between molecules, thereby increasing enzyme activity. At temperatures near the maximum growth temperature, growth ceases, presumably due to inactivation of enzymes.

The effect of temperature on bacteria can be determined by measuring the population growth rate. The phases of growth of a bacterial population are shown in the figure on page 140. During log phase, the cells are growing at the fastest rate possible under the conditions provided. **Generation** or **doubling time** is the time it takes for one cell to divide into two cells—or the time required for the population of cells to double. The shorter the generation time, the faster the growth. The growth of a bacterial population can be determined by inoculating a growth medium with a few cells and counting the cells over time. However, bacteria in suspension in a broth scatter light and cause the transparent broth to appear turbid. This turbidity can be measured with a spectrophotometer (Appendix C) to determine bacterial growth. Although turbidity is not a direct measure of bacterial numbers, increasing turbidity does indicate growth.

Changes in the logarithmic absorbance scale on the spectrophotometer correspond to changes in the number of cells. Consequently, a growth curve can be obtained by graphing absorbance (Y-axis) vs. time (X-axis). The rate of growth is indicated by the slope of the lines; faster growth produces a steeper (higher-number) slope (Figure 20.2).

In this exercise, we will plot growth curves for a bacterium at different temperatures to determine the preferred temperature range of this species.

Materials

Flask containing nutrient broth

Spectrophotometer tubes (2)

Sterile 5-ml pipettes (11)

Sterile 10-ml pipette

Spectrophotometer

Culture

Escherichia coli

Techniques Required

Aseptic technique, Exercise 10

Pipetting, Appendix A

Spectrophotometry, Appendix C

Graphing, Appendix D

Procedure

Each student group is assigned a temperature: 15°C, room temperature, 35°C, 45°C, or 55°C.

1. Aseptically transfer 2–4 ml of nutrient broth to one of the spectrophotometer tubes. This is your control to *standardize* the spectrophotometer. Read Appendix C.
2. Aseptically inoculate a flask of nutrient broth with 10 ml of *E. coli*.

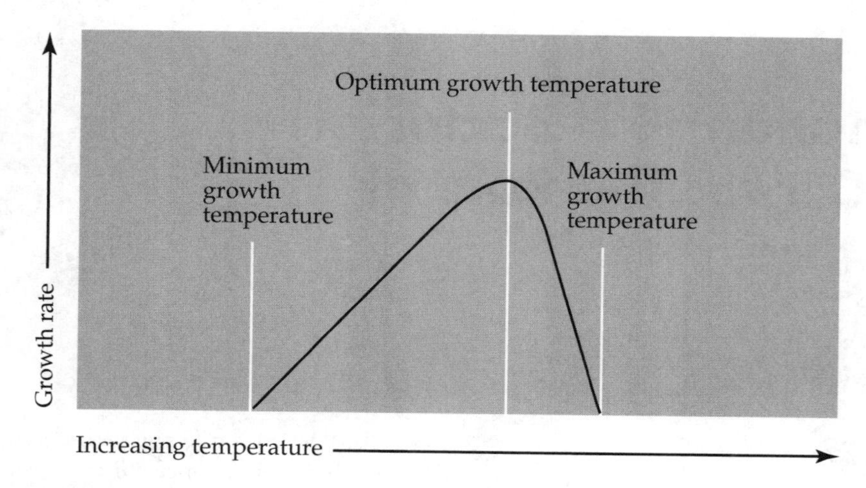

Figure 20.1
Growth response of bacteria within their growing temperature range.

3. Swirl to mix the contents. Transfer 2–4 ml from the flask to the second spectrophotometer tube. Wipe the surface of the spectrophotometer tube with a low-lint, nonabrasive paper such as a Kimwipe. Place the tube in the spectrophotometer, wait about 45 sec, and measure the absorbance (Abs.) on the spectrophotometer.
4. Record all measurements.

After taking a reading, empty the spectrophotometer tube into disinfectant. Rinse the spectrophotometer tube with distilled water and discard the water into disinfectant.

5. Place the flask at your *assigned* temperature.
6. Record the absorbance every 10 minutes for 60 to 90 minutes. Remove the flask from the water bath for the minimum time possible to take each sample. To take a sample, aseptically pipette 2–4 ml into the second spectrophotometer tube.
7. Graph the data you obtained. *Read* Appendix D before drawing your graph.
8. Determine the generation time. Select two points, (*a*) and (*b*), in the log phase during which the absorbance doubled and determine the number of minutes required for the culture to go from (*a*) to (*b*). See Figure 20.2.

Time (min)	Temp A	Temp B
0	0.011	0.012
48	0.022	*a* 0.045
58	0.035	0.081
68	*a* 0.047	0.056
78	0.062	0.075
88	0.081	0.078
97	*b* 0.099	0.078
108	0.109	0.071
118	0.137	0.088
128	0.149	*b* 0.089
138	0.167	0.098
148	0.174	0.106
158	0.190	0.109
168	0.207	0.114
178	0.222	0.117
188	0.297	0.118
198	0.288	0.126
600	0.800	0.125
1470	0.864	0.130
Slope between *a* and *b*	0.0018	0.0004
Generation time (min)	29	80

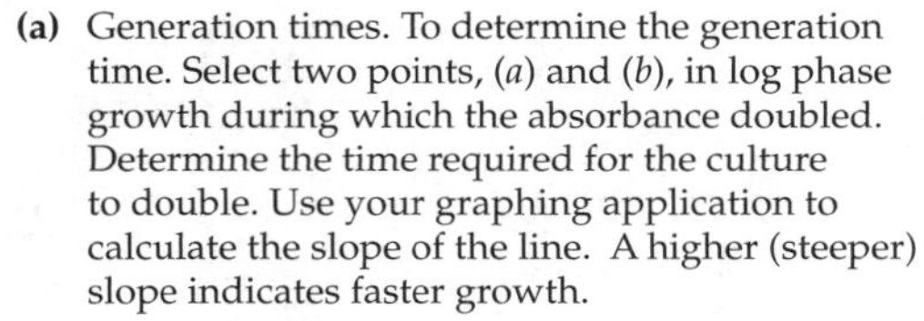

(a) Generation times. To determine the generation time. Select two points, (*a*) and (*b*), in log phase growth during which the absorbance doubled. Determine the time required for the culture to double. Use your graphing application to calculate the slope of the line. A higher (steeper) slope indicates faster growth.

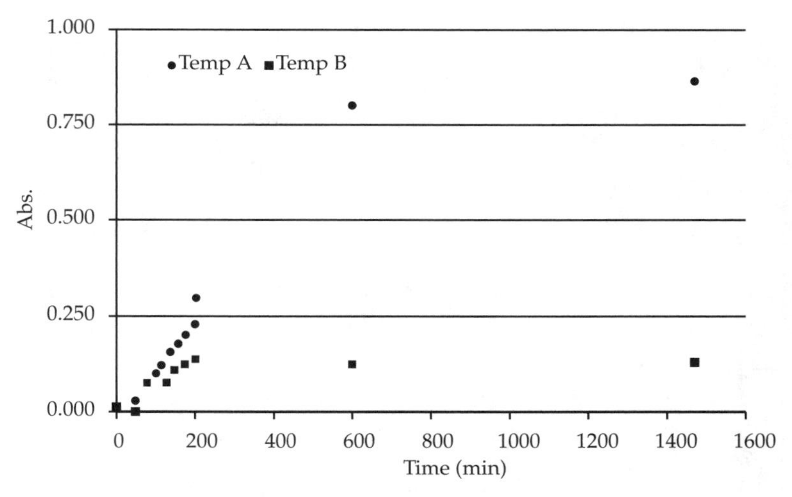

(b) A graph of all the data shows logarithmic and stationary phases.

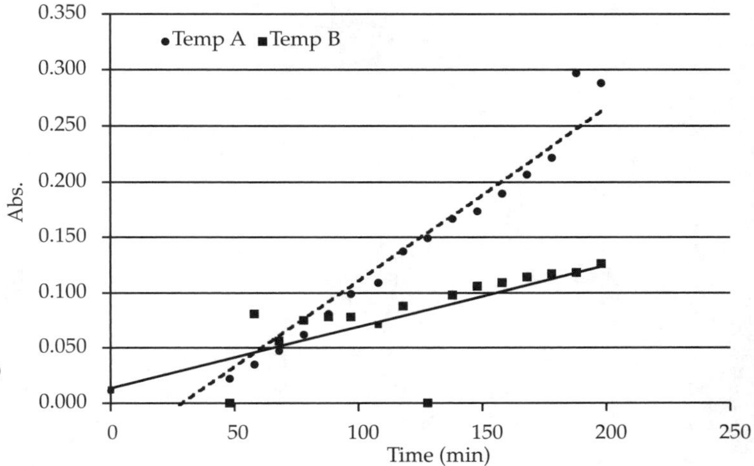

(c) A graph of the first three hours expands the log phase. You can compare the growth by comparing the slopes of the best-fit lines.

Figure 20.2

Showing bacterial growth using absorbance values. **(a)** The generation time is the number of minutes required for the absorbance to double. **(b)** Plotting all the data shows when the cultures reach stationary phase. **(c)** This graph is used to examine log phase. The best-fit lines show the rate at which the population is growing.

Exercise 20

LABORATORY REPORT

Determination of a Bacterial Growth Curve: The Role of Temperature

NAME ______________________

DATE ______________________

LAB SECTION ______________________

Purpose

____

____

____

Data

Record your data below or in a computer application spreadsheet.

Temperature: __________

Time (min)	Absorbance	Time (min)	Absorbance
0			

Plot your data on graph paper or use a computer graphing application. Mark absorbance on the Y-axis and time on the X-axis. Obtain data for other temperatures from your classmates. Plot these data on the same graph using different lines or colors. Label the phases of growth.

Use your data and that of your classmates to fill in this table:

Temperature	Doubling Time (min)

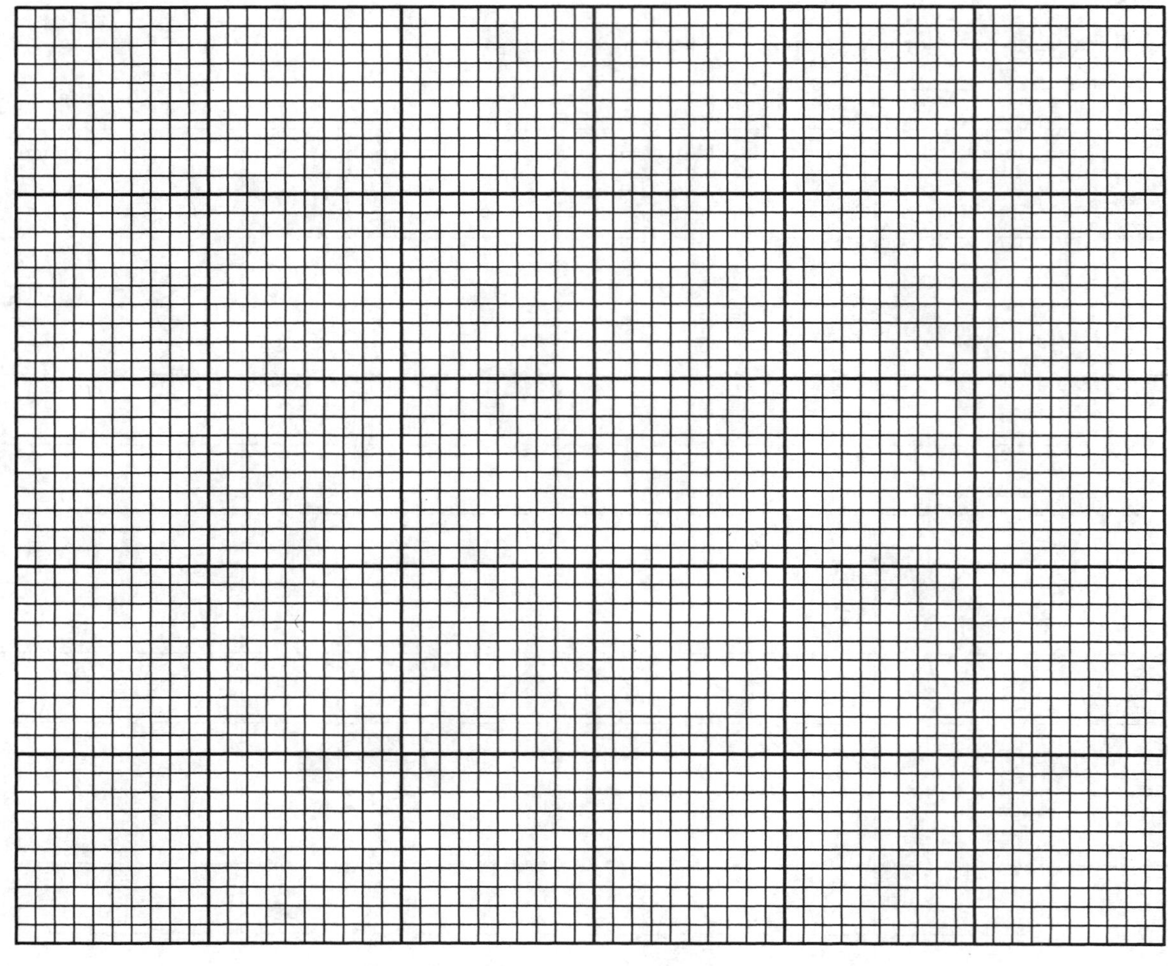

Conclusions

1. Summarize the effect of temperature on growth of *E. coli*. ______________________________

____

____

____

2. From the class data, what is *E. coli*'s optimum temperature? ________ Does this agree with your textbook?________

   If not, provide a brief explanation. ______________________________

Questions

1. Why aren't you likely to get lag and death phases in this experiment? ______________________________

____

____

2. What is the optimum growth temperature for human pathogens? ______________________________

3. Where in nature would you most likely find a thermophile? ______________________________

   A psychrophile? ______________________________

4. What is the effect of temperature on enzymes? ______________________________

____

   Can you use any examples from your experiments? ______________________________

Critical Thinking

1. The following graph shows a likely relationship between bacterial growth and oxygen use in glucose broth. Explain the relationship.

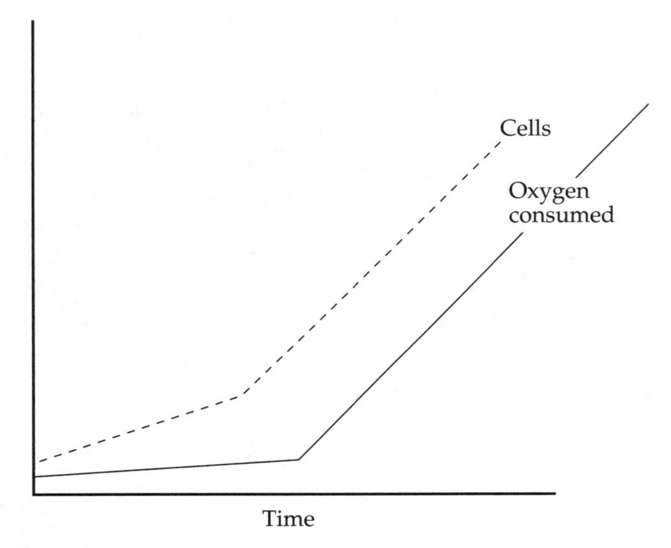

2. The following data were obtained for a log phase culture using plate counts and spectrophotometry:

No. of Cells/ml	Absorbance
5×10^4	0.04
5×10^5	0.26
6×10^7	0.72

Assume that you inoculated a flask yesterday and now you need a broth culture containing 5×10^6 cells/ml to perform an experiment. The concentration of cells in the 24-hour culture is much too high; its absorbance is greater than 1. What is the easiest way to get the desired cell concentration from your 24-hour flask?

3. You have isolated a bacterium that produces an antibiotic. The bacterium grows well at 37°C but doesn't produce the antibiotic at that temperature. Antibiotic production is best at 30°C, but the cells grow so poorly at that temperature that you can't get enough antibiotic. The relationships are shown in the following graphs. Design a procedure to maximize the biomass (total cells) and the yield of the antibiotic.

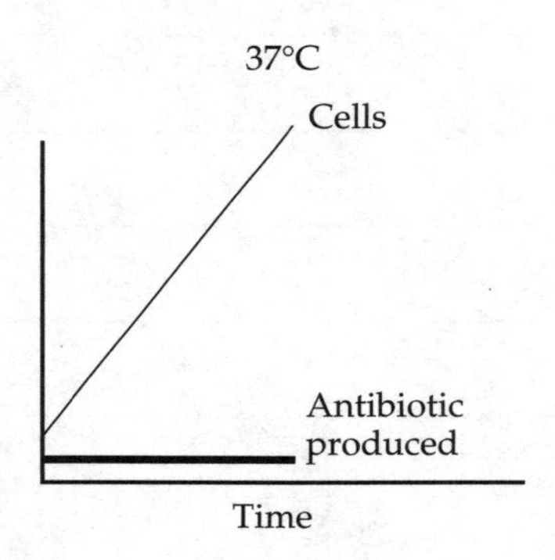

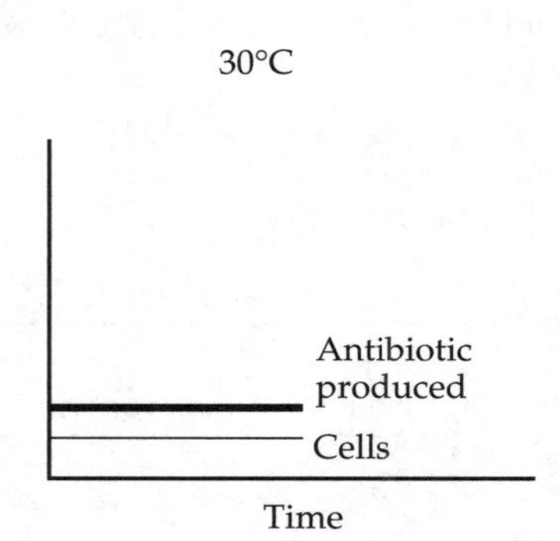

4. Why is turbidity not an accurate measurement of viable bacteria in a culture?

Exercise 21

EXERCISE 21

Biofilms

Objectives

After completing this exercise, you should be able to:

1. Define biofilm.
2. Explain the importance of biofilms in clinical medicine.
3. Demonstrate biofilm growth.

Background

Biofilms are populations or communities of microorganisms that attach and grow on a solid surface that has been exposed to water. These microorganisms are usually encased in an extracellular polysaccharide that they synthesize. As the polysaccharide enlarges, other microbes may adhere to it so several different species of bacteria and fungi may be in one biofilm. Biofilms may be found on essentially any environmental surface in which sufficient moisture is present. Biofilms may also float. The pellicle that you've seen on some liquid culture media is an example (see Color Plate X.1). This biofilm at the air–water interface allows the bacteria to float near the surface and get oxygen while getting nutrients from the liquid. Individual cells would have to expend too much energy swimming to stay near oxygen.

As bacterial cells grow, gene expression in bacterial cells changes in a process called quorum sensing. **Quorum sensing** is the ability of bacteria to communicate and coordinate behavior. Bacteria that use quorum sensing produce and secrete a signaling chemical called an *inducer*. As the inducer diffuses into the surrounding medium, other bacterial cells move toward the source and begin producing inducer. The concentration of inducer increases with increasing cell numbers. This, in turn, brings more cells and causes synthesis of more inducer.

Biofilms are not just layers of cells. A biofilm is usually composed of pillars and channels through which water can flow, bringing nutrients and taking away wastes (Figure 21.1).

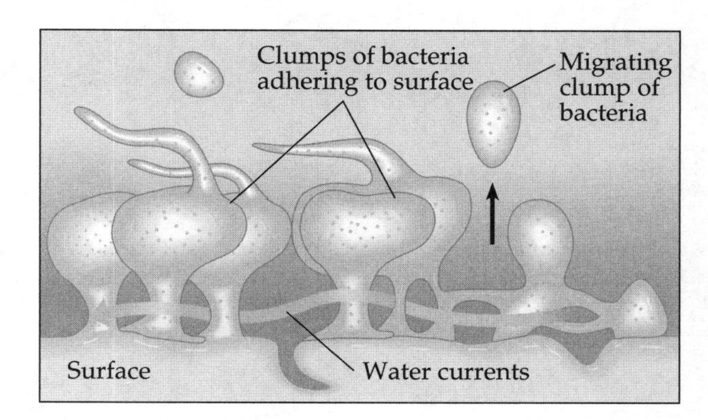

Figure 21.1
Biofilm. Water currents move through the pillars of slime formed by the growth of bacteria attached to a solid surface.

Increasing evidence indicates that biofilms play a key role in disease. Healthy bronchioles are usually sterile; however, *Pseudomonas aeruginosa* establishes a permanent infection in cystic fibrosis patients by forming a biofilm. In healthy individuals, bacteria can form biofilms on implanted prostheses. Bacteria that break off biofilms of indwelling medical devices such as catheters or protheses are a continuous source of infection in patients. The microbes in biofilms are generally more resistant to antibiotics and more difficult for the immune system to destroy.

Materials

Microscope slides (3)

Coplin jar or beaker and slide rack

Methylene blue (third day)

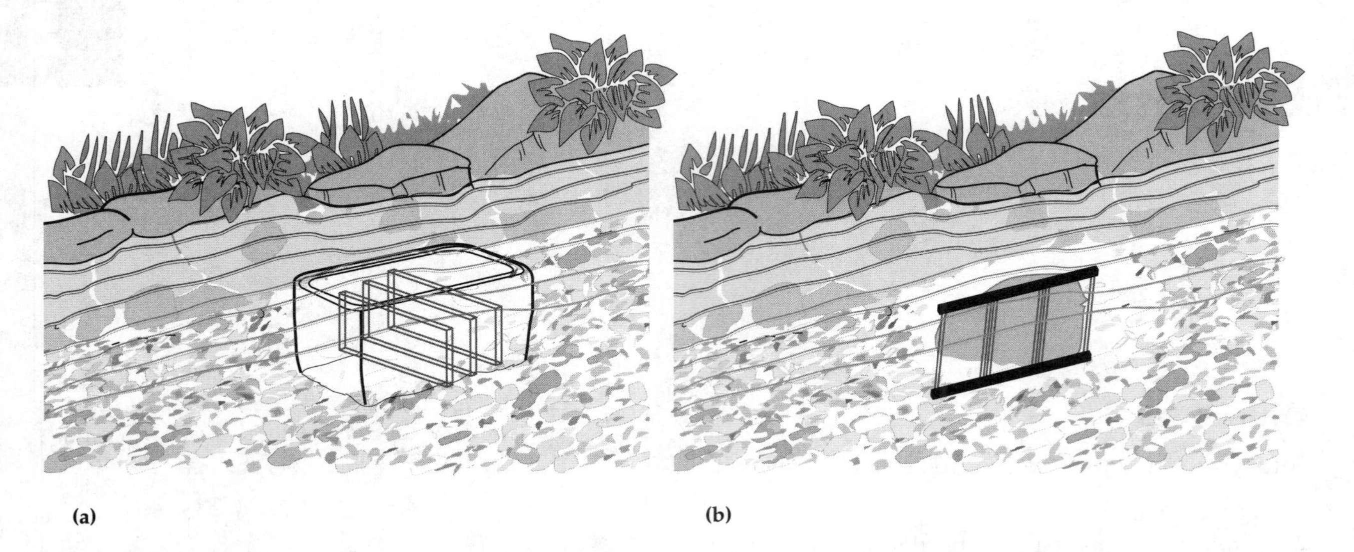

Figure 21.2
In the environment. Set slides in a pond or stream where they won't be disturbed by using **(a)** an open staining box or **(b)** a holder made from a plastic report holder.

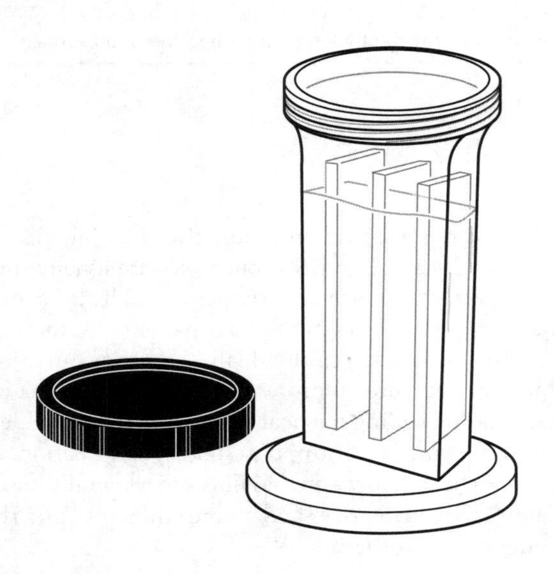

Figure 21.3
In the laboratory. Set slides in the Coplin jar. Fill the jar with liquid, leaving a few millimeters of the slides exposed above the liquid.

Cultures

Hay infusion incubated 3–5 days in the light *or*

Pond or stream water

Techniques Required

Simple staining, Exercise 3

Procedure

1. Establish a biofilm by using one of the following methods:
   a. In the environment
      1. Place three slides in a horizontal staining dish or a plastic slide rack.
      2. Place the slide dish or rack in a pond or stream in an area that is easily accessible and where the slides will not be displaced (Figure 21.2).
      3. Mark the location and note the location in your lab report.
   b. In the laboratory
      1. Place the three slides in a Coplin jar (Figure 21.3).
      2. Add enough of the culture to the jar to nearly cover the slides; leave a few millimeters.

3. Place the cap loosely on the jar and incubate the jar at room temperature.

2. Collect your biofilm.
   a. At the next laboratory period, remove one slide and label it "1."
   b. Wipe one side of the slide clean, leaving the biofilm on the other side.
   c. Air-dry and heat-fix the slide. Save the slide in your drawer for staining in step 3.
   d. Repeat steps 2a–c at the next two laboratory periods, labeling the next slide "2" and the last slide "3."

3. Observe your biofilm.
   a. On the third laboratory period, compare the appearance of the slides without a microscope. Then stain the three slides with methylene blue (Figure 3.3).
   b. Examine the stained slides microscopically using the oil immersion objective. Record your observations.

Exercise 21

Laboratory Report

Biofilms

NAME ____________________

DATE ____________________

LAB SECTION ____________________

Purpose

____

Data

Where did you set your slides? ____________________

	Slide 1	Slide 2	Slide 3
Incubated (days)			
Appearance of the slide with an unaided eye			
Microscopic appearance of the slide			

Conclusions

Compare the slides. Did any changes occur over successive days? ____________________

What different types of microorganisms did you see? What was the most abundant? ____________________

Questions

1. Are the cells evenly distributed on your slides? Provide an explanation for their distribution? ______________

____

____

2. Is a natural biofilm a pure culture? ______________________

____

____

3. How is a biofilm beneficial to bacteria? ______________________

____

____

4. What cell structures help bacteria attached to a solid surface? ______________________

____

____

5. What is a quorum in law? ______________________

   How is this related to quorum sensing in bacteria? ______________________

6. Why is quorum sensing important for biofilm formation? ______________________

____

____

Critical Thinking

1. Bacteria are the second most common cause of artificial implant failures in humans. In these cases, why isn't an infection usually detected from cultures of blood or tissue?

2. Suggest a reason why dental researchers were the first to become aware of the importance of biofilms to disease.

3. Infection-control personnel in a hospital isolated *Mycobacterium chelonae* from 27 patients. All the patients had undergone an endoscopic procedure. Endoscopes were processed after each use with an automated disinfector, which washed the endoscopes with a detergent solution, disinfected them with 2.0% glutaraldehyde for 10 minutes, and rinsed them with sterile water. What was the source of the infection?

4. A 3-year-old boy was treated with amoxicillin for a middle ear infection. He was better for a few days after completing antibiotic therapy when the infection recurred. Amoxicillin was again prescribed because the *S. pneumoniae* isolated from ear fluid was amoxicillin sensitive. Again the boy was better after completing the antibiotic therapy, but the infection recurred. What is the underlying cause of the recurrent infections?

Part Six Control of Microbial Growth

EXERCISES

The destruction of microbes by heat was employed by Lazzaro Spallanzani in the 1760s. He heated nutrient broth to kill preexisting life in his attempts to disprove the concept of spontaneous generation. In the 1860s, Pasteur heated broth in specially designed flasks and ended the debate over spontaneous generation (see the illustration on the next page). In Exercise 22, we will examine the effectiveness of heat for killing microbes.

The surgeon Joseph Lister was greatly influenced by Pasteur's demonstrations of the omnipresence of microorganisms and his proof that microorganisms cause decomposition of organic matter. Lister had observed the disastrous consequences of compound bone fractures (in which the skin is broken) compared to the relative safety of simple bone fractures. He had heard of the treatment of sewage with carbolic acid (phenol) to prevent diseases in villages that had sewage spills. In 1867, Lister wrote:

> *It appears that all that is requisite is to dress the wound with some material capable of killing these septic germs**

to prevent disease and death due to microbial growth.

---

*Quoted in W. Bulloch. *The History of Bacteriology*. New York: Dover Publications, 1974, p. 46.

In Part Six, we will examine current methods of controlling microbial growth, with special attention focused on the use of heat (Exercise 22), ultraviolet radiation (Exercise 23), disinfectants (Exercise 24), antimicrobials (Exercise 25), and hand scrubbing (Exercise 26).

Pasteur's experiment that disproved spontaneous generation involved boiling nutrient broth in an S-neck flask. Microorganisms did not appear in the cooled broth even after a long period of time.

Exercise 22

Physical Methods of Control: Heat

The successful man lengthens his stride when he discovers that the signpost has deceived him; the failure looks for a place to sit down.

John R. Rogers

Objectives

After completing this exercise, you should be able to:

1. Compare the bactericidal effectiveness of dry heat and moist heat.
2. Evaluate the heat tolerance of microbes.
3. Define and provide a use for each of the following: incineration, hot-air oven, pasteurization, boiling, and autoclaving.

Background

The use of extreme temperature to control the growth of microbes is widely employed. Generally, if heat is applied, microbes are killed; if cold temperatures are used, microbial growth is inhibited.

Bacteria exhibit different tolerances to the application of heat. Heat sensitivity is genetically determined and is partially reflected in the *optimal* growth ranges, which are **psychrophilic** (about 15°C), **psychrotrophic** (20°C to 30°C), **mesophilic** (25°C to 40°C), **thermophilic** (45°C to 65°C), **hyperthermophilic** (about 80°C or higher), and by the presence of heat-resistant endospores (Figure 22.1). Overall, bacteria are more heat resistant than most other forms of life. Heat sensitivity of organisms can be affected by container size, cell density, moisture content, pH, and medium composition.

Heat can be applied as dry or moist heat. **Dry heat,** such as that in hot-air ovens or incineration (for example, flaming loops), denatures enzymes, dehydrates microbes, and kills by oxidation effects. A standard application of dry heat in a hot-air oven is 170°C for 2 hours. The heat of hot air is not readily transferred to a cooler body such as a microbial cell. Moisture transfers heat energy to the microbial cell more efficiently than dry air, resulting in the denaturation of enzymes. **Moist heat** methods include pasteurization, boiling, and autoclaving. In **pasteurization** the temperature is maintained at 63°C for 30 minutes or 72°C for 15 seconds to kill designated organisms that are pathogenic or cause spoilage. **Boiling** (100°C) for 10 minutes will kill

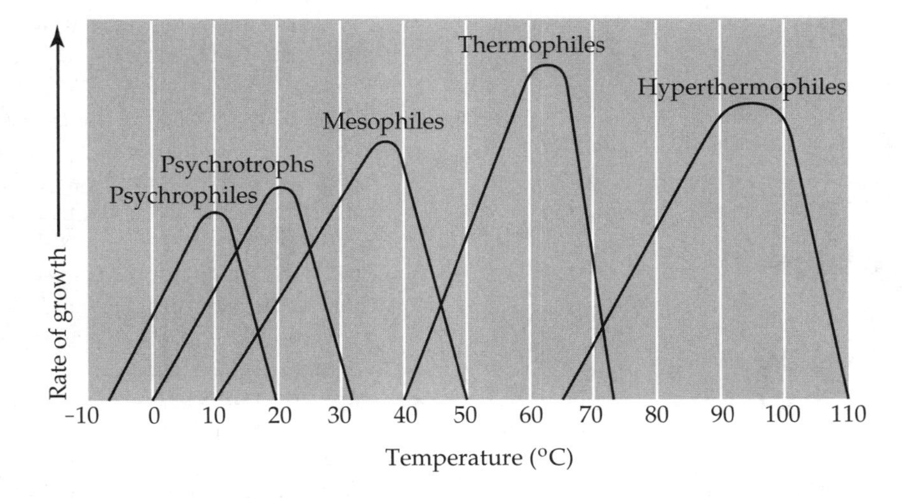

Figure 22.1

Typical growth responses of different types of microorganisms to temperature.

Table 22.1
Relationship Between Pressure and Temperature of Steam

Pressure (pounds per square inch, psi, in excess of atmospheric pressure)	Temperature (°C)
0 psi	100°C
5 psi	110°C
10 psi	116°C
15 psi	121°C
20 psi	126°C
30 psi	135°C

Source: G. J. Tortora, B. R. Funke, and C. L. Case. *Microbiology: An Introduction*, 9th ed. San Francisco, CA: Benjamin Cummings, 2007.

vegetative bacterial cells; however, endospores are not inactivated. The most effective method of moist heat sterilization is **autoclaving,** the use of steam under pressure. Increased pressure raises the boiling point of water and produces steam with a higher temperature (Table 22.1). Standard conditions for autoclaving are 15 psi, at 121°C, for 15 minutes. This is usually sufficient to kill endospores and render materials sterile.

There are two different methods of measuring heat effectiveness. **Thermal death time (TDT)** is the length of time required to kill all bacteria in a liquid culture at a given temperature. The less common **thermal death point (TDP)** is the temperature required to kill all bacteria in a liquid culture in 10 minutes.

Materials

Petri plates containing nutrient agar (2)

Thermometer

Empty tube

Beaker

Hot plate or tripod and asbestos pad

Ice

Cultures (as assigned)

Group A:

Old (48 to 72 hours) *Bacillus subtilis*
Young (24 hours) *Bacillus subtilis*

Group B:

Staphylococcus epidermidis
Escherichia coli

Group C:

Young (24 hours) *Bacillus subtilis*
Escherichia coli

Group D:

Mold (*Penicillium*) spore suspension
Old (48 to 72 hours) *Bacillus subtilis*

Demonstration

Autoclaved and dry-heated soil

Techniques Required

Inoculating loop technique, Exercise 10

Aseptic technique, Exercise 10

Plate streaking, Exercise 11

Graphing, Appendix D

Procedure

Each pair of students is assigned two cultures and a temperature.

Group	Group
A: 63°C ________	A: 72°C ________
B: 63°C ________	B: 72°C ________
C: 63°C ________	C: 72°C ________
D: 63°C ________	D: 72°C ________

You can share beakers of water as long as the effect of the same temperature is being evaluated.

1. Divide two plates of nutrient agar into five sections each. Label the sections "0," "30 sec," "2 min," "5 min," and "15 min."
2. Set up a water bath in the beaker, with the water level higher than the level of the broth in the tubes. Do not put the broth tubes into the water bath at this time. Carefully put the thermometer in a test tube of water in the bath.
3. Streak the assigned organisms on the "0" time section of the appropriate plate. Why are we using "old" and "young" *Bacillus* cultures? ____________
   ____
4. Raise the temperature of the bath to the desired temperature and maintain that temperature. Use ice to adjust the temperature. Why was 63°C selected as one of the temperatures? ____________
   ____

5. Place the broth tubes of your organism into the bath when the temperature is at the desired point. After 30 seconds, remove the tubes, resuspend the culture, streak a loopful on the corresponding sections, and return the tubes to the water bath. Repeat at 2, 5, and 15 minutes. What is the longest time period that any microbe is exposed to heat?

   ___

6. When you are done, clean the beaker and return the materials. Incubate the plates, inverted, at 35°C until the next lab period. Record your results and the results for the other organisms tested: (−) = no growth, (+) = minimum growth, (2+) = moderate growth, (3+) = heavy growth, and (4+) = maximum growth.
7. Examine the demonstration plates and record your observations. (Refer to Color Plate V.3.) Collect results from your classmates to complete the data table in your Laboratory Report.

Exercise 22

Physical Methods of Control: Heat

LABORATORY REPORT

NAME ______________________

DATE ______________________

LAB SECTION ______________________

Purpose

____

____

____

Data

Record growth on a scale from (−) to (4+).

Organism	Temperature/Time									
	63°C					72°C				
	0	30 sec	2 min	5 min	15 min	0	30 sec	2 min	5 min	15 min
Old *Bacillus subtilis*										
Young *Bacillus subtilis*										
Staphylococcus epidermidis										
Escherichia coli										
Mold (*Penicillium*) spores										

Demonstration Plates

	Control	Autoclaved	Dry-Heated
Number of colonies			
Number of different colonies			

Use a computer graphing application to graph the effect of heating on each organism or draw your graphs below.

Graph your cultures at ____°C.

Organism ______________________

Growth 4 3 2 1 0

Time (min)

Organism ______________________

Growth 4 3 2 1 0

Time (min)

Conclusions

____

Questions

1. Compare the heat sensitivity of fungal spores to that of bacterial endospores. ______________________

2. Compare the effectiveness of autoclaving and dry heat. ______________________

3. Give an example of an application (use) of thermal death time. ______________________

4. In the exercise, was the thermal death time or thermal death point determined? ______________________

____

5. Give an example of a nonlaboratory use of each of the following methods to control microbial growth:

   a. Incineration: ____

   b. Pasteurization: ____

   c. Autoclaving: ____

6. Define pasteurization. What is the purpose of pasteurization? ______________________

____

____

____

Critical Thinking

1. Explain why fungi and *Bacillus* sometimes grow better after heat treatment.

2. The decimal reduction time (DRT) is the time it takes to kill 90% of cells present. Assume that a DRT value for autoclaving a culture is 1.5 minutes. How long would it take to kill all the cells if 10^6 cells were present? What would happen if you stopped the heating process at 9 minutes?

3. Indicators are used in autoclaving to ensure that sterilization is complete. One type of chemical indicator turns color when it has reached a specific temperature; the other type turns color when it has reached a specified temperature and been exposed to steam. Which type of indicator should be used?

4. A biological indicator used in autoclaving is a vial containing 10^9 *Geobacillus stearothermophilus* cells that is placed in the autoclave with the material to be sterilized. After autoclaving, the vial is incubated and examined for growth. Why is this species used as opposed to *E. coli* or *Bacillus subtilis*?

[illegible]

[illegible]

[illegible]

[illegible]

Critical Thinking

1. [illegible]

2. [illegible]

3. [illegible]

4. [illegible]

Exercise 23

Physical Methods of Control: Ultraviolet Radiation

Objectives

After completing this exercise, you should be able to:

1. Examine the effects of ultraviolet radiation on bacteria.
2. Explain the method of action of ultraviolet radiation and light repair of mutations.

Background

Radiant energy comes to the Earth from the Sun and other extraterrestrial sources, and some is generated on Earth from natural and human-made sources. The **radiant energy spectrum** is shown in Figure 23.1. Radiation differs in wavelength and energy. The shorter wavelengths have more energy. X rays and gamma rays are forms of **ionizing radiation.** Their principal effect is to ionize water into *highly reactive free radicals* (with unpaired electrons) that can break strands of DNA. The effect of radiation is influenced by many variables, such as the age of the cells, media composition, and temperature.

Some **nonionizing** wavelengths are essential for biochemical processes. The main absorption wavelengths for green algae, green plants, and photosynthetic bacteria are shown in Figure 23.1a. Animal cells synthesize vitamin D in the presence of light around 300 nm. Nonionizing radiation between 15 and 400 nm is called **ultraviolet (UV).** Wavelengths below 200 nm are absorbed by air and do not reach living organisms. The most lethal wavelengths, sometimes called biocidal, are in the **UVC** range, 200–290 nm. These wavelengths correspond to the optimal absorption wavelengths of DNA (Figure 23.1b). **UVB** wavelengths (290–320 nm) can also cause damage to DNA. **UVA** wavelengths (320–400 nm) are not as readily absorbed and are therefore less active on living organisms.

Ultraviolet light induces *thymine dimers* in DNA, which result in a mutation. Mutations in critical genes may result in the death of the cell unless the damage is repaired. When thymine dimers are exposed to visible light, *photolyases* are activated; these enzymes split the dimers, restoring the DNA to its undamaged state. This is called **light repair** or **photoreactivation.** Another repair mechanism, called **dark repair,** is independent of light. Dimers are removed by endonuclease, DNA polymerase replaces the nucleotides, and DNA ligase seals the sugar–phosphate backbone.

As a sterilizing agent, ultraviolet radiation is limited by its poor penetrating ability. It is used to sterilize some heat-labile solutions, to decontaminate hospital operating rooms and food-processing areas, and to disinfect wastewater.

In this exercise, we will investigate the penetrating ability of ultraviolet light and light repair by using lamps of the desired wavelength. (See Color Plate IV.1.)

Materials

Petri plates containing nutrient agar (3)

Sterile cotton swabs (3)

Covers (choose one): Gauze; 3, 6, or 12 layers of paper; cloth; aluminum foil; clear glass; sunglasses; or plastic

Ultraviolet lamp (265 nm)

Plastic safety glasses

Cultures (as assigned)

Bacillus subtilis

Staphylococcus epidermidis

Micrococcus luteus

Techniques Required

Aseptic technique, Exercise 10

Procedure

1. Swab the surface of each plate with *one* of the cultures; to ensure complete coverage, swab the surface in two directions. Label the plates "A," "B," and "C."
2. Remove the lid of an inoculated plate and cover one-half of the plate with one of the covering materials (Figure 23.2). Cover one-half of each of the remaining plates with the same material.

Do not look at the ultraviolet light, and do not leave your hand exposed to it. Wear safety glasses.

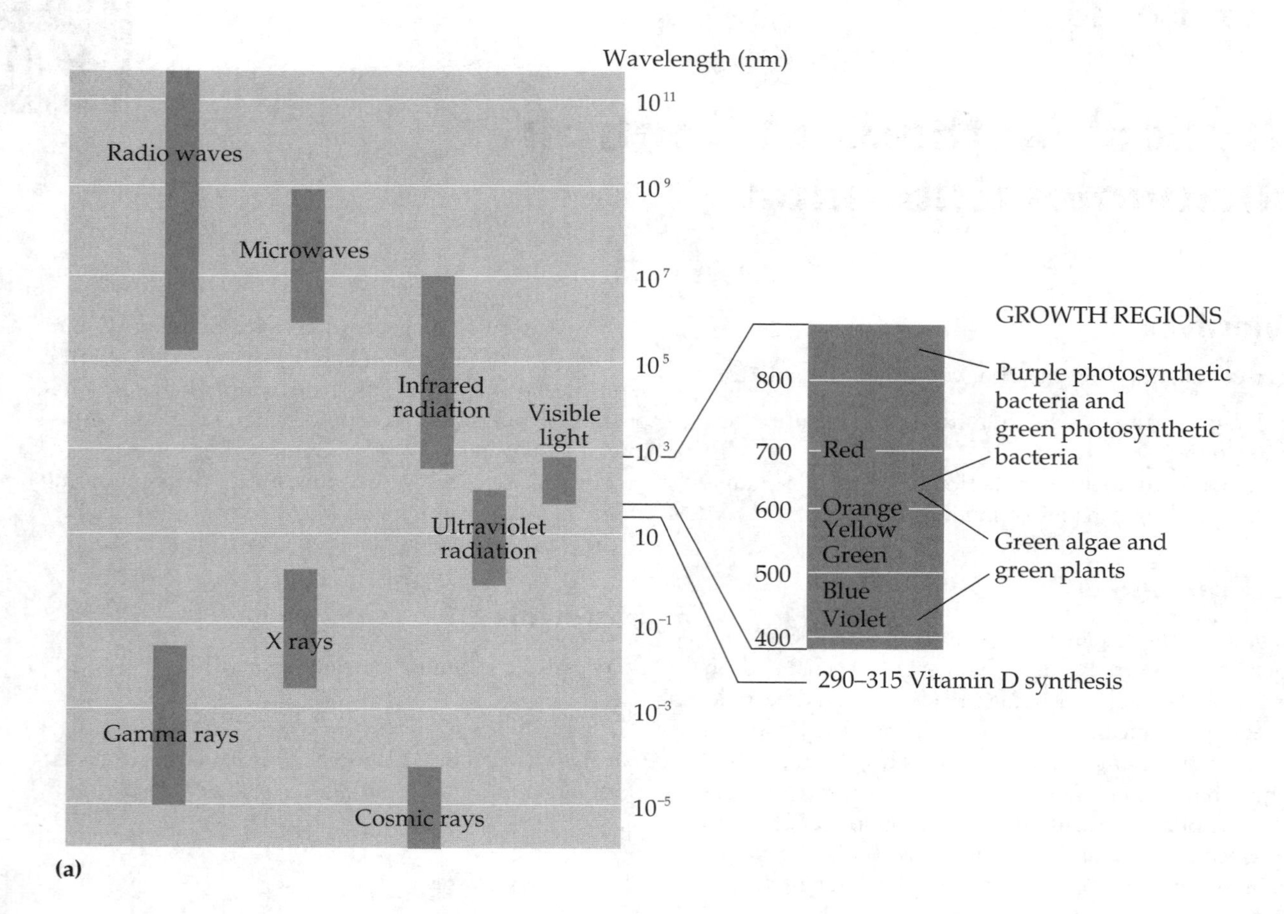

(a)

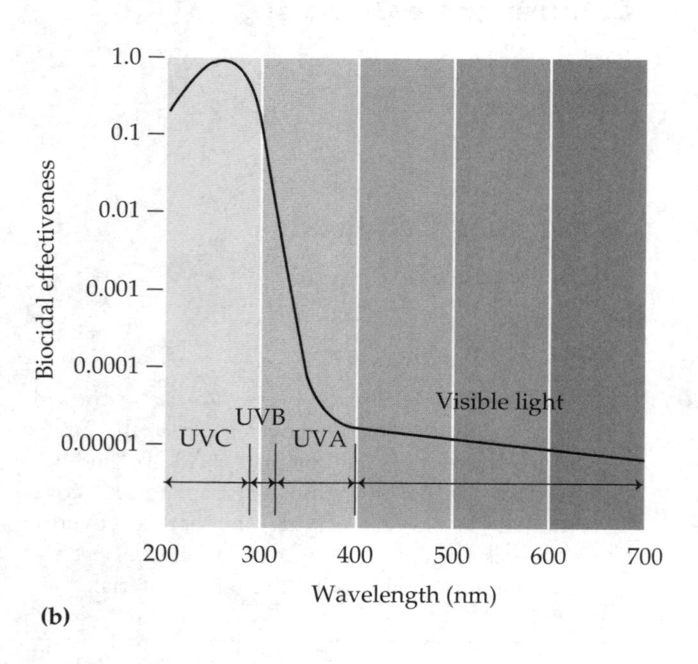

(b)

Figure 23.1

Radiant energy. **(a)** Radiant energy spectrum and absorption of light for growth. **(b)** Biocidal effectiveness of radiant energy between 200 and 700 nanometers (nm) (from UV to visible red light).

3. Place each plate directly under the ultraviolet light 24 cm from the light with the *lid off*, agar-side up, with the covering material on one-half of the plate. Why should the lid be removed? ______________
____
____

   Plate A: Expose to UV light for 30 seconds. Remove the covering material, replace the lid, and incubate in a dark environment at room temperature.
   Plate B: Expose to UV light for 30 seconds. Remove the covering material and replace the Petri plate lid. Incubate in sunlight or under lights at room temperature.
   Plate C: Expose for 60 seconds. Remove covering material, replace the lid, and incubate in a dark environment at room temperature.
4. Incubate all three plates at room temperature until the next period.
5. Examine all plates and record your results. Observe the results of students using the other organisms.

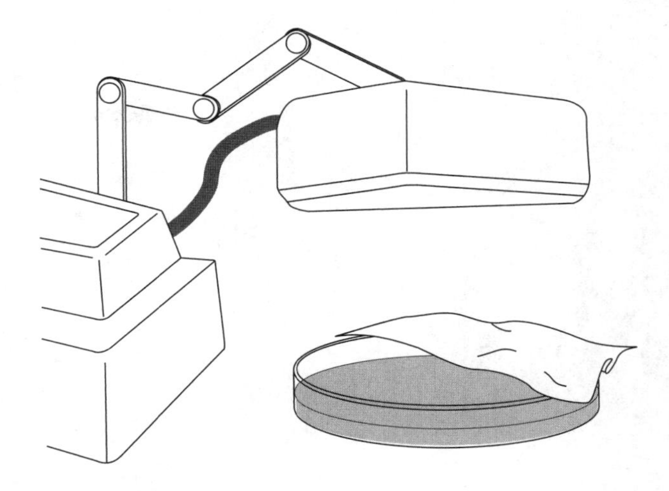

Figure 23.2
With the lid removed, cover one-half of an inoculated Petri plate with one of the covering materials.

Exercise 23

LABORATORY REPORT

Physical Methods of Control: Ultraviolet Radiation

NAME ______________________

DATE ______________________

LAB SECTION ______________________

Purpose

____

____

____

Data

What organism did you use? ______________________

What did you use to cover one-half of each plate? ______________________

Sketch your results. Note any pigmentation.

A

B

C

Classmates' results:

Bacteria						
Plate	A	B	C	A	B	C
Amount of growth on UV-exposed area						
Pigment						

Conclusions

____

____

____

Questions

1. If the *Bacillus* had sporulated before exposure to radiation, would that affect the results? ______________________

____

____

2. What are the variables in ultraviolet radiation treatment? ______________________

____

____

3. Many of the microorganisms found on environmental surfaces are pigmented. Of what possible advantage is the pigment? ______________________

____

____

Critical Thinking

1. Can dark repair be a factor in this experiment?

2. Why are there still some colonies growing in the areas exposed to ultraviolet light?

3. How might the results differ if a UVA lamp were used? A UVB lamp?

4. Considering your results, discuss the possible effects of UV radiation on the ecology of a lake if the UV radiation has the same effect on the lake bacteria as it did on the bacteria in your experiment.

Exercise 24

Chemical Methods of Control: Disinfectants and Antiseptics

One nineteenth-century method of avoiding cholera: Wear a pouch of foul-smelling herbs around your neck. If the odor is bad enough, disease carriers will spare you the trouble of avoiding them.

ANONYMOUS

Objectives

After completing this exercise, you should be able to:

1. Define the following terms: disinfectant and antiseptic.
2. Describe the use-dilution test.
3. Evaluate the relative effectiveness of various chemical substances as antimicrobial agents.

Background

A wide variety of chemicals called **antimicrobial agents** are available for controlling the growth of microbes. Chemotherapeutic agents are used internally and will be evaluated in another exercise. **Disinfectants** are chemical agents used on inanimate objects to lower the level of microbes on their surfaces; **antiseptics** are chemicals used on living tissue to decrease the number of microbes. Disinfectants and antiseptics affect bacteria in many ways. Those that result in bacterial death are called **bactericidal agents.** Those causing temporary inhibition of growth are **bacteriostatic agents.**

No single chemical is the best to use in all situations. Antimicrobial agents must be matched to specific organisms and environmental conditions. Additional variables to consider in selecting an antimicrobial agent include pH, solubility, toxicity, organic material present, and cost. In evaluating the effectiveness of antimicrobial agents, the concentration, length of contact, and whether it is lethal (*-cidal*) or inhibiting (*-static*) are the important criteria. The standard method for measuring the effectiveness of a chemical agent is the **American Official Analytical Chemist's use-dilution test.** For most purposes, three strains of bacteria are used in this test: *Salmonella enterica* Choleraesuis, *Staphylococcus aureus*, and *Pseudomonas aeruginosa*. To perform a use-dilution test, metal rings are dipped into standardized cultures of the test bacteria grown in liquid media, removed, and dried. The rings are then placed into a solution of the disinfectant at the concentration recommended by the manufacturer for 10 minutes at 20°C. The rings are then transferred to a nutrient medium to permit the growth of any surviving bacteria. The effectiveness of the disinfectant can then be determined by the amount of resulting growth. The use-dilution test is limited to bactericidal compounds and cannot be used to evaluate bacteriostatic compounds.

In this exercise, we will perform a modified use-dilution test.

Materials

Petri plates containing nutrient agar (2)

Sterile water

Sterile tubes (3)

Sterile 5-ml pipettes (2)

Sterile 1-ml pipettes (2)

Test substance: chemical agents such as bathroom cleaner, floor cleaner, mouthwash, lens cleaner, and acne cream. Bring your own.

Culture

Staphylococcus aureus

Techniques Required

Inoculating loop technique, Exercise 10

Aseptic technique, Exercise 10

Pipetting, Appendix A

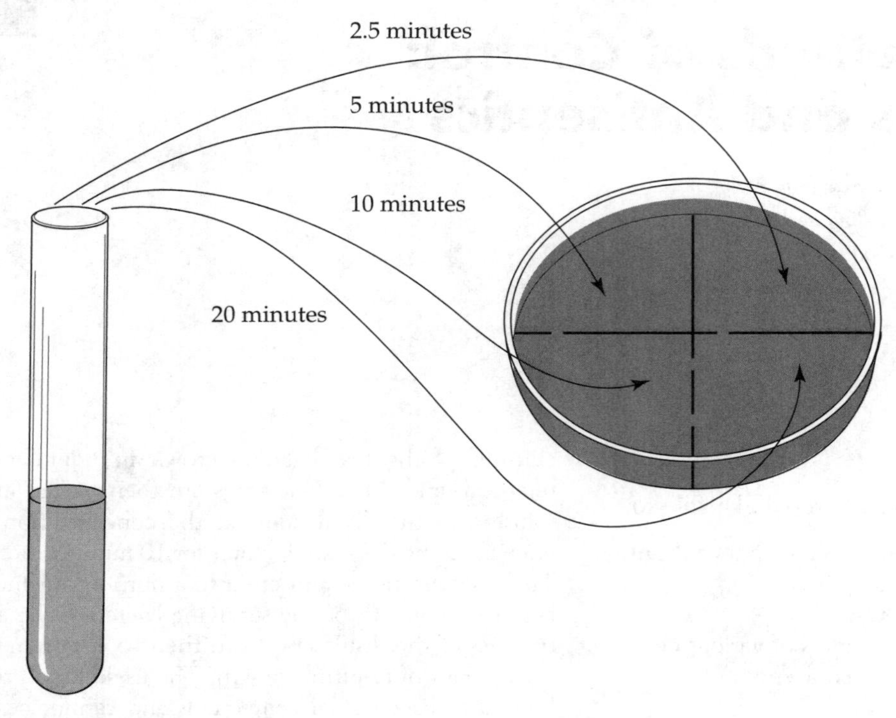

Figure 24.1

Transfer a loopful from the tube containing the chemical and *Staphylococcus* to the appropriate sector at the time intervals shown. Repeat the procedure with a loopful from the tube containing the test substance and *Staphylococcus* onto the appropriate sector of the second nutrient agar plate.

Procedure

1. Using sterile water, prepare a dilution of the test substance in a sterile tube, diluted to the strength at which it is normally used. If it is a paste, it must be suspended in sterile water.
2. Transfer 5 ml of the diluted test substance to a sterile tube. If the test substance is normally used at full strength, then don't dilute it for this experiment. Label the tube. Add 5 ml of your laboratory disinfectant to another sterile tube. What is the disinfectant you use to disinfect your lab bench?

   ____

   Label the tube.
3. Divide one plate of nutrient agar into five sections. Label the sections "0," "D-2.5," "D-5," "D-10," and "D-20." The D stands for laboratory disinfectant.
4. Label the other nutrient agar plate for the other chemical and divide it into four sections. Label the sections "2.5," "5," "10," and "20."
5. Inoculate the 0 sector with a loopful of *S. aureus*.
6. Aseptically add 0.5 ml of the *S. aureus* culture to each tube prepared in step 2.
7. Transfer one loopful from each tube to a corresponding sector at 2.5 minutes, 5 minutes, 10 minutes, and 20 minutes (Figure 24.1).
8. Incubate the plates, inverted, at 35°C until the next lab period. (Discard the chemical/bacteria mixtures in the To Be Autoclaved area.)
9. Observe the plates for growth. Record the growth as (−) = no growth, (+) = minimum growth, (2+) = moderate growth, (3+) = heavy growth, and (4+) = maximum growth.

Exercise 24

LABORATORY REPORT

Chemical Methods of Control: Disinfectants and Antiseptics

NAME ______________________

DATE ______________________

LAB SECTION ______________________

Purpose ______________________

Data

Time of Exposure (min)	Amount of Growth		
	Control	Lab Disinfectant	Chemical: ________
0			
2.5			
5			
10			
20			

Conclusions ______________________

Questions

1. Was this a fair test? Is it representative of the effectiveness of the test substance? ______________________

2. Read the label of the preparation you tested. What is (are) the active ingredient(s)? ____________________

____

Using your textbook or another reference, find the method of action of the active ingredient(s) in the test substance.

____

____

____

3. What is the use-dilution method? ____

____

____

Critical Thinking

1. How could the procedures used in this experiment be altered to measure bacteriostatic effects?

2. In the use-dilution test, a chemical is evaluated by its ability to kill 10^6 to 10^8 dried *Clostridium sporogenes* or *Bacillus subtilis* endospores. Why is this considered a stringent test?

3. The effectiveness of disinfectants can be measured in DRT values. DRT, or decimal reduction time, is the length of time it takes to kill 90% of a test population of bacteria. The DRT values for contact lens disinfectants against *Serratia marcescens* are as follows:

Disinfectant	DRT Value (min)	Disinfectant	DRT Value (min)
Chlorhexidine, 0.005%	2.8	Thimerosal, 0.002%	138.9
Hydrogen peroxide, 3%	3.1	Polyquaternium-1, 0.001%	383.3

Which disinfectant is most effective? ____________________ What is the minimum time that lenses with 10^2 bacteria should be soaked in chlorhexidine? ____

In polyquaternium-1? ____________________ What if the lenses are contaminated with *Staphylococcus* or *Acanthamoeba*? ____

____

Why isn't a higher concentration of disinfectant used? ____

Exercise 25

Chemical Methods of Control: Antimicrobial Drugs

The aim of medicine is to prevent disease and prolong life; the ideal of medicine is to eliminate the need of a physician.

WILLIAM JAMES MAYO

Objectives

After completing this exercise, you should be able to:

1. Define the following terms: antibiotic, antimicrobial agent, and MIC.
2. Perform an antibiotic sensitivity test.
3. Provide the rationale for the agar diffusion technique.

Background

The observation that some microbes inhibited the growth of others was made as early as 1874. Pasteur and others observed that infecting an animal with *Pseudomonas aeruginosa* protected the animal against *Bacillus anthracis*. Later investigators coined the word **antibiosis** (against life) for this inhibition and called the inhibiting substance an **antibiotic.** In 1928, Alexander Fleming observed antibiosis around a *Penicillium* mold growth on a culture of staphylococci. He found that culture filtrates of *Penicillium* inhibited the growth of many gram-positive cocci and *Neisseria* spp. In 1940, Selman A. Waksman isolated the antibiotic streptomycin, produced by an actinomycete. This antibiotic was effective against many bacteria that were not affected by penicillin. Actinomycetes remain an important source of antibiotics. Today, research investigators look for antibiotic-producing actinomycetes and fungi in soil and have synthesized many antimicrobial substances in the laboratory. Antimicrobial chemicals absorbed or used internally, whether natural (antibiotics) or synthetic, are called **antimicrobial agents.**

A physician or dentist needs to select the correct antimicrobial agent intelligently and administer the appropriate dose in order to treat an infectious disease; then the practitioner must follow that treatment in order to be aware of resistant forms of the organism that might occur. The clinical laboratory isolates the **pathogen** (disease-causing organism) from a clinical sample and determines its sensitivity to antimicrobial agents.

In the **disk-diffusion method,** a Petri plate containing an agar growth medium is inoculated uniformly over its entire surface. Paper disks impregnated with various antimicrobial agents are placed on the surface of the agar. During incubation, the antimicrobial agent *diffuses* from the disk, from an area of high concentration to an area of lower concentration. An effective agent will inhibit bacterial growth, and measurements can be made of the size of the **zones of inhibition** around the disks. The concentration of antimicrobial agent at the edge of the zone of inhibition represents its **minimum inhibitory concentration (MIC).** The MIC is determined by comparing the zone of inhibition with MIC values in a standard table (Table 25.1). The MIC values are determined by doing a broth dilution test in a laboratory by using a test bacterium. The zone size is affected by such factors as the diffusion rate of the antimicrobial agent and the growth rate of the organism. To minimize the variance between laboratories, the standardized **Kirby-Bauer test** for agar diffusion methods is performed in many clinical laboratories with strict quality controls. This test uses *Mueller-Hinton agar.* Mueller-Hinton agar allows the antimicrobial agent to diffuse freely.

In this exercise, we will evaluate antimicrobial agents by the disk-diffusion method.

Materials

Petri plate containing Mueller-Hinton agar

Sterile cotton swab

Dispenser and antimicrobial disks

Forceps

Alcohol

Ruler (second period)

Cultures (as assigned)

Staphylococcus aureus broth

Escherichia coli broth

Pseudomonas aeruginosa broth

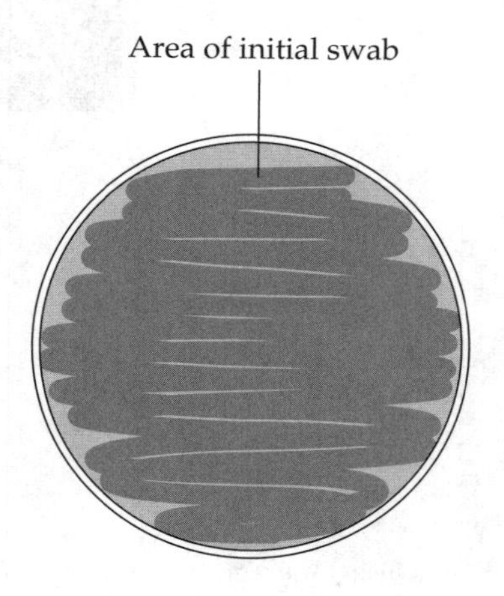

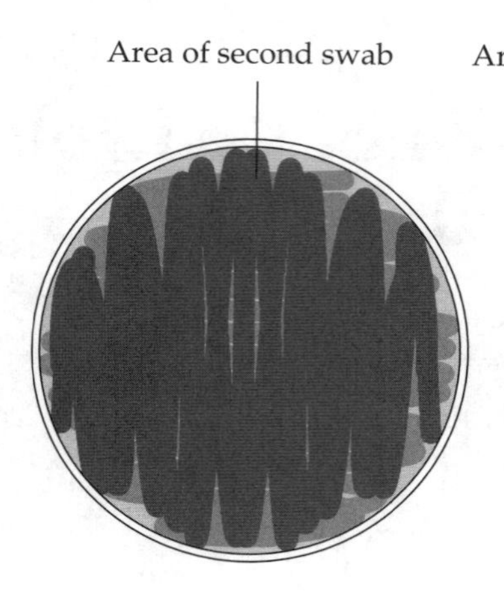

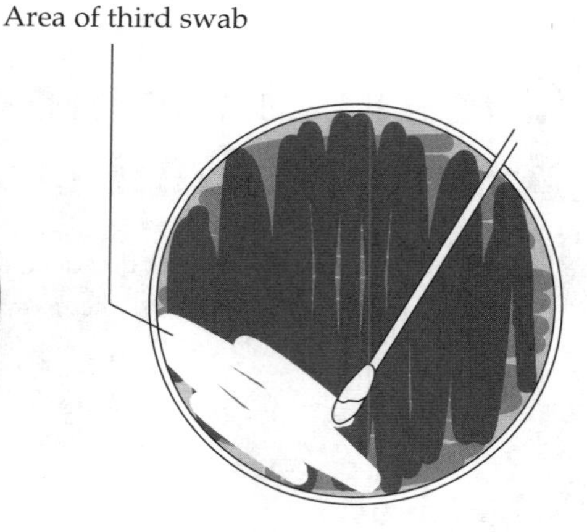

Figure 25.1

Dip a cotton swab in the culture to be tested and swab across the surface of the agar without leaving any gaps. Using the same swab, swab the agar in a direction perpendicular to the first inoculum. Repeat, swabbing the agar at a 45° angle to the first inoculum.

Techniques Required

Inoculating loop technique, Exercise 10

Aseptic technique, Exercise 10

Procedure

1. Aseptically swab the assigned culture onto the appropriate plate. Swab in three directions to ensure complete plate coverage (Figure 25.1). Why is complete coverage essential? ______________

   ____

   Let stand at least 5 minutes.
2. Follow procedure a or b.
   a. Place the antimicrobial-impregnated disks by pushing the dispenser over the agar. Sterilize your loop and touch each disk with the sterile inoculating loop to ensure better contact with the agar. Record the agents and the disk codes in your Laboratory Report. Circle the corresponding chemicals in Table 25.1.
   b. Sterilize forceps by dipping them in alcohol and burning off the alcohol.

While it is burning, hold the forceps pointed down. Keep the beaker of alcohol away from the flame.

Obtain a disk impregnated with a antimicrobial agent and place it on the surface of the agar (Figure 25.2a). Gently tap the disk with the forceps to ensure better contact with the agar. Repeat, placing five to six different disks the same distance apart on the Petri plate. See the location of the disks in Figure 25.2b. Record the agents and the disk codes in your

(a) Place disks impregnated with antimicrobial agents on an inoculated culture medium with sterile forceps to get the pattern shown in **(b)**.

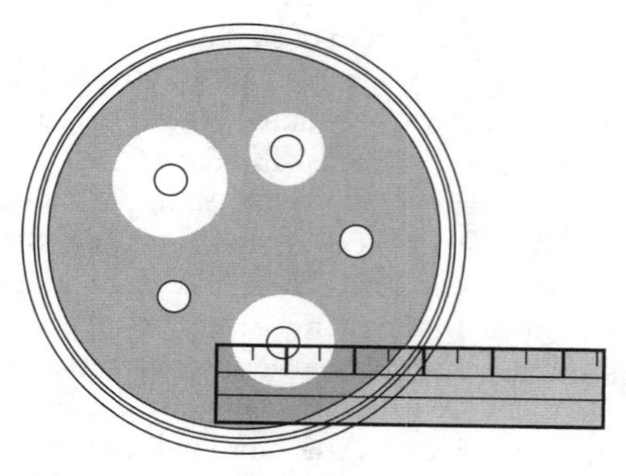

(b) After incubation, measure the diameters of zones of inhibition.

Figure 25.2

Disk-diffusion method.

Table 25.1

Interpretation of Inhibition Zones of Test Cultures

Disk Symbol	Antimicrobial Agent	Disk Content	Diameter of Zones of Inhibition (mm)		
			Resistant	Intermediate	Susceptible
AM	Ampicillin when testing gram-negative bacteria	10 µg	<13	14–16	>17
	Ampicillin when testing gram-positive bacteria	10 µg	<28	—	>29
C	Chloramphenicol	30 µg	<12	13–17	>18
CAZ	Ceftazidime	30 µg	<14	15–17	>18
CB	Carbenicillin	100 µg	<19	—	>23
	Carbenicillin when testing *Pseudomonas*	100 µg	<13	—	>17
CF	Cephalothin	30 µg	<14	—	>18
CIP	Ciprofloxacin	5 µg	<15	16–20	>21
E	Erythromycin	15 µg	<13	14–22	>23
Fox	Cefoxitin (Mefoxin)	30 µg	<14	—	>18
G	Sulfisoxazole (Gantrisin)	300 µg	<12	13–16	>17
GM	Gentamicin	10 µg	<12	13–14	>15
IPM	Imipenem	10 µg	<13	14–15	>16
P	Penicillin G when testing staphylococci	10 units	<28	—	>29
	Penicillin G when testing other bacteria	10 units	<14	—	>15
R	Rifampin	5 µg	<16	17–19	>20
S	Streptomycin	10 µg	<11	12–14	>15
SxT	Trimethoprim/ Sulfamethoxazole	1.25 µg/ 23.75 µg	<10	11–15	>16
Te	Tetracycline	30 µg	<14	15–18	>19
VA	Vancomycin	30 µg	<9	10–11	>12
	Vancomycin when testing enterococci	30 µg	<14	15–16	>17

Source: National Committee for Clinical Laboratory Standards. Performance Standards for Antimicrobial Disk Susceptibility Tests.

Laboratory Report. Circle the corresponding chemicals in Table 25.1.

3. Incubate the plate, inverted, at 35°C until the next period. Measure the zones of inhibition in millimeters, using a ruler on the underside of the plate (Figure 25.2b). If the diameter is difficult to measure, the radius from the center of the disk to the edge of the zone can be measured. Multiply the radius by 2 to get the diameter of the zone. Record the zone size and, based on the values in Table 25.1, indicate whether the organism is susceptible, intermediate, or resistant. Record the results of students using the other two bacteria. (See Color Plates VIII.1 and VIII.2.)

Exercise 25

LABORATORY REPORT

Chemical Methods of Control: Antimicrobial Drugs

NAME ______________________

DATE ______________________

LAB SECTION ______________________

Purpose

____

____

____

Data

Antimicrobial Agent	Disk Code	*Staphylococcus aureus*		*Escherichia coli*		*Pseudomonas aeruginosa*	
		Zone Size	S, I, or R*	Zone Size	S, I, or R*	Zone Size	S, I, or R*
1.							
2.							
3.							
4.							
5.							
6.							
7.							
8.							

*S = susceptible; I = intermediate; R = resistant.

Conclusions

Which antimicrobial agents were most effective against each organism? ______________________

____

____

Questions

1. Is the disk-diffusion technique measuring bacteriostatic or bactericidal activity? Briefly explain.

____

____

2. In which growth phase is an organism most sensitive to an antimicrobial agent? ______________________

3. Why is the disk-diffusion technique not a perfect indication of how the drug will perform in vivo? What other factors are considered before using the antimicrobial agent in vivo? ______________________

____

____

4. Using your textbook or other references, match each of the antimicrobial agents listed in Table 25.1 with its type and method of action.

Type of Antimicrobial Agent

a. Aminoglycosides
b. β-lactams
c. Carbapenems
d. Cephalosporins
e. Glycopeptides
f. Macrolides
g. Monobactams
h. Quinolones
i. Sulfonamides
j. Tetracyclines
k. None of the above

Method of Action

1. Inhibit enzyme activity
2. Inhibit cell wall synthesis
3. Inhibit protein synthesis
4. Inhibit nucleic acid synthesis

	Type	Method of Action
Ampicillin		
Carbenicillin		
Cefoxitin		
Ceftazidime		
Cephalothin		
Chloramphenicol		
Ciprofloxacin		
Erythromycin		
Gentamicin		
Imipenem		
Penicillin G		
Rifampin		
Streptomycin		
Sulfisoxazole		
Tetracycline		
Trimethoprim/Sulfamethoxazole		
Vancomycin		

5. From your results, which type of antimicrobial agent was most effective against gram-negative bacteria?

____

Against gram-positive bacteria? ______________________________

Critical Thinking

1. What effect would the presence of tetracycline in the body have on penicillin therapy?

2. The following results were obtained from a disk-diffusion test against a bacterium:

Antibiotic	Zone of Inhibition (mm)
A	6
B	18
C	11
D	18

Which drug should be used to treat an infection caused by this bacterium? Briefly explain.

3. The broth dilution test can be used to determine the effectiveness of an antibiotic. In this test, serial dilutions of the antibiotic were set up in the wells of a microtiter plate. Equal amounts of broth culture of *Staphylococcus aureus* were added to each well. After incubation, the wells were examined for bacterial growth. Wells with no growth were subcultured in nutrient broth without the antibiotic. Results were recorded as (+) for growth and (−) for no growth.

Antibiotic	Dilution	Growth	Growth in Subculture
A	1:10 through 1:70	−	−
	1:80	−	−
	1:90	−	+
	1:100	−	+
	1:200 through 1:500	+	+
B	1:10 through 1:150	−	−
	1:160	−	+
	1:170	+	+
	1:180	+	+
	1:190 through 1:500	+	+

What is the minimum bactericidal concentration of each antibiotic? ____________________

What is the minimum bacteriostatic concentration? ____________________

Which antibiotic is more effective against *S. aureus*? ____________________

Which antibiotic is more effective against *Salmonella enterica*? ____________________

Exercise 26

Effectiveness of Hand Scrubbing

People are sick because they are poor; they become poorer because they are sick, and they become sicker because they are poorer.

ANONYMOUS

Objectives

After completing this exercise, you should be able to:

1. Evaluate the effectiveness of handwashing and a surgical scrub.
2. Explain the importance of aseptic technique in the hospital environment.

Background

The skin is sterile during fetal development. After birth, a baby's skin is colonized by many bacteria for the rest of its life. As an individual ages and changes environments, the microbial population changes to match the environmental conditions. The microorganisms that are more or less permanent are called **normal microbiota.** Microbes that are present only for days or weeks are referred to as **transient microbiota.**

Discovery of the importance of handwashing in disease prevention is credited to Ignaz Semmelweis at Vienna General Hospital in 1846. He noted that the lack of aseptic methods was directly related to the incidence of puerperal fever and other diseases. Medical students would go directly from the autopsy room to the patient's bedside and assist in child delivery without washing their hands. Less puerperal sepsis occurred in patients attended by midwives, who did not touch cadavers. Semmelweis established a policy for the medical students of handwashing with a chloride of lime solution that resulted in a drop in the death rate due to puerperal sepsis from 12% to 1.2% in one year. Guidelines from the Centers for Disease Control and Prevention state that "handwashing is the single most important procedure for preventing nosocomial infections," yet recent studies in hospitals show handwashing rates as low as 31%.

A layer of oil and the structure of the skin prevent the removal of all bacteria by handwashing. Soap helps remove the oil, and scrubbing will maximize the removal of bacteria. Hospital procedures require personnel to wash their hands before attending a patient, and a complete surgical scrub—removing the transient and many of the resident microbiota—is done before surgery. Transient microbiota are usually removed after 10 to 15 minutes of scrubbing with soap. The surgeon's skin is never sterilized. Only burning or scraping it off would achieve that.

In this exercise, we will examine the effectiveness of washing skin with soap and water. Only organisms capable of growing aerobically on nutrient agar will be observed. Because organisms with different nutritional and environmental requirements will not grow, this procedure will involve only a minimum number of the skin microbiota.

Materials

Petri plates containing nutrient agar (2)

Scrub brush

Bar soap or liquid soap (bring one from home)

Waterless hand cleaner

Techniques Required

Colony morphology, Exercise 9

Procedure

1. Select two nutrient agar plates.
   a. Divide one nutrient agar plate into four quadrants. Label the sections 1 through 4. Label the plate "Water." ________
   b. Divide the other nutrient agar plate into five sections. Label the sections 1 through 5. Label the plate "Soap." Which soap will you use? ________
2. Use the "Water" plate first. Touch section 1 with your fingers, wash well *without* soap, shake off excess water, and, while your hands are still wet, touch section 2. Do not dry your fingers with a towel. Wash

again, and, while your hands are still wet, touch section 3. Wash a final time, and touch section 4. Touch the same fingers to the plate each time.

3. Use your same hand on the plate labeled "Soap." Wash well with soap, rinse, shake off the excess water, and then touch section 1.
4. Wash again with soap, rinse, shake off the excess water, and then touch section 2.
5. Using a brush and soap, scrub your hand for 2 minutes, rinse, and shake off the excess water; then touch section 3.
6. Repeat the soap-and-brush scrub for 4 minutes, rinse, and shake off the excess water; then touch section 4.
7. Use a waterless hand-cleaning product; then touch section 5. What are the active ingredients in the product? ______________________________
8. Incubate the plates, inverted, at 35°C until the next period.
9. Speculate on your expected results, and record them in your Laboratory Report.
10. Record the growth as (−) = no growth, (+) = minimum growth, (2+) = moderate growth, (3+) = heavy growth, and (4+) = maximum growth.

Exercise 26

LABORATORY REPORT

Effectiveness of Hand Scrubbing

NAME ______________________

DATE ______________________

LAB SECTION ______________________

Purpose

____

____

____

Expected Results

Before the next lab period, indicate the relative amounts of growth you *expect* in each quadrant.

Section	Water Alone	Soap
1.	(No washing)	
2.		
3.		
4.		
5.		(Waterless hand cleaner)

Data

Indicate the relative amounts of growth in each quadrant.

Section	Water Alone	Soap (type: ____________)
1.	(No washing)	
2.		
3.		
4.		
5.		(Waterless hand cleaner)

Conclusions

______

______

Questions

1. Did your results differ from your expected results? ______

   Briefly explain why or why not. ______

2. What is a surgeon trying to accomplish with a 10-minute scrub with a brush followed by an antiseptic?

   ______

   ______

3. How do normal microbiota and transient microbiota differ? ______

   ______

4. Using your classmates' data, compare the results from bar soap and liquid soap. ______

   ______

   ______

Critical Thinking

1. If most of the normal microbiota and transient microbiota aren't harmful, then why must hands be scrubbed before surgery?

2. The following data were collected from soaps after 1 week of use at a hospital nurses' handwashing station. Neither bacteria nor fungi were isolated from any of the products before use.

   Isolation of aerobic bacteria from 25 soap products. Data are expressed as percentage of soap products contaminated.

		Liquid Soap: Type of Closure			
Organisms	Bar Soap	Screw Top	Slit/Flip	Flip/Pump	Pump
Total bacteria	95%	71%	39%	10%	0%
Gram-positive cocci	95%	71%	39%	10%	0%
Gram-negative rods	12%	1%	1%	1%	0%

   What conclusions can you draw from these data?

Part Seven Microbial Genetics

EXERCISES

All of the characteristics of bacteria—including growth patterns, metabolic activities, pathogenicity, and chemical composition—are inherited. These traits are transmitted from parent cell to offspring through genes. **Genetics** is the study of genes: how they carry information, how they are replicated and passed to the next generation or to another organism, and how their information is expressed within an organism to determine the particular characteristics of that organism (see the illustration on the next page).

The information stored in genes is called the **genotype.** All of the genes may not be expressed. **Phenotype** refers to the actual expressed characteristics, such as the ability to perform certain biochemical reactions (e.g., synthesis of a capsule or pigment).

The control of gene expression is explained by the operon model (Exercise 27). This model was first proposed by François Jacob and Jacques Monod to explain why *E. coli* makes lactose-catabolizing enzymes in the presence, but not in the absence, of lactose. Bacteria are invaluable to the study of genetics because large numbers can be cultured inexpensively, and their relatively simple genetic composition facilitates studying the structure and function of genes.

Bacteria undergo genetic change because of mutation or recombination. These genetic changes result in a change in the genotype. When a change occurs, in most instances we can expect that the product encoded by certain genes will be changed. An enzyme encoded by a changed gene may become inactive. This might be disadvantageous or even lethal if the cell loses a phenotypic trait it needs. Some genetic changes may be beneficial.

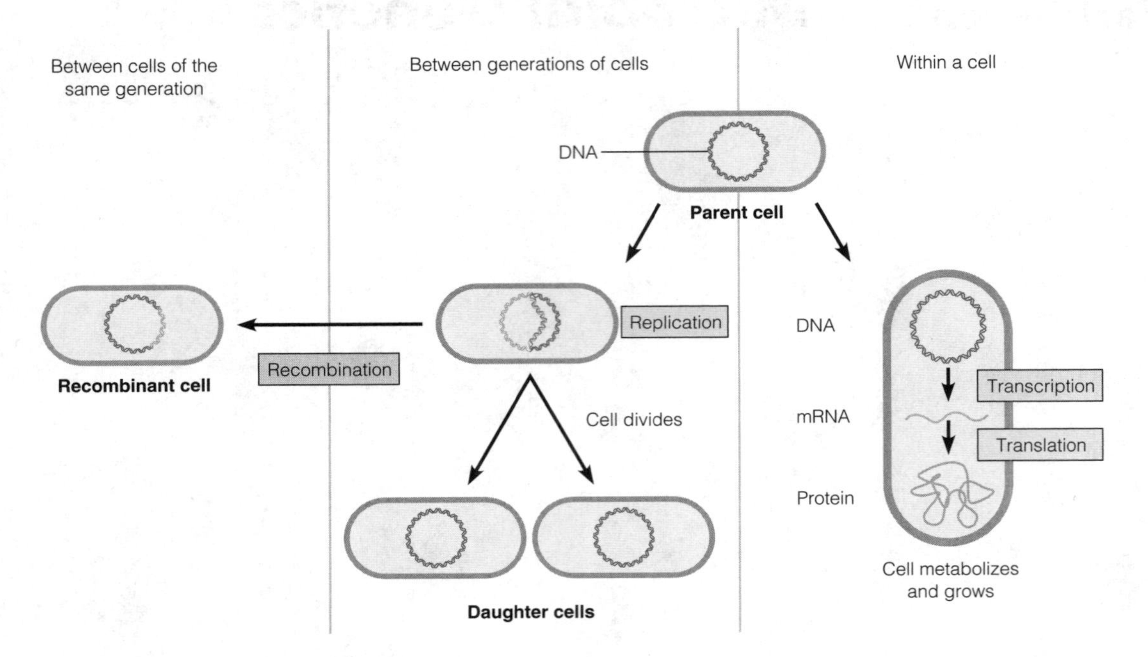

The flow of genetic information. Genetic information can be transferred between generations of cells, through DNA replication. Occasionally, genetic information can be transferred between cells of the same generation, through recombination. Genetic information is also used within a cell to produce the proteins the cell needs to function, through transcription and translation.

For instance, if an altered enzyme encoded by a changed gene has new enzymatic activity, the cell may be able to grow in new environments.

Genetic change in microbial populations is relatively easy to observe. In 1949, Howard B. Newcombe wrote:

> *Numerous bacterial variants are known which will grow in environments unfavorable to the parent strain, and to explain their occurrence two conflicting hypotheses have been advanced. The first assumes that the particular environment produces the observed change in some bacteria exposed to it, whereas the second assumes that the variants arose spontaneously during growth under normal conditions.**

Salvador Luria and Max Delbrück first demonstrated the latter **spontaneous mutation** hypothesis in 1943.

Bacteria can be mutated to get a different product from an altered gene (Exercise 28). New genes can be inserted into bacteria by transformation (Exercise 29). The desired gene can be isolated and identified by DNA fingerprinting (Exercise 30). Artificial manipulation of genes is known as **genetic engineering.** In genetic engineering, genes can be inserted into a vector and incorporated into a bacterium by transformation (Exercise 31).

Cancer is usually the result of mutations in the nucleic acid of a chromosome. We will perform a test using principles of microbial genetics for identification of possible cancer-inducing substances in Exercise 32.

---

*Quoted in H. A. Lechevalier and M. Solotorovsky. *Three Centuries of Microbiology*. New York: Dover Publications, 1974, p. 502.

Exercise 27

EXERCISE 27

Regulation of Gene Expression

Objectives

After completing this exercise, you should be able to:

1. Define operon, induction, and repression.
2. Determine the inducers for the enzymes tested in this exercise.

Background

Many genes are expressed constantly throughout the life of a cell. However, some proteins may be needed only during a particular growth phase or in a particular environment. A cell can conserve energy by making only those proteins needed at a particular time.

The operon model was proposed to explain why *Escherichia coli* made the enzyme β-galactosidase in the presence of lactose but not in its absence. An operon consists of a promoter, the region where RNA polymerase binds; the operator, which signals whether transcription will occur; and the structural genes they control. The structural genes code for the peptides. A regulatory gene encodes a regulatory protein, which binds to the operator to activate or inhibit transcription. In an **inducible operon,** transcription is activated or induced by the presence of a particular substance—in this case, lactose. The regulatory protein has an allosteric site to which the inducer can bind. The inducer then inactivates the repressor. A **repressible**

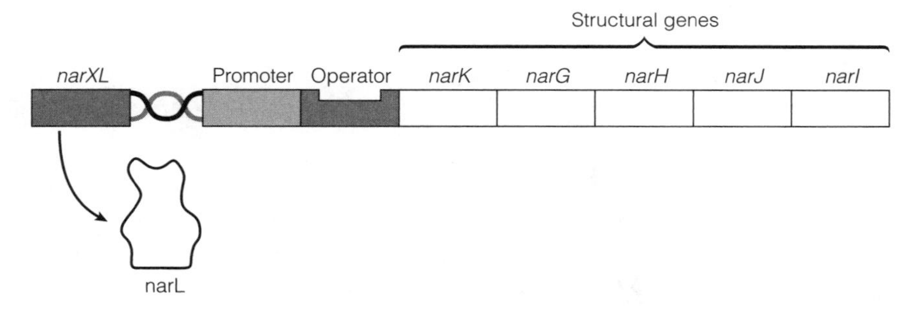

(a) The *nar* operon. The regulatory gene (*narXL*) encodes the regulatory protein narL.

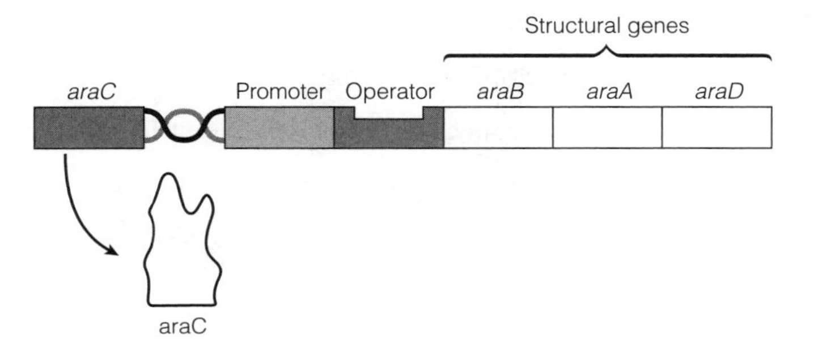

(b) The *ara* operon. The regulatory gene (*araC*) encodes the regulatory protein araC.

Figure 27.1
An operon consists of a promoter, an operator, and the structural genes they control.

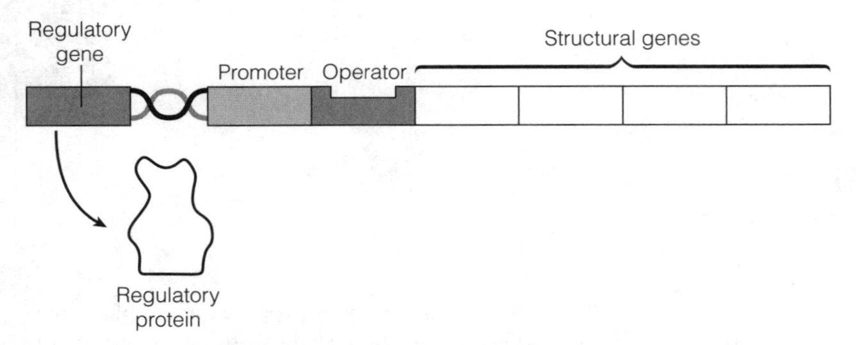

(a) A regulatory gene codes for a regulatory protein.

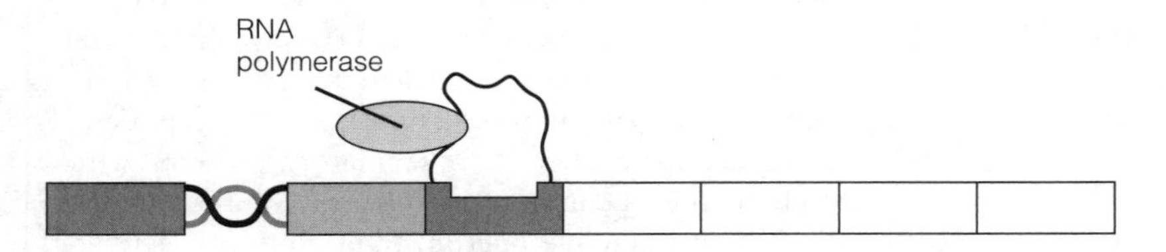

(b) The regulatory protein binds to the operator and prevents transcription of the structural genes.

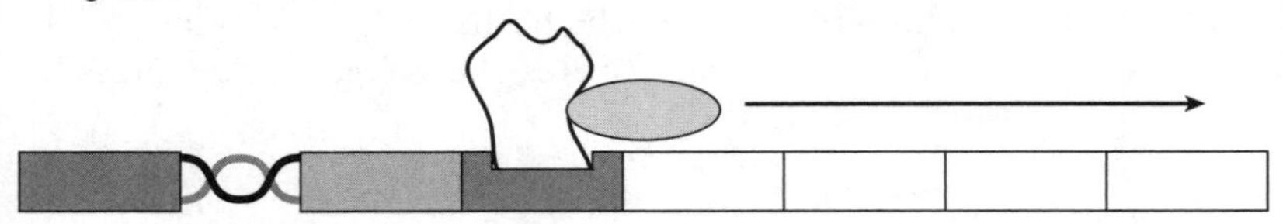

(c) A change in shape of the repressor protein activates transcription of the structural genes.

Figure 27.2

The regulatory proteins that control the *nar* and *ara* operons can inhibit or activate transcription.

operon is inhibited when a specific small molecule is present. The synthesis of the amino acid tryptophan in *E. coli* is inhibited by tryptophan. If tryptophan binds to the regulatory protein, the tryptophan–protein complex acts as a repressor to inhibit synthesis of more tryptophan.

In *E. coli* the genes for nitrate reductase are encoded by the ***nar* operon** (Figure 27.1a), and the enzymes for the catabolism of arabinose are encoded by the ***ara* operon** (Figure 27.1b). The regulatory genes for these operons produce a protein that binds to the operator to inhibit transcription or to promote binding of RNA polymerase, depending on the shape of the regulatory proteins (Figure 27.2).

In this experiment, the *ara* structural genes have been replaced with a gene called pGLO* that encodes green fluorescent protein (see Color Plate II.1). The pGLO gene, isolated from the jellyfish *Aequoria victoria*, is widely used to study gene expression. You will investigate the *ara* or *nar* operons in *E. coli* in this exercise. Before beginning, write your hypothesis in your Laboratory Report.

*pGLO is a trademark of Bio-Rad Laboratories.

Materials

Small test tubes (2)

1-ml pipettes (3)

Nitrate solution

Brewer anaerobic jar

Petri plate containing glucose-nutrient agar

Petri plate containing glucose-arabinose-nutrient agar

Petri plate containing arabinose-nutrient agar

Spreading rod

Alcohol

Nitrate reagents A and B

Ultraviolet lamp (second period)

Cultures

Escherichia coli for nitrate reductase

Escherichia coli pGLO for arabinose catabolism

Techniques Required

Anaerobic culture techniques, Exercise 19

Nitrate reduction test, Exercise 17

Pipetting, Appendix A

Procedure

Nitrate Reductase

1. Label two small tubes "Aerobic" and "Anaerobic." Add 0.5 ml *E. coli* to each tube. Then add 0.2 ml nitrate (NO_3^-) solution to each tube.
2. Immediately place the "anaerobic" tube in the anaerobic Brewer jar and activate the GasPak (Figure 19.2). Begin shaking the "aerobic" tube. Do not splash the contents out of the tube, but shake it enough to aerate the entire culture.
3. After 15 minutes, test for the presence of nitrite ions (NO_2^-) by adding 5 drops of nitrate reagent A and 5 drops of nitrate reagent B to each tube. Be sure to test an uninoculated nitrate broth tube for nitrate ion. Let the tubes stand at room temperature for 15 to 20 minutes. The presence of a pink color indicates nitrite ions. Observe the tubes for a color change and record your results. (See Color Plate III.12)

Arabinose Catabolism

1. Aseptically transfer 0.1 ml of the *E. coli* pGLO culture to the surface of the glucose-nutrient agar, glucose-arabinose-nutrient agar, and the arabinose-nutrient agar.
2. Disinfect a spreading rod by dipping it in alcohol, quickly igniting the alcohol in a Bunsen burner flame, and letting the alcohol burn off.

While it is burning, hold the spreading rod pointed *down*. Keep the beaker of alcohol away from the flame.

Let the spreading rod cool.

Figure 27.3
Inoculation using a spreading rod.

3. Spread the liquid on the surface of each plate. Let the Petri plates sit undisturbed until the liquid diffuses into the agar (Figure 27.3). Disinfect the spreading rod and return it.
4. Incubate the plates in an inverted position at 35°C for 24 to 48 hours.
5. Observe the plates using an ultraviolet lamp. Are any of the colonies producing a fluorescent protein?

____

Do not look at the ultraviolet light, and do not leave your hand exposed to it.

Exercise 27

LABORATORY REPORT

Regulation of Gene Expression

NAME ______________________

DATE ______________________

LAB SECTION ______________________

Purpose ______________________

____

____

Hypothesis ______________________

____

____

Results

Nitrate Reductase

	Color After NO_3^- Reagents	NO_3^- Reduction
Control		
E. coli, aerobic		
E. coli, anaerobic		

Arabinose Catabolism

	Growth	Fluorescence
Glucose-nutrient agar		
Arabinose-nutrient agar		
Glucose-arabinose-nutrient agar		

Conclusions

Did you accept or reject your hypothesis? Briefly explain. ______________________________

____

____

____

Questions

1. Write the chemical reaction catalyzed by nitrate reductase.

2. What is the inducer for *nar*? ______________________________

3. Of what value is nitrate reductase to *E. coli*? ______________________________

   ____

   ____

4. Chemically what is arabinose? ______________________________

5. What is the inducer for *ara*? ______________________________

6. What effect did the presence of glucose have on *ara*? Briefly explain. ______________________________

   ____

   ____

Critical Thinking

1. Several bacteria including *Bacillus subtilis* reduce nitrate to ammonia as the first step in amination. This is called assimilatory nitrate reduction. *E. coli*'s transformation of nitrate is called dissimilatory nitrate reduction. How do assimilatory and dissimilatory nitrate reduction differ?

2. What is the valence of N in NO_3^-? In NO_2^-? What is gaining electrons in nitrate reduction?

3. Assume you replaced the *ara* gene with the gene for amylase in the *ara* operon. What would be produced in the presence of arabinose?

4. *E. coli* cells do not normally fluoresce in the presence of arabinose. What do they normally do?

5. Why did the *E. coli* cells fluoresce in the presence of arabinose in this experiment? Why not in the absence of arabinose?

Exercise 28

Isolation of Bacterial Mutants

Though coli may bother and vex us,
It's hard to believe they outsex us.
They accomplish seduction
by viral transduction
Foregoing the joys of amplexus.

SEYMOUR GILBERT

Objectives

After completing this exercise, you should be able to:

1. Define the following terms: prototroph, auxotroph, and mutation.
2. Differentiate between direct and indirect selection.
3. Isolate bacterial mutants by replica plating.

Background

For practical purposes, genes and the characteristics for which they code are stable. However, when millions of bacterial progeny are produced in just a few hours of incubation, genetic variants may occur. A variant is the result of a **mutation**—that is, a change in the sequence of nucleotide bases in the cell's DNA.

Metabolic mutants can be easily identified and isolated. Catabolic mutations could result in a bacterium that is deficient in the production of an enzyme needed to utilize a particular substrate. Anabolic mutations result in bacteria that are unable to synthesize an essential organic chemical. "Wild-type," or nonmutated, bacteria are called **prototrophs,** and mutants are called **auxotrophs.** Auxotrophs have been isolated that either cannot catabolize certain organic substrates or cannot synthesize certain organic chemicals, such as amino acids, purine or pyrimidine bases, or sugars.

The cultures used in this exercise are capable of synthesizing all of their growth requirements from glucose–minimal salts medium (Table 28.1). Mutations will be induced by exposing the cells to the mutagenic wavelengths of ultraviolet light (Exercise 23.) After exposure to ultraviolet radiation, auxotrophs that cannot grow on the glucose–minimal salts medium will be identified.

Because of the low rate of appearance of recognizable mutants, special techniques have been developed to select for desired mutants. **Direct selection** is used to pick out mutant cells while rejecting the unmutated parent cells. Direct selection will be used in Exercise 31.

Table 28.1

Composition of Glucose–Minimal Salts Agar

Glucose	0.1 g
Dipotassium phosphate	0.7 g
Monopotassium phosphate	0.2 g
Sodium citrate	0.05 g
Magnesium sulfate	0.01 g
Ammonium sulfate	0.15 g
Agar	1.5 g
Water	100 ml

We will use **indirect selection** with the **replica-plating technique** in this exercise.

In replica plating, mutated bacteria are grown on a nutritionally complete solid medium. An imprint of the colonies is made on velveteen-covered or rubber-coated blocks and transferred to a glucose–minimal salts solid medium and to a complete solid medium. Colonies that grow on the complete medium but not on the minimal medium are auxotrophs. Auxotrophs are identified *indirectly* because they will not grow.

Materials

Petri plates containing nutrient agar (complete medium) (3)

99-ml water dilution blanks (3)

Sterile 1-ml pipettes (4)

Ultraviolet lamp, 260 nm

Spreading rod

Alcohol

Safety goggles

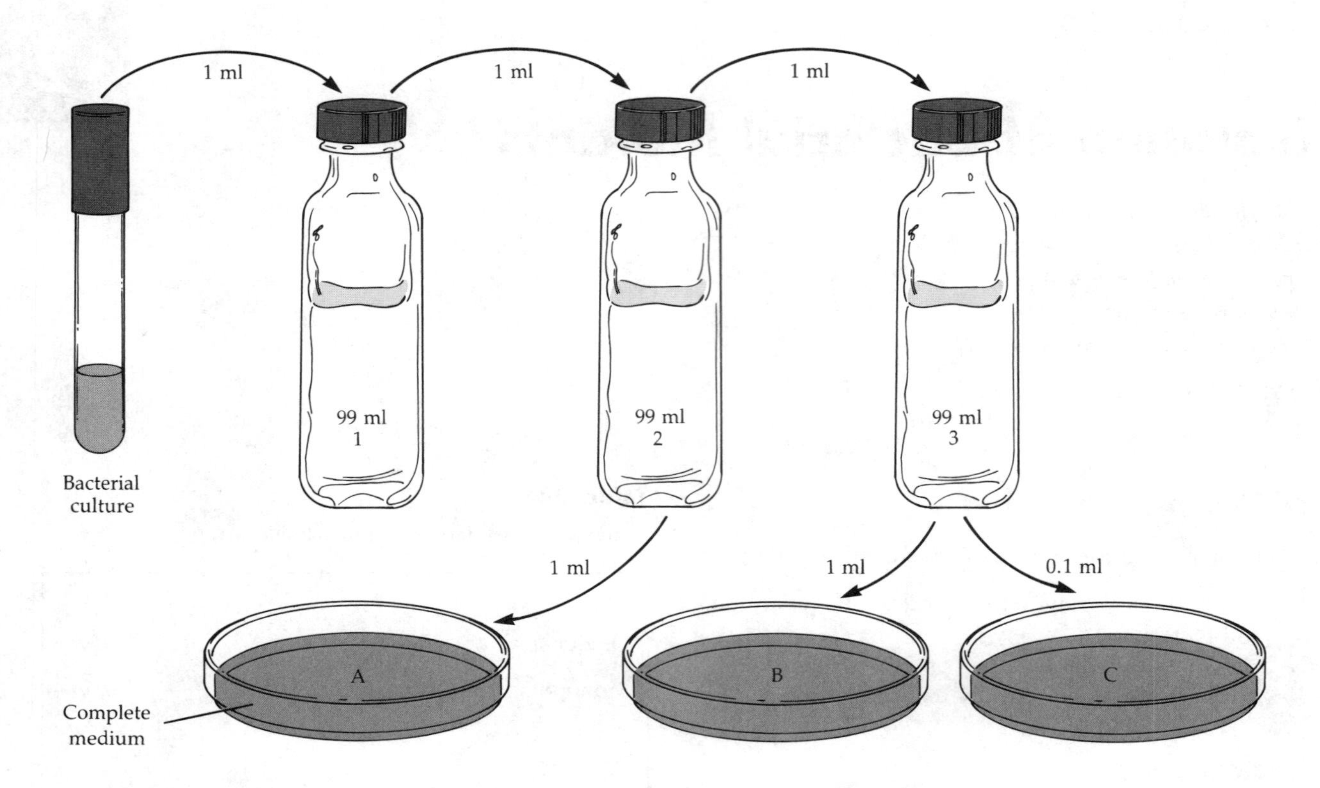

Figure 28.1
Dilution procedure.

Second Period

Petri plate containing glucose–minimal salts agar (minimal medium)

Petri plate containing nutrient agar (complete medium)

Sterile replica-plating block

Culture

Serratia marcescens

Techniques Required

Spreading rod technique, Exercise 27

Pipetting, Appendix A

Serial dilution technique, Appendix B

Procedure*

First Period

1. Label the nutrient agar plates "A," "B," and "C" and the dilution blanks "1," "2," and "3."
2. Aseptically pipette 1 ml of the assigned broth culture to dilution blank 1 and mix well.
3. Using another pipette, transfer 1 ml from dilution blank 1 to dilution blank 2, as shown in Figure 28.1, and mix well. What dilution is in bottle 2?

   ____

   ____
4. Using another pipette, transfer 1 ml from dilution blank 2 to bottle 3 and 1 ml to the surface of plate A.
5. Mix bottle 3, and transfer 1 ml to the surface of plate B with a sterile pipette and 0.1 ml to the surface of plate C.
6. Disinfect a spreading rod by dipping it in alcohol, quickly igniting the alcohol in a Bunsen burner flame, and letting the alcohol burn off.

While it is burning, hold the spreading rod pointed *down*. Keep the beaker of alcohol away from the flame.

Let the spreading rod cool.

7. Spread the liquid on the surface of plate C over the entire surface. Do the same with plates B and A (Figure 27.3). Let the Petri plates sit undisturbed until the liquid diffuses into the agar.

*Adapted from C. W. Brady. "Replica Plate Isolation of an Auxotroph." Unpublished paper. Whitewater, WI: University of Wisconsin, n.d.

8. Disinfect the spreading rod and return it. Why is it not necessary to disinfect the spreading rod between each plate when proceeding from plate C to B to A?

   ____

   ____

9. Put the plate with the cover off, agar-side up, about 30 cm from the ultraviolet lamp. Position the plate directly under the lamp. Turn the lamp on for 30 to 60 seconds, as assigned by your instructor. Wear your safety goggles.

Do not look at the ultaviolet light, and do not leave your hand exposed to it. Wear safety goggles.

10. Incubate the plates at 25°C until the next period.

Second Period

1. Select a plate with 25 to 50 isolated colonies. Mark the bottom of the plate with a reference mark (Figure 28.2a). This is the master plate.
2. Mark the uninoculated complete and minimal media with a reference mark on the bottom of each plate.
3. If your replica-plating block is already assembled, proceed to step 4. If not, follow these instructions. Carefully open the package of velveteen. Place the replicator block on the center of the velveteen. Pick up the four corners of the cloth and secure them tightly on the handle with a rubber band.
4. Inoculate the sterile media by either step a **or** step b, as follows:
   a. Hold the replica-plating block by resting it on the table with the rough surface, or velveteen, up (Figure 28.2b). Invert the master plate selected in step 1 on the block, and allow the master plate agar to lightly touch the block. Remove the lid from the minimal medium, align the reference marks, and touch the uninoculated minimal agar with the inoculated replica-plating block. Replace the lid. Remove the lid from the uninoculated complete medium and inoculate with the replica-plating block, keeping the reference marks the same.
   b. Place the master plate and the uninoculated plates on the table. Place the reference marks in a 12 o'clock position and remove the lids. Touch the replica-plating block to the master plate; then, without altering its orientation, gently touch it to the minimal medium, then to the complete medium (Figure 28.2c).
5. Replace the lids, and incubate as before. Refrigerate the master plate.
6. After incubation, compare the plates and record your results. (See Color Plate V.4.)

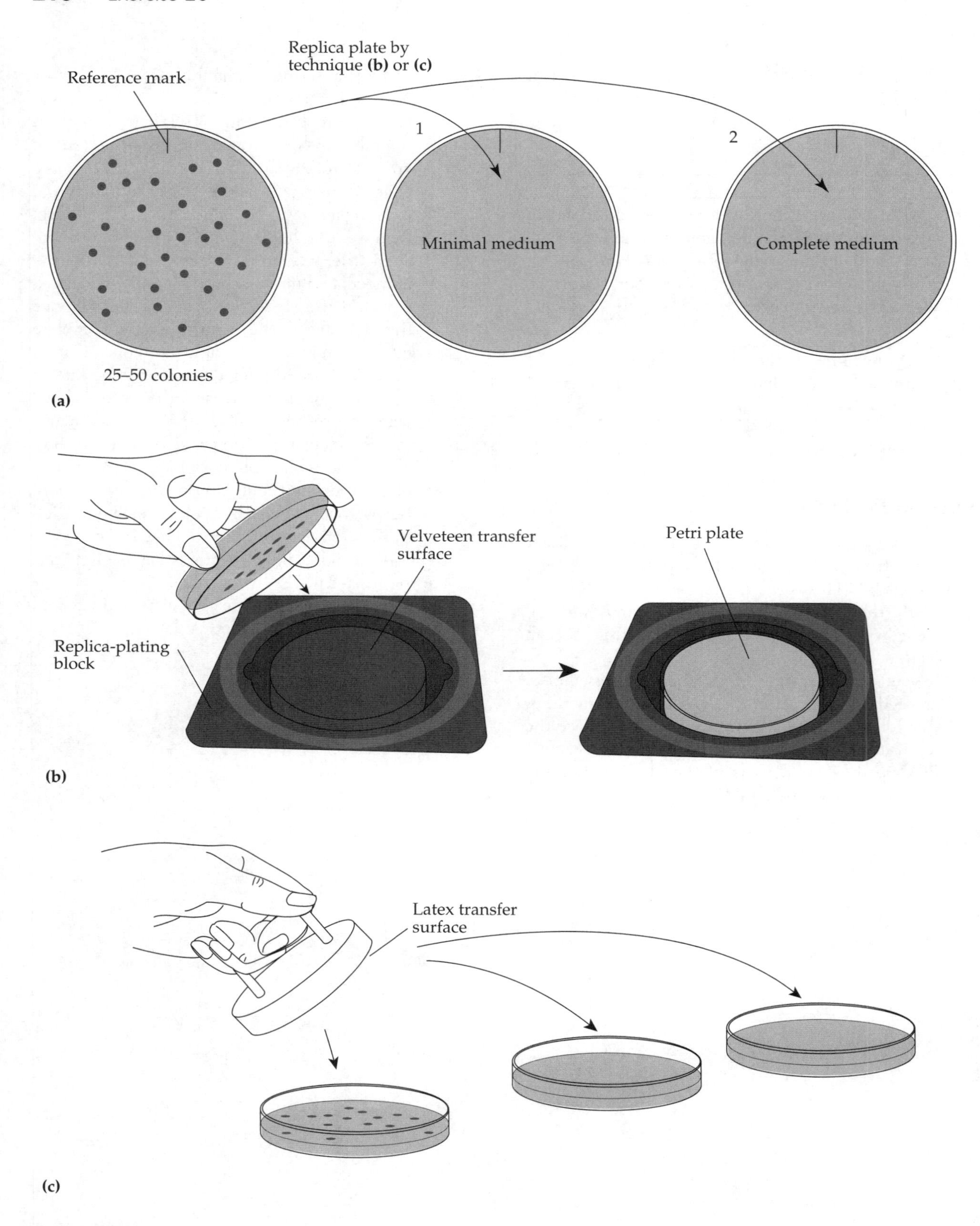

Figure 28.2

Replica plate. **(a)** Schematic diagram. **(b)** Transfer with a stationary replica-plating block. **(c)** Transfer by moving the latex transfer surface from plate to plate.

Exercise 28

Isolation of Bacterial Mutants

LABORATORY REPORT

NAME ____________________

DATE ____________________

LAB SECTION ____________________

Purpose

____

Data

Mark the location of colonies, and note any changes in pigmentation. Circle the auxotrophs on the diagram of the complete medium.

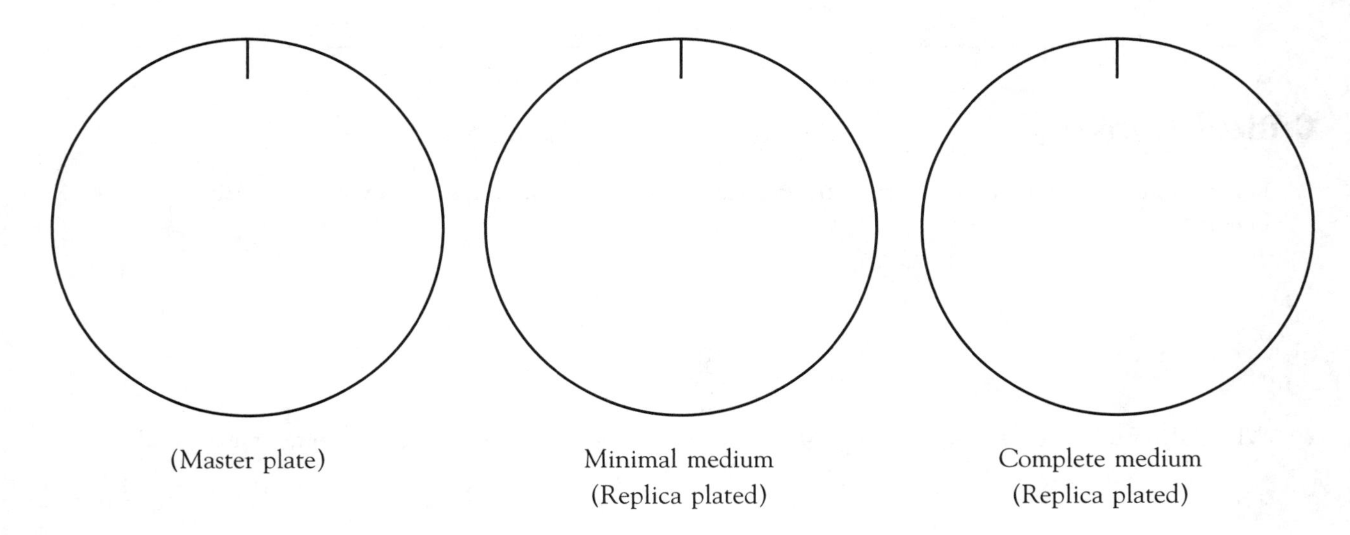

(Master plate)

Minimal medium (Replica plated)

Complete medium (Replica plated)

Class Data

	S. marcescens	
Students	UV Exposure (sec)	Number of Auxotrophs

Conclusions

What is the effect of longer exposure to ultraviolet light? ______________________

Questions

1. Why was the complete medium, as well as the minimal medium, replica-plated? ______________________

2. Why is replica plating usually used for indirect selection of mutants? ______________________

Critical Thinking

1. How would you isolate a mutant from the replica plate? Does your technique involve a catabolic or anabolic mutant?

2. Why do you think bacteria such as *Staphylococcus aureus* that survive on surfaces are often pigmented?

3. How would you identify the growth factor(s) needed by an auxotroph?

4. How would you use the replica-plating technique to isolate a mutant that is sensitive to an antibiotic?

Exercise 29

EXERCISE 29

Transformation of Bacteria

Objectives

After completing this exercise, you should be able to:

1. Explain the process of isolating DNA.
2. Define transformation and auxotroph.
3. Isolate DNA and use it to transform an auxotroph.

Background

Scientists from many different disciplines have used the technique of genetic engineering to genetically alter bacteria and eukaryotic organisms. In **genetic engineering,** genes are isolated from one organism and incorporated by manipulation in a laboratory into other bacterial or eukaryotic cells.

The first step in genetic engineering is extraction of DNA from the cell that possesses the desired gene. To extract DNA from bacteria, we disrupt the cells by chemical or mechanical means or by osmotic pressure.

The resuspension buffer used in this experiment contains lysozyme and RNase. What is the purpose of each of these enzymes? ______________________

____

Cell debris is precipitated with SDS (sodium dodecyl sulfate).

The 5 M NaCl denatures proteins such as DNase and stabilizes the DNA by forming an Na^+ shell around the negatively charged phosphates of the DNA. The DNA is precipitated as a viscous, translucent mass with cold alcohol.

Transformation is a rare event involving acquisition by a *recipient* bacterium of small pieces of DNA released from a dead *donor* bacterium. The acquired pieces of DNA can give the recipient bacterium new characteristics. In this experiment, you will extract DNA from an auxotrophic mutant strain of *Bacillus subtilis*. This strain, designated met$^-$, cannot synthesize the amino acid methionine. You will use its DNA to transform a different *B. subtilis* mutant that cannot synthesize the amino acid phenylalanine (phe−). Neither of these bacteria can grow on media that lack the required amino acids. Glucose–minimal salts agar (GMSA) is such a medium (Table 28.1). GMSA contains glucose as a carbon and energy source; the other macronutrients, nitrogen, phosphate, and sulfur, are provided as inorganic salts. However, this medium does not provide amino acids so only bacteria that can synthesize their own amino acids will be able to grow on it.

Materials

Centrifuge tube

Resuspension buffer

SDS

Proteinase

5 M NaCl

95% ethyl alcohol, ice cold

Glass rod

Graduated cylinder

Sterile test tube

Petri plates containing glucose–minimal salts agar (minimal medium) (3)

Spreading rod

Alcohol

Micropipette and tip (10–100 μl)

Sterile 1-ml pipettes (2)

Pipette bulb or propipette

Water baths, 37°C and 60°C

Centrifuge

Cultures

Bacillus subtilis met$^-$, phe$^+$

B. subtilis met$^+$, phe$^-$

Techniques Required

Aseptic technique, Exercise 10

Spreading rod technique, Exercise 27

Pipetting, Appendix A

Procedure*

Isolation of DNA

1. Centrifuge the *Bacillus* met^-, phe^+ culture for 10 minutes at 3,000 rpm, to collect cells in the bottom of the tube.

Balance the centrifuge by placing a tube containing 10 ml of water opposite your culture tube.

Discard the supernatant (liquid portion) by carefully pouring it into a container of disinfectant.

2. Add 5 ml of resuspension buffer to the centrifuge tube and mix to resuspend the pellet. Incubate the tube in a 37°C water bath for 30 minutes.
3. Add 0.1 ml of SDS and 50 μl of proteinase. Why did the appearance of the suspension change? ____
4. Incubate the tube in a 60°C water bath for 30 minutes. Remove the tube from the water bath and let it cool to room temperature.
5. Add 1 ml of 5 M NaCl and mix gently.
6. Gently add 20 ml of cold ethyl alcohol by letting it run down the side of the tube. You should have two distinct layers. Which layer (alcohol or water) is on top? ____
7. Let the tube stand in a test-tube rack for 10 minutes. You should see a precipitate form at the interface of the two liquids. What is the precipitate? ____
8. Gently mix the two layers with a glass rod using a circular motion and spool the fibers of DNA onto the rod. (See Color Plate II.2.) Describe the appearance of your spooled DNA.
9. Place the rod with the spooled DNA into a sterile test tube.

Transformation

1. Label three minimal media plates: "phe^- cells," "met^- DNA," and "DNA + phe^- cells."
2. Pipette 0.1 ml of the *Bacillus* met^+, phe^- onto the appropriate two plates and spread with a sterile spreading rod. Disinfect a spreading rod by dipping it in alcohol, quickly igniting the alcohol in a Bunsen burner flame, and letting the alcohol burn off.

While it is burning, hold the spreading rod pointed *down*. Keep the beaker of alcohol away from the flame.

3. Spread your spooled DNA over the surface of the appropriate two plates (Figure 27.3). Let the plates stand for about 15 minutes. Disinfect the spreading rod and return it.
4. Incubate the plates at 37°C for 48 hours. Which plates are the controls? ____
5. Observe the plates and record your results.

*Adapted from N. Kapp. Unpublished manuscript. San Bruno, CA: Skyline College, n.d.

Exercise 29

Laboratory Report

Transformation of Bacteria

NAME ____________________

DATE ____________________

LAB SECTION ____________________

Purpose

Expected Results

Answer the following before you complete the experiment.

	Inoculum		
	Bacillus phe$^-$	met$^-$ DNA	DNA + *Bacillus* phe$^-$
Do you expect growth?			

Results

Describe the appearance, color, texture, and shape of your spooled DNA.

	Inoculum		
	Bacillus phe$^-$	met$^-$ DNA	DNA + *Bacillus* phe$^-$
Number of colonies			

Conclusions

Did you transform *Bacillus*? How do you know?

Questions

1. How were the *Bacillus* cells disrupted in this exercise?

2. What physical characteristics of DNA allow it to be spooled onto a glass rod? Why is it not possible to spool out precipitated proteins? ____

____

____

3. Why do you suspect that your spooled product is DNA and not RNA? ______________________

____

____

4. What new gene were you looking for in the transformed *Bacillus*? ______________________

Critical Thinking

1. If growth occurred on the "DNA + *Bacillus* phe$^-$" plate, how could you rule out contamination? Mutation?

2. Design an experiment to determine whether *Bacillus* DNA could be used to transform any other species.

3. It appears that *N. gonorrhoeae* acquired an antibiotic resistance gene (*tet*M) from *Streptococcus*. How did the *Neisseria* do this? How could you prove it was from *Streptococcus*?

Exercise 30

DNA Fingerprinting

Objectives

After completing this exercise, you should be able to:

1. Define restriction enzyme and RFLP.
2. Perform agarose gel electrophoresis.
3. Identify your unknown.

Background

Restriction enzymes recognize specific short nucleotide sequences in double-stranded DNA and cleave both strands of the molecule (Table 30.1). Restriction enzymes are bacterial enzymes that restrict the host range of bacteriophages. The enzymes were discovered in laboratory experiments when phages were used to infect bacteria other than their usual hosts. Restriction enzymes in the new host destroyed the phage DNA. Today, more than 300 restriction enzymes have been isolated and purified for use in DNA research. Each enzyme recognizes a different nucleotide sequence in DNA. The enzymes are named with three-letter abbreviations for the bacteria from which they were isolated.

DNA cleaved with restriction enzymes produces **restriction fragment length polymorphisms (RFLPs).** The size and number of the pieces is determined by **agarose gel electrophoresis** (Appendix G). The digested DNA is placed near one end of a thin slab of agarose and immersed in a buffer to allow current to flow through the agarose. Electrodes are attached to both ends of the gel, and a current is applied. Each piece of DNA then migrates toward the positive electrode at a rate determined by its size. The DNA fragments are visualized by staining with methylene blue or ethidium bromide.

These enzymes can be used to characterize DNA because a specific restriction enzyme will cut a molecule of DNA everywhere a specific base sequence occurs. When the DNA molecules from two different microorganisms are treated with the same restriction enzyme, the restriction fragments that are produced can be separated by electrophoresis. A comparison of the number and sizes of restriction fragments produced from different organisms provides information about their genetic similarities and differences—the more similar the patterns, the more closely related the organisms are expected to be.

In forensic science, analyses of DNA can be used to determine the father of a child or the perpetrator of a crime. DNA samples are digested to produce RFLPs that are separated by gel electrophoresis. The DNA fragments are transferred to a membrane filter by blotting so the fragments on the paper are in the same positions as the fragments on the gel. The paper is then heated to produce single strands of DNA. Small pieces of radioactively labeled single-stranded DNA called probes are used to distinguish core sequences of nucleotides. When the probe is added to the gel, it will anneal to its complementary strand. The gel can then be placed on photographic film, and the radioactive label will expose the film to produce an *autoradiograph*.

In this exercise, we will compare plasmids from an unknown bacterial species to plasmids from known bacteria in order to identify the unknown.

Materials

Digestion of DNA

DNA samples

Restriction enzyme

Restriction buffer

Table 30.1

Recognition Sequences of Some Restriction Endonucleases

Enzyme	Bacterial Source	Recognition Sequence
*Eco*RI	*Escherichia coli*	G↓AATTC CTTAA↑G
*Hae*III	*Haemophilus aegyptius*	GG↓CC CC↑GG

Micropipette, 1–10 μl

Micropipette tips (13)

Microcentrifuge tubes (5)

Electrophoresis of DNA Samples

Electrophoresis buffer

Casting tray and comb

Agarose, 0.8%

Tracking dye

Electrophoresis chamber and power supply

Ethidium bromide or methylene blue

Transilluminator or light box

Safety goggles

Techniques Required

Pipetting, Appendix A

Electrophoresis, Appendix G

Procedure

Every pair of students will perform five digests: One from the unknown and one each from four known bacteria.

Digestion of DNA

1. Label five microcentrifuge tubes "U," "1," "2," "3," and "4."
2. To each tube add 5 μl of restriction buffer. Using a new pipette tip, add 4 μl of restriction enzyme to each tube.
3. Add the DNA samples to each tube as follows. Use a different pipette tip for each sample. Why?

   ____

Tube	DNA
U	Unknown, 4 μl
1	Species 1, 4 μl
2	Species 2, 4 μl
3	Species 3, 4 μl
4	Species 4, 4 μl

4. Centrifuge the tubes for 1 to 2 seconds to mix. Be sure to balance the centrifuge. Then incubate the tubes in a 37°C water bath for 45 minutes. After incubation, the tubes can be frozen if electrophoresis will be done during another lab period.

Electrophoresis of DNA Samples

1. Read about electrophoresis in Appendix G. Use the melted agarose to pour a gel, as described in Appendix G.
2. If necessary, defrost the tubes of digested DNA by holding them in your hand or placing them in a 37°C water bath.
3. Add 1 μl of tracking dye to each tube. Centrifuge the tubes for 1 to 2 seconds to mix. *Be sure the centrifuge is balanced.*
4. Load 14 μl of one of your samples prepared previously into a well. Using a different pipette tip, load your other samples.
5. Once the wells have been filled, apply power (125 V) to the chamber. During the run, you will see the tracking dye migrate. The tracking dye will separate into its two component dyes during electrophoresis. Turn off the power before the faster dye runs off the gel. Good separation of the DNA fragments occurs when the two dyes have separated by 4 or 5 cm. Run the gel until the dye is near the end of the gel; then turn off the current and remove the gel.
6. To stain the gel, use either step a or step b as follows. (See Color Plate II.3.)
   - **a.** Transfer the gel to the ethidium bromide staining tray for 5 to 10 minutes.

Wear safety goggles. Do not touch the ethidium bromide. It is a mutagen.

Transfer the gel to tap water to destain for 5 minutes. (Chlorine in tap water will inactivate residual ethidium.) Place your gel on the transilluminator and close the plastic lid.

Do not look directly at the UV light. Do not turn the UV light on until the plastic lid is down.

Ethidium bromide that is bound to DNA does not wash off and will fluoresce with UV light.

   - **b.** Transfer the gel to the methylene blue staining tray for 30 minutes to 2 hours. Destain by placing the gel in water for 30 minutes to overnight. Place your gel on a light box. Methylene blue that is bound to DNA does not wash off.

7. Carefully draw the location of the bands or photograph your gel using the transilluminator camera.

Exercise 30

DNA Fingerprinting

Laboratory Report

Name ______________________

Date ______________________

Lab Section ______________________

Purpose

____

Data

Draw your gel so you have all five digests represented. Label the contents of each lane.

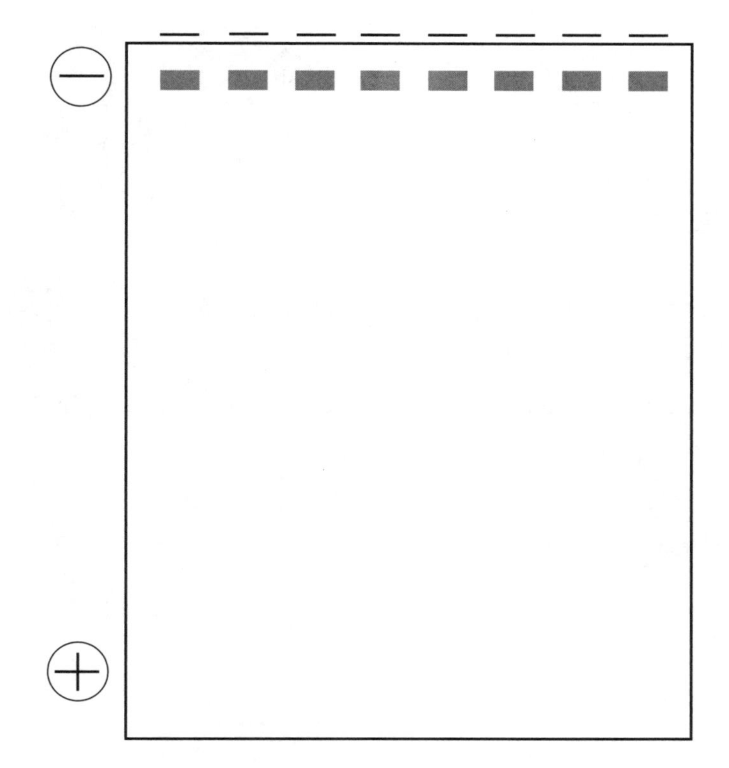

Conclusions

What is the identity of your unknown? How can you tell? ______________________

Questions

1. What restriction enzyme did you use? ______________________

   What nucleotide sequence does it recognize? ______________________

2. Using this map of pLAB30 (18 kb), give the number of restriction fragments and their lengths that would result from digesting pLAB30 with *Eco*RI, *Bam*HI, and both enzymes together.

Enzyme	Fragments Number	Size
*Eco*RI	________	________
*Bam*HI	________	________
*Eco*RI + *Bam*HI	________	________

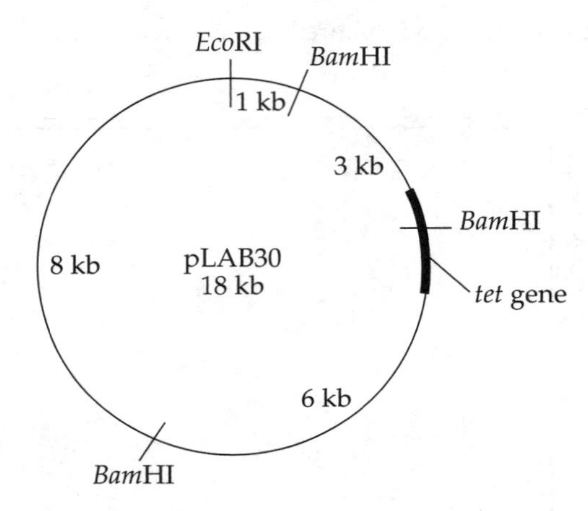

Which enzyme will give the smallest piece containing the tetracycline-resistance gene? ______________________

Critical Thinking

1. Differentiate between a gene and an RFLP.

2. How could you use this technique to trace the source of *E. coli* O157:H7?

Why would you want to trace *E. coli*?

Of what advantage is this technique over conventional biochemical tests?

3. The following agarose gel electrophoresis patterns were obtained from *Eco*RI digests of DNA from different isolates. The gels were developed with a DNA probe for a particular guanine–cytosine sequence. Which appear to be most closely related? Do any appear to be the same species? Briefly explain how you arrived at your conclusions.

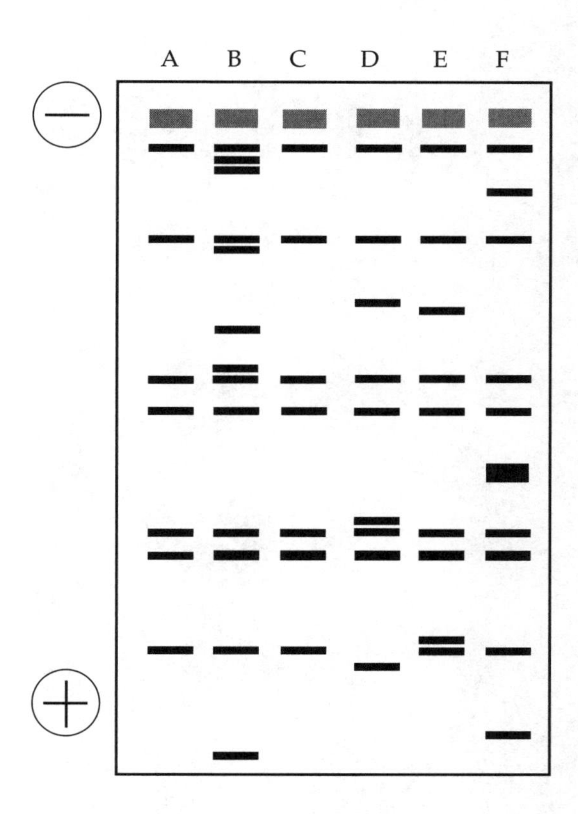

Exercise 31

Genetic Engineering

Objectives

After completing this exercise, you should be able to:

1. Define transformation, restriction enzyme, and ligation.
2. Analyze DNA by electrophoresis.
3. Accomplish genetic change through transformation.

Background

For the genetic engineering of a bacterium by using transformation, the desired gene is first inserted into a plasmid. Next, the recombinant plasmid is acquired by the bacterium. If the gene is expressed in the cell, the cell will produce a new protein product.

Genetic engineering is made possible by restriction enzymes. A **restriction enzyme** is an enzyme that recognizes and cuts only one particular sequence of nucleotide bases in DNA, and it cuts this sequence in the same way each time. These enzymes are divided into two major classes, based on their manner of cleavage. Class I enzymes bind to specific sequences but catalyze cleavage at sites up to thousands of base pairs from the binding sequence. Class II enzymes recognize specific sequences and catalyze cleavage within or next to these sequences. The recognition sequences usually include inverted repeats (*palindromes*) so that the same sequence is cut at the same point in both strands. Some of these enzymes catalyze cleavage of the two strands so that the ends of the resulting fragments are fully base-paired (**blunt ends**). Other enzymes catalyze a staggered cleavage of the two DNA strands so that resulting fragments have unpaired single strands (**sticky ends**) (Figure 31.1).

We will use the restriction enzymes *Hind*III and *Bam*HI. *Hind*III, an enzyme isolated from *Haemophilus influenzae*, cuts double-stranded DNA at the arrows in the sequence

A↓AGCTT
TTCGA↑A

*Bam*HI, isolated from *Bacillus amyloliquefaciens*, cuts double-stranded DNA at the arrows in the sequence

G↓GATCC
CCTAG↑G

In this exercise, we will use a plasmid containing an ampicillin-resistance gene (pAMP) and another plasmid with a kanamycin-resistance gene (pKAN) to transform an antibiotic-sensitive strain of *E. coli* (amp^s/kan^s). We will cut the two circular plasmids using restriction enzymes (Figure 31.2). The DNA fragments will be ligated using DNA ligase, and the ligated DNA will be used to transform the *E. coli*. *E. coli* does not normally take up new DNA; its cell wall will be damaged by treatment with $CaCl_2$ so DNA can enter. We will use agarose gel electrophoresis to determine whether the restriction enzymes cut the plasmids.

Materials

Plasmid Digestion

Sterile microcentrifuge tubes (2)

Micropipette, 1–10 µl

Sterile micropipette tips (4)

pAMP plasmid

pKAN plasmid

C-C-G-**G-A-A-T-T-C**-C-C-C-A-C-**G-G-C-C**-G-A
G-G-C-**C-T-T-A-A-G**-G-G-G-T-G-**C-C-G-G**-C-T

(a) Before cleavage

C-C-G-**G** | **A-A-T-T-C**-C-C-C-A-C-**G-G** | **C-C**-G-A
G-G-C-**C-T-T-A-A** | **G**-G-G-G-T-G-**C-C** | **G-G**-C-T

(b) After cleavage

Figure 31.1

*Eco*RI recognizes G↑AATTC and leaves sticky ends. *Hae*III recognizes GG↑CC to produce blunt ends.

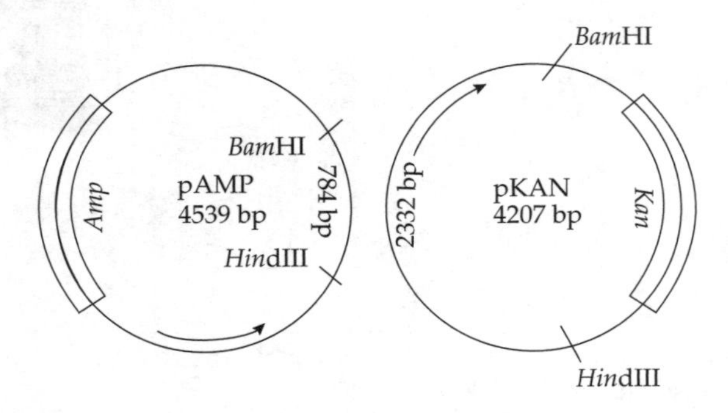

Figure 31.2

Restriction maps of plasmids showing locations of antibiotic-resistance markers. DNA is measured in the number of base pairs (bps).

Restriction enzyme buffer

*Bam*HI and *Hin*dIII enzymes

37°C water bath

Electrophoresis of Plasmids

Microcentrifuge tubes (3)

Micropipette, 1–10 μl

Sterile micropipette tips (7)

Casting tray and comb

Agarose, 0.8%

pAMP plasmid

Electrophoresis buffer

Tracking dye

Electrophoresis chamber and power supply

Ethidium bromide or methylene blue stain

Transilluminator or light box

Safety goggles

Ligation of DNA Pieces

Sterile microcentrifuge tubes (2)

Micropipette, 1–10 μl

Sterile micropipette tips (5)

Ligation buffer

65°C water bath

Ice

Sterile distilled water (dH_2O)

DNA ligase

Transformation of Competent amp^s/kan^s E. coli Cells

Plate containing nutrient agar

Sterile microcentrifuge tubes (2)

Micropipettes, 1–10 μl and 200–1000 μl

Sterile micropipette tips (4)

50 mM of $CaCl_2$

Ice

43°C water bath

35°C water bath

Sterile nutrient broth

Culturing of Transformed Cells

Micropipette, 10–100 μl

Sterile micropipette tips (2)

Plates containing nutrient agar (3)

Plates containing ampicillin-nutrient agar (3)

Plates containing kanamycin-nutrient agar (3)

Plates containing ampicillin-kanamycin-nutrient agar (2)

Spreading rod and alcohol

Culture

E. coli amp^s/kan^s

Techniques Required

Inoculating loop technique, Exercise 10

Spreading rod, Exercise 27

Pipetting, Appendix A

Electrophoresis, Appendix G

Procedure (Figure 31.3)

A. Plasmid Digestion

Keep all reagents ice cold.

1. Label two sterile microcentrifuge tubes "1" and "2."
2. Aseptically add reagents as shown below.

Tube	pAMP	pKAN	Restriction Buffer	*Bam*HI/ *Hin*dIII
1	5.5 μl	—	7.5 μl	2 μl
2	—	5.5 μl	7.5 μl	2 μl

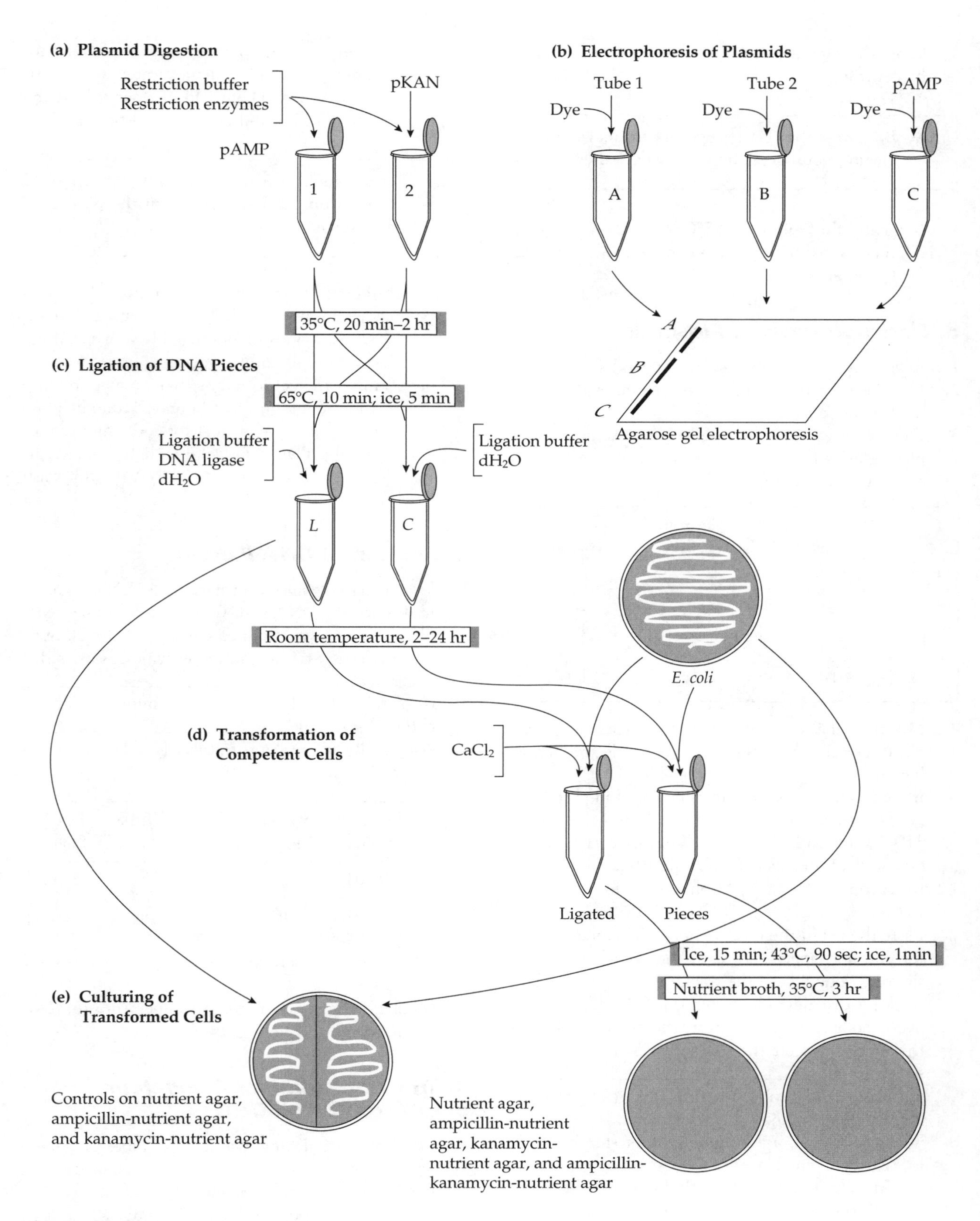

Figure 31.3

Procedure for plasmid digestion, electrophoresis, ligation, and transformation of *E. coli*.

3. Close the caps and mix the tubes for 1 to 2 seconds by centrifuging.

> **Be sure the centrifuge is balanced by placing your tubes opposite each other.**

4. Incubate the tubes at 35°C for 20 minutes to 2 hours, as instructed. Tubes can be frozen until the next lab period.

B. Electrophoresis of Plasmids

1. Read about electrophoresis in Appendix G.
2. Use the melted agarose to pour a gel, as described in Appendix G.
3. Prepare three tubes as listed below. Remove samples aseptically from tubes 1 and 2. Freeze tubes 1 and 2 for procedure part C.

Tube	Uncut pAMP	Sample from Part A Tube #1	Sample from Part A Tube #2	Tracking Dye
A	—	5 μl	—	1 μl
B	—	—	5 μl	1 μl
C	5 μl	—	—	1 μl

4. Mix the tubes by centrifuging for 1 to 2 seconds.
5. Fill the electrophoresis chamber with electrophoresis buffer.
6. If necessary, defrost the tubes prepared in part A by holding them in your hand or placing them in a 37°C water bath. Load 5 μl from tube A into the first well of your gel. Load 5 μl from tube B into the second well and 5 μl from tube C into the third well.
7. Place the gel in the electrophoresis chamber, attach the lid, and apply 125 V to the chamber. During the run, you will see the tracking dye migrate. The tracking dye migrates faster than the DNA. The dye is added so you can turn the power off before the DNA runs off the gel. Why do the two dyes separate? ______________________
   Turn off the current when the two dyes have separated by 4 or 5 cm.
8. To stain the gel, use either step a or step b as follows. (See Color Plate II.3.)
   a. Transfer the gel to the ethidium bromide staining tray for 5 to 10 minutes.

> **Wear safety goggles. Do not touch the ethidium bromide. It is a mutagen.**

   Transfer the gel to tap water to destain for 5 minutes. (Chlorine in tap water will inactivate residual ethidium.) Place your gel on the transilluminator and close the plastic lid.

> **Do not look directly at the UV light. Do not turn on the UV light until the plastic lid is down.**

   Ethidium bromide that is bound to DNA does not wash off and will fluoresce with UV light. Draw the bands or photograph your gel using the transilluminator camera.
   b. Transfer the gel to the methylene blue staining tray for 30 minutes to 2 hours. Destain by placing the gel in water for 30 minutes to overnight. Place your gel on a light box. Methylene blue that is bound to DNA did not wash off. Draw the bands.

C. Ligation of DNA Pieces

1. Label two sterile microcentrifuge tubes "L" for "ligated" to describe the DNA and "C" for control (unligated).
2. Heat tubes 1 and 2 from part A at 65°C for 10 minutes to destroy the restriction enzymes. Why is this necessary? ______________________
3. Cool tubes 1 and 2 on ice for 5 minutes. Why is cooling necessary? ______________________
4. Aseptically add reagents to tubes L and C, as shown below.

Tube	Ligation Buffer	Sterile dH_2O	Tube #1	Tube #2	DNA Ligase
L	10 μl	3 μl	3 μl	3 μl	1 μl
C	10 μl	4 μl	3 μl	3 μl	—

5. Close caps securely and centrifuge the tubes for 1 to 2 seconds.
6. Incubate the tubes at room temperature for 2 to 24 hours.

D. Transformation of Competent amp^s/kan^s *E. coli* Cells

1. Label two sterile microcentrifuge tubes "Ligated" and "Pieces." Aseptically pipette 250 μl of ice-cold $CaCl_2$ into each tube. Keep the tubes on ice.
2. With a sterile loop, transfer a loopful of *E. coli* to each tube. Mix to make a suspension.
3. Aseptically add 10 μl from tube L to the tube labeled "Ligated." Save tube L for part E. Add 10 μl from tube C to the "Pieces" tube.

4. Incubate the tubes on ice for 15 minutes to allow the DNA to settle on the cells.
5. *Heat-shock* the cells by placing them in a 43°C water bath for 90 seconds so the DNA will enter the cells.
6. Return the tubes to the ice for 1 minute, then add 700 μl sterile nutrient broth. Incubate the tubes in a 35°C water bath for at least 3 hours.

E. Culturing of Transformed Cells

1. Prepare the following controls. Inoculate one-half of a nutrient agar plate with a loopful of the amp^s/kan^s *E. coli* cells. Inoculate the other half with a loopful of the ligated plasmid DNA preparation (tube L). Similarly, inoculate an ampicillin-nutrient agar plate and a kanamycin-nutrient agar plate. How many plates do you have? __________
2. Add 100 μl from the "Ligated" tube to each of the following plates:

   Nutrient agar

   Ampicillin-nutrient agar

   Kanamycin-nutrient agar

   Ampicillin-kanamycin-nutrient agar

3. Disinfect a spreading rod by dipping it in alcohol, quickly igniting the alcohol, and letting the alcohol burn off.

While it is burning, hold the spreading rod pointed *down*. Keep the beaker of alcohol away from the flame.

   Let the spreading rod cool.

4. Spread cells over one plate with the sterile spreading rod (Figure 27.3). Disinfect the spreading rod before spreading the cells over each of the remaining plates.
5. Repeat steps 2 through 4 to inoculate four plates from the "Pieces" tube.
6. Label the plates and incubate them for 24 hours at 35°C.
7. Observe the plates and record your results.

Exercise 31

Genetic Engineering

Laboratory Report

Name ____________________

Date ____________________

Lab Section ____________________

Purpose

____

____

____

Data

Sketch your electrophoresis results. Note the approximate sizes of the DNA fragments.

Controls:

Inoculum	Growth on		
	Nutrient Agar	Ampicillin-Nutrient Agar	Kanamycin-Nutrient Agar
amp^s/kan^s *E. coli*			
DNA			

Transformation:

Inoculum	Number of Colonies			
	Nutrient Agar	Ampicillin-Nutrient Agar	Kanamycin-Nutrient Agar	Ampicillin-Kanamycin-Nutrient Agar
Ligated				
Pieces				

Conclusions

1. Did you cut the plasmids? How do you know? ____________________

2. What do the results of your transformation experiment indicate?

Questions

1. What is the purpose of each control? ____________________

2. If there was no growth on the DNA + *E. coli* plates, what went wrong? ____________________

3. Why were the *E. coli* bacteria treated with $CaCl_2$ before the plasmids were added? ____________________

4. Do the enzymes used in this experiment produce blunt or sticky ends? Why was this type of enzyme used?

Critical Thinking

1. If you got growth on the DNA + *E. coli* plates, how could you rule out contamination? Mutation?

2. How could you prove ligation of the two plasmids occurred and not two separate transformation events?

3. Using the following gel, determine whether recombination between the plasmid and chromosome occurred. Briefly explain how you can tell it is transformation.
Lane A: *Eco*RI digest of cell's DNA
Lane B: Undigested plasmid
Lane C: *Eco*RI digest of plasmid
Lane D: *Eco*RI digest of cell's DNA after a transformation experiment
Lane E: *Hind*III digest of cell's DNA
Lane F: *Hind*III digest of plasmid
Lane G: *Hind*III digest of cell's DNA after a transformation experiment

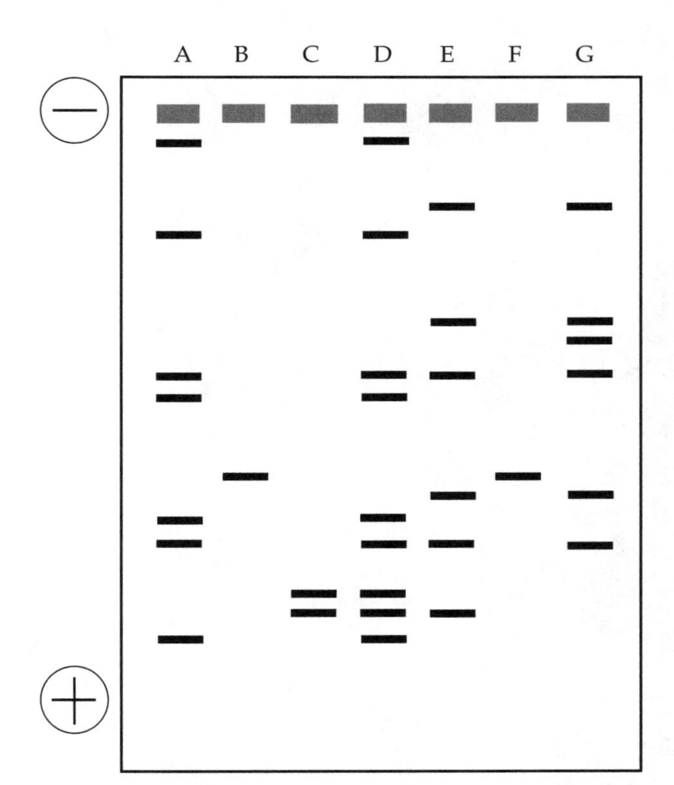

Exercise 32

Ames Test for Detecting Possible Chemical Carcinogens

The requirements of health can be stated simply. Those fortunate enough to be born free of significant congenital disease or disability will remain well if three basic needs are met: they must be adequately fed; they must be protected from a wide range of hazards in the environment; and they must not depart radically from the pattern of personal behavior under which man evolved, for example, by smoking, overeating, or sedentary living.

THOMAS MCKEOWN

Objectives

After completing this exercise, you should be able to:

1. Differentiate between the terms mutagenic and carcinogenic.
2. Provide the rationale for the Ames test.
3. Perform the Ames test.
4. Differentiate between a revertant and a mutant.

Background

Every day we are exposed to a variety of chemicals, some of which are **carcinogens**—that is, they can induce cancer. Historically, animal models have been used to evaluate the carcinogenic potential of a chemical, but the procedures are costly and time consuming and result in the inadequate testing of some chemicals. Many chemical carcinogens induce cancer because they are **mutagens** that alter the nucleotide base sequence of DNA.

Bruce Ames and his coworkers at the University of California at Berkeley developed a fast, inexpensive assay for mutagenesis using a *Salmonella* auxotroph. Most chemicals that have been shown to cause cancer in animals have proven mutagenic in the **Ames test.** Considering not all mutagens induce cancer, the Ames test is a screening technique that can be used to identify high-risk compounds that must then be tested for carcinogenic potential. The Ames test uses auxotrophic strains of *Salmonella enterica* Typhimurium, which cannot synthesize the amino acid histidine (his$^-$). The strains are also defective in *dark excision* repair of mutations (***uvrB***), and an ***rfa*** mutation eliminates a portion of the lipopolysaccharide that coats the bacterial surface. The *rfa* mutation prevents the *Salmonella* from growing in the presence of sodium desoxycholate or crystal violet and increases the cell wall permeability; consequently, more mutagens enter the cell. The *uvrB* mutation minimizes repair of mutations; as a result, the bacteria are much more sensitive to mutations. To grow the auxotroph, histidine and biotin (because of the *uvrB*) must be added to the culture media.

In the Ames spot test, a small sample of the test chemical (a suspected mutagen) is placed on the surface of glucose–minimal salts agar (GMSA, Table 28.1) seeded with a lawn of the *Salmonella* auxotroph. (A small amount of histidine allows all the cells to go through a few divisions.) If certain mutations occur, the bacteria may *revert* to a wild type, or prototroph, and grow to form a colony. Only bacteria that have mutated (reverted) to his$^+$ (able to synthesize histidine) will grow into colonies. In theory, the number of colonies is proportional to the mutagenicity of the chemical. (See Color Plate II.4.)

In nature many chemicals are neither carcinogenic nor mutagenic, but they are metabolically converted to mutagens by liver enzymes. The original Ames test could not detect the mutagenic potential of these in vivo (in-the-body) conversions. In 1973, Ames and his colleagues modified the spot test to add mammalian liver enzymes and the *Salmonella* auxotroph to melted soft agar (containing histidine) that is overlaid on GMSA. The mutagens to be tested are then placed on the agar overlay. *Salmonella* is exposed to the "activated" chemicals to test for mutagenicity. The treat-and-plate method is another modification in which the test chemical and liver enzymes are incubated in an aerobic environment. Then a sample of the activited chemicals is added to soft agar containing the *Salmonella* auxotroph and histidine and poured onto GMSA.

In this exercise, you will use the spot test with liver enzymes. The liver enzymes are called S9 because slurry of liver cells that have been broken apart is centrifuged at 9000×*g**. The supernatant contains the cytoplasm including enzymes.

Salmonella enterica **Typhimurium is a potential pathogen.**

Materials

Petri plates containing glucose–minimal salts agar (2)

Tubes containing 4 ml of soft agar (glucose–minimal salts + 0.05 mM histidine and 0.05 mM biotin) (2)

Tube containing rat liver enzymes

Sterile filter paper disks

Sterile 1-ml pipettes (3)

Forceps and alcohol

Paper disks soaked with suspected mutagens:

- Benzo(α)pyrene
- Ethidium bromide
- Nitrosamine
- 2-aminofluorene
- Cigarette smoke condensates
- Food coloring
- Hair dye
- Household compound (your choice)
- Maraschino cherries
- Hot dog
- Cosmetic or personal care item (your choice)

Culture

Salmonella enterica Typhimurium his⁻, *uvrB*, *rfa*

Techniques Required

Pipetting, Appendix A

Aseptic technique, Exercise 10

Procedure

1. Label one glucose–minimal salts agar plate "Liver Enzymes" and the other "No Enzymes."
2. Aseptically pipette 0.1 ml *Salmonella* into one of the soft agar tubes, add 0.5 ml of liver enzymes, and quickly pour it over the surface of the "Liver Enzyme" plate (Figure 32.1a). Tilt the plate back and forth gently to spread the agar evenly. Let it harden.
3. Obtain another soft agar tube. Aseptically pipette 0.1 ml of *Salmonella* into the soft agar and quickly pour it over the surface of the "No Enzymes" plate. Why does the soft agar contain histidine? ______
____

Dip the forceps in alcohol, and with the tip pointed *down*, burn off the alcohol. Keep the beaker of alcohol away from the flame.

4. Aseptically place three to five disks saturated with the various suspected mutagens on the surface of the soft agar overlay of each plate (Figure 32.1b). If the chemical is crystalline, place a few crystals directly on the agar. Label the bottom of the Petri plates with the name of each chemical. Place one more sterile disk on each plate for a control. Why is this a control? ______________________

Be very careful. These are potentially dangerous compounds.

5. Incubate both plates, right-side up, at 35°C for 48 hours.
6. Describe your results. As the chemical diffuses into the agar, a concentration gradient is formed. A mutagenic chemical will give rise to a ring of revertant colonies surrounding a disk (Color Plate II.4, Figure 32.1c). If a compound is toxic, a zone of inhibition will also be observed around the disk.

---

*Precise centrifugation conditions are specified in terms of units of gravity (×g) because centrifuges with different size rotating heads produce different centrifugal force, even at the same speed.

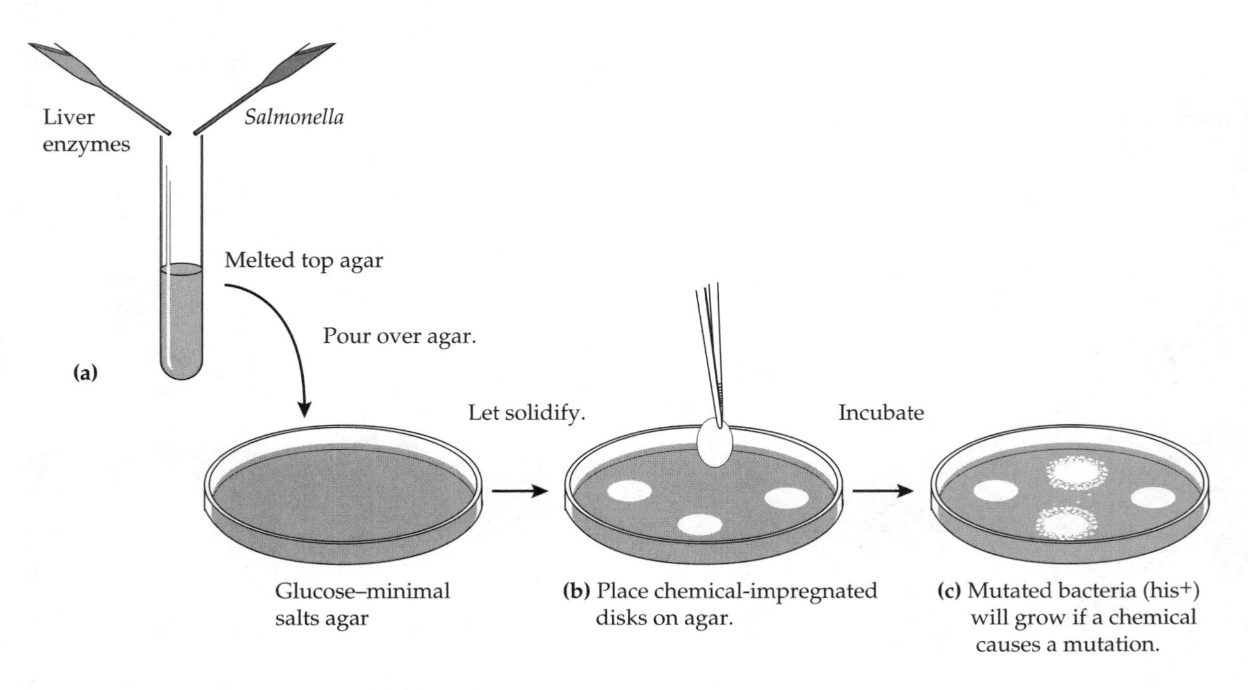

Figure 32.1

The Ames test. **(a)** *Salmonella* his$^-$ and liver enzymes are mixed in melted top agar which is poured over glucose–minimal salts agar. **(b)** Disks impregnated with suspected mutagens are placed on the agar. **(c)** Only bacteria that have mutated (reverted) to his$^+$ (able to synthesize histidine) will grow into colonies.

Exercise 32

LABORATORY REPORT

Ames Test for Detecting Possible Chemical Carcinogens

NAME ______________________

DATE ______________________

LAB SECTION ______________________

Purpose ______________________

____

____

Data

Show the location of paper disks and bacterial growth on the plates. Number the disks to correspond to the table below.

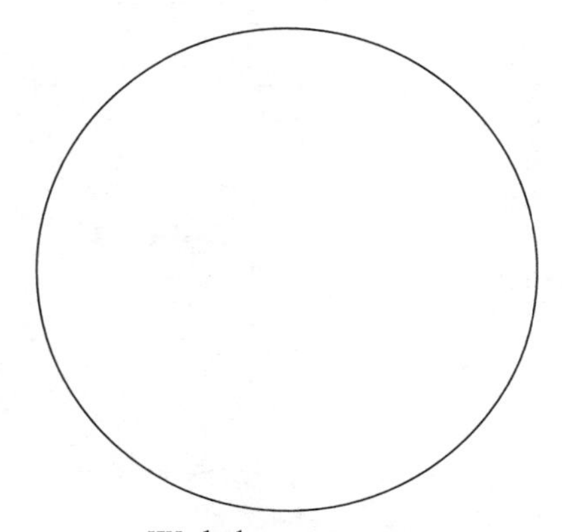

With liver enzymes

Without liver enzymes

Complete the following table.

Compound	With Liver Enzymes		Without Liver Enzymes		Mutagenic
	Growth (large colonies)	Toxicity (zone of inhibition, mm)	Growth (large colonies)	Toxicity (zone of inhibition, mm)	
1 (Control)					
2					
3					
4					
5					
6					

Conclusions

Did the liver enzymes affect the mutagenicity of any of the chemicals tested? ______________________

____

____

Questions

1. Why are mutants used as test organisms in the Ames test? ______________________

   ____

   ____

2. What is the advantage of this test over animal tests? Disadvantages? ______________________

   ____

   ____

Critical Thinking

1. Does this technique give a minimum or maximum mutagenic potential? Briefly explain.

2. Is it possible for a chemical to be negative in the Ames test yet be a carcinogen? Briefly explain.

3. In the following case, the Ames test was used to determine the mutagenic capability of 2-aminofluorene. What can you conclude from these data? Write your answer below the table.

Test Substance	Relative Amounts of Growth in the Ames Test
2-Aminofluorene	+
2-Aminofluorene activated by:	
Liver enzymes	++++
Intestinal enzymes	++
Bacteroides fragilis enzymes	++
Intestinal enzymes + *B. fragilis* enzymes	+++

Part Eight The Microbial World

EXERCISES

Organisms with eukaryotic cells include algae, protozoa, fungi, and higher plants and animals. The eukaryotic cell is typically larger and structurally more complex than the prokaryotic cell. The DNA of a eukaryotic cell is enclosed within a membrane-bounded **nucleus.** In addition, eukaryotic cells contain membrane-bounded **organelles,** which have specialized structures and perform specific functions (see the illustration on the next page).

Mushrooms, molds (Exercise 34), and yeasts are fungi. Yeasts are possibly the best known microorganisms. They are widely used in commercial processes and can be purchased in the supermarket for baking. Yeasts are unicellular fungi (Exercise 34).

Van Leeuwenhoek was the first to observe the yeast responsible for fermentation in beer:

> *I have made divers observations of the yeast from which beer is made and I have generally seen that it is composed of globules floating in a clear medium (which I judged to be the beer itself).**

Many algae are only visible through the microscope, while others can be a few meters long. Algae are important producers of oxygen and food for protozoa and other organisms. A few unicellular algae, such as the agents of "red tides,"

---

*Quoted in H. A. Lechevalier and M. Solotorovsky. *Three Centuries of Microbiology.* New York: Dover Publications, 1974, p. 502.

Alexandrium catanella and related species, are toxic to animals, including humans, when ingested in large numbers.

Algae and cyanobacteria (Exercise 36) are photoautotrophs; fungi and protozoa are all chemoheterotrophs. Protozoa, originally called "infusoria," were of interest to early investigators. In 1778, Friedrich von Gleichen studied food vacuoles by feeding red dye to his infusoria. A refinement of this experiment will be done in Exercise 37. The compound light microscope is invaluable for identification of molds, algae, cyanobacteria, and protozoa; however, many yeasts and bacteria look alike through the microscope. These organisms are identified by biochemical and genetic testing. In Exercise 36, you will use biochemical tests to identify a bacterium.

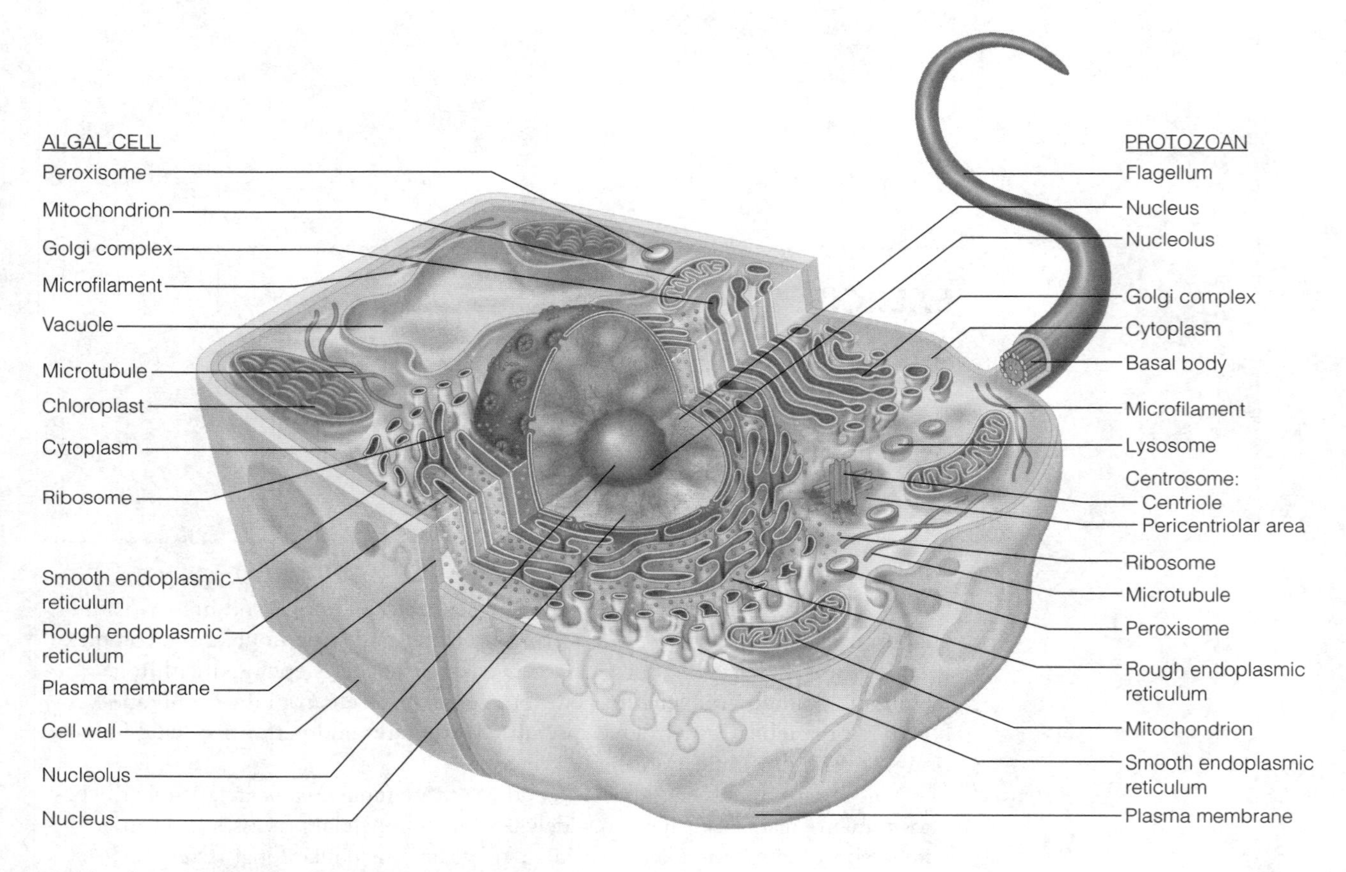

Highly schematic diagram of a composite eukaryotic cell, half algal and half protozoan.

Exercise 33

EXERCISE 33

Unknown Identification and *Bergey's Manual*

The strategy of discovery lies in determining the sequence of choice of problems to solve. Now it is in fact very much more difficult to see a problem than to find a solution to it. The former requires imagination, the latter only ingenuity.

JOHN BERNAL

Objectives

After completing this exercise, you should be able to:

1. Explain how bacteria are characterized and classified.
2. Use *Bergey's Manual.*
3. Identify an unknown bacterium.

Background

In microbiology, a system of classification must be available to allow the microbiologist to categorize and classify organisms. Communication among scientists would be very limited if no universal system of classification existed. Until recently, the **taxonomy** (grouping) of bacteria was difficult because few definite anatomical or visual differences exist. With these limitations, most bacteria are identified through evaluation of primary characteristics, such as morphology and growth patterns, and secondary characteristics, such as metabolism and serology. A characteristic that is critical for distinguishing one bacterial group from another may be irrelevant for identification of other bacteria.

The most important reference for bacterial taxonomy is ***Bergey's Manual,**** in which bacteria are classified according to similarities in their ribosomal RNA (rRNA). The Domains Bacteria and Archaea are included in *Bergey's Manual*. Bacteria and Archaea with similar rRNA sequences are grouped together into taxa. The taxa used for the Bacteria and Archaea are phylum, class, order, family, genus, and species. Although the characteristics of a given group are relatively constant, through repeated laboratory culture, atypical bacteria will be found. This variability, however, only heightens the fun of classifying bacteria.

You will be given an unknown heterotrophic bacterium to characterize and identify. By using careful deduction and systematically compiling and analyzing data, you should be able to identify the bacterium.

Keys to species of bacteria are provided in Appendix H, Figures H.1 and H.2. These keys are an example of a dichotomous classification system—that is, a population is repeatedly divided into two parts until a description identifies a single member. Such a key is sometimes called an "artificial key" because there is no single correct way to write one. You may want to check your conclusion with the species description given in *Bergey's Manual.*

To begin your identification, ascertain the purity of the culture you have been given and prepare stock and working cultures. Avoid contamination of your unknown. Note growth characteristics and Gram-stain appearance for clues about how to proceed. After culturing and staining the unknown, many bacterial groups can be eliminated. Final determination of your unknown will depend on carefully selecting the relevant biochemical tests and weighing the value of one test over another in case of contradictions. (See Color Plate XIV) Enjoy!

Materials

Petri plates containing Trypticase soy agar (2)

Trypticase soy agar slant (2)

All stains, reagents, and media previously used

Culture

Unknown bacterium # ______________________

Techniques Required

Compound light microscopy, Exercise 1

Hanging-drop procedure, Exercise 2

**Bergey's Manual of Systematic Bacteriology*, 2nd ed., 5 vols. (2005), is the reference for classification. *Bergey's Manual of Determinative Bacteriology*, 9th ed. (1994), is used for laboratory identification of culturable bacteria and archaea.

Wet-mount technique, Exercise 2

Negative staining, Exercise 4

Gram staining, Exercise 5

Acid-fast staining, Exercise 6

Endospore, capsule, and flagella staining, Exercise 7

Inoculating loop and needle technique, Exercise 10

Aseptic technique, Exercise 10

Plate streaking, Exercise 11

OF test, Exercise 13

Starch hydrolysis, Exercise 13

MRVP tests, Exercise 14

Fermentation tests, Exercise 14

Citrate test, Exercise 14

Protein catabolism, Exercises 15 and 16

Catalase test, Exercise 17

Nitrate reduction test, Exercise 17

Oxidase test, Exercise 17

Procedure

1. Streak your unknown onto the agar plates for isolation. Incubate one plate at 35°C and the other at room temperature for 24 to 48 hours. Note the growth characteristics (Figure 9.2) and the temperature at which each one grows best.
2. Aseptically inoculate two Trypticase soy agar slants from a colony on your streak plate. Incubate them for 24 hours. Describe the resulting growth (Figure 10.7). Keep both slant cultures in the refrigerator. One is your stock culture; the other is your working culture. Subculture your stock culture onto another slant when your working culture is contaminated or not viable. Keep the working culture in the refrigerator when it is not in use.
3. Use your working culture for all identification procedures. When a new slant is made and its purity demonstrated, discard the old working culture. What should you do if you think your culture is contaminated? ______________________________

   ____
4. Read the keys in Appendix H, Figures H.1 and H.2, to develop ideas on how to proceed. Perhaps determining staining characteristics might be a good place to start. What shape is it? ______________ What can be eliminated? ______________
5. After determining its staining and morphologic characteristics, determine which biochemical tests you will need. Do not be wasteful. Inoculate *only* what is needed. It is not necessary to repeat a test—do it once accurately. Do not perform unnecessary tests.
6. If you come across a new test—one not previously done in this course—determine whether it is essential. Can you circumvent it? ______________ If not, consult your instructor.
7. Record your results in the Laboratory Report and identify your unknown.

Exercise 33 **LABORATORY REPORT**

Unknown Identification and *Bergey's Manual*

NAME ______________________

DATE ______________________

LAB SECTION ______________________

Purpose ______________________

____

____

Data

Write "not tested" (NT) next to tests that were not performed. Unknown # __________

Morphological, Staining, and Cultural Characteristics	Microscopic Examination Sketches: Label and Give Magnification (_____×)
The cell Staining characteristics Gram ________ Age ________ Other ________ Age ________ Shape ________ Size ________ Arrangement ________ Endospores (position) ________ ________ Motility ________ Determined by ________ Colonies on Trypticase soy agar Diameter ________ Appearance ________ Color ________ Elevation ________ Margin ________ Consistency ________ Agar slant Age ________ Amount of growth ________ Pattern ________ Color ________	

Results: Essential Biochemical Characteristics

	Time (hr)			Temp.____ °C
	24	48	. .	
Glucose				
Lactose				
Mannitol				
Catalase				
Oxidase				
H_2S				
Nitrate reduction				
Indole				
Methyl red				
V–P				
Citrate				

Abbreviations:
A = Acid
G = Gas
a = slight acid
alk = alkaline
+ = positive
− = negative
ng = no growth

Other characteristics, special media, etc. ______________________

____

____

Conclusion

What organism was in unknown # __________ ? ____

Questions

1. On a separate sheet of paper, write your rationale for arriving at your conclusion.

2. Why is it necessary to complete the identification of a bacterium based on its physiology rather than its morphology? ____

____

____

____

3. Place the following organisms next to their description in the key shown below:

Actinomyces
Bacillus
Bacteroides
Campylobacter
Clostridium
Corynebacterium
Cyanobacteria
Escherichia
Lactobacillus
Mycobacterium
Mycoplasma
Neisseria
Nitrosomonas
Nocardia
Pseudomonas
Purple sulfur bacteria
Staphylococcus
Streptococcus
Streptomyces
Thiobacillus
Treponema

I. Gram-positive
  A. Rods
    1. Produce conidiospores
      a. Aerobic ____________________
      b. Anaerobic ____________________
    2. Endospore-forming
      a. Obligate anaerobe ____________________
      b. Facultative anaerobe ____________________
    3. Nonendospore-forming
      a. Acid-fast
        (1) Mycelia formed ____________________
        (2) Mycelia not formed ____________________
      b. Not acid-fast
        (1) Regular ____________________
        (2) Club-shaped; stain irregularly ____________________

B. Cocci

1. Oxidative; catalase-positive ______________________
2. Fermentative; catalase-negative ______________________

C. No cell wall ______________________

II. Gram-negative

A. Heterotrophic

1. Helical or vibroid
   a. Possess axial filament ______________________
   b. Possess flagella ______________________
2. Rods
   a. Aerobic ______________________
   b. Facultatively anaerobic ______________________
   c. Anaerobic ______________________
3. Cocci ______________________

B. Autotrophic

1. Photoautotrophic
   a. Produce O_2 ______________________
   b. Do not produce O_2 ______________________
2. Chemoautotrophic
   a. Oxidize N-containing inorganic compounds ______________________
   b. Oxidize S-containing inorganic compounds ______________________

EXERCISE 34

Exercise 34

Fungi: Yeasts

Objectives

After completing this exercise, you should be able to:

1. Culture and identify yeasts.
2. Differentiate between yeasts and bacteria.

Background

Fungi possess eukaryotic cells and can exist as unicellular or multicellular organisms. They are chemoheterotrophic and obtain nutrients by absorbing dissolved organic material through their cell walls and plasma membranes. Fungi (with the exception of yeasts) are aerobic. Unicellular yeasts, multicellular molds, and macroscopic species such as mushrooms are included in the Kingdom Fungi. Compared to bacteria, fungi generally grow better in more acidic conditions and tolerate higher osmotic pressure and lower moisture. They are larger than bacteria, with more cellular and morphologic detail. In contrast to bacterial characterization, primary characteristics, such as morphology and cellular detail, are used to identify fungi, with little attention given to secondary characteristics, such as metabolism and antigenic composition. Fungi are structurally more complex than bacteria but are less diverse metabolically.

Yeasts are nonfilamentous, unicellular fungi that are typically spherical or oval in shape. Yeasts are widely distributed in nature and frequently found on fruits and leaves as a white, powdery coating. When budding yeasts reproduce asexually, cell division is uneven. A new cell forms as a protuberance (bud) from the parent cell (Figure 34.1a). In some instances, when buds fail to detach themselves, a short chain of cells called a **pseudohypha** forms (Figure 34.1b). When yeasts reproduce sexually, they may produce one of several types of sexual spores. The type of sexual spore produced by a species of yeast is used to classify the yeast to a phylum. Metabolic activities are also used to identify genera of yeasts.

Yeasts are facultative anaerobes. Their metabolic activities are used in many industrial fermentation processes. Yeasts are used to prepare many foods, including bread, and beverages, such as wine and beer.

In the laboratory, **Sabouraud agar,** a selective medium, is commonly used to isolate yeast. Sabouraud agar has very simple nutrients (glucose and peptone) and a low pH, which inhibits the growth of most other organisms. Many of the techniques useful in working with bacteria can be applied to yeasts.

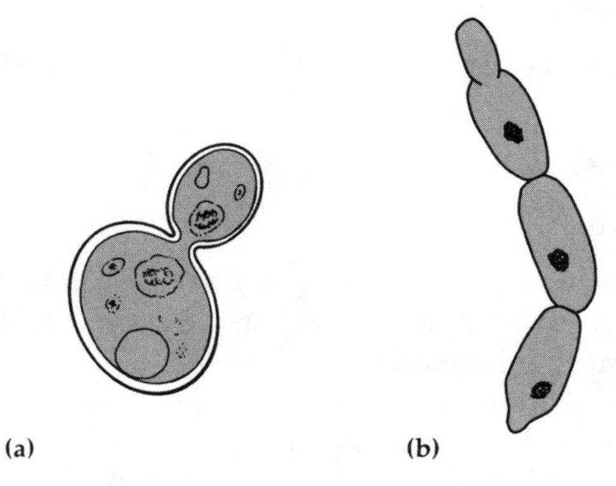

Figure 34.1

Budding yeasts. **(a)** A bud forming from a parent cell. **(b)** Pseudohyphae are short chains of cells formed by some yeasts.

Materials

Glucose fermentation tubes (2)

Sucrose fermentation tubes (2)

Petri plates containing Sabouraud agar (2)

Bottle containing glucose–yeast extract broth

Sterile cotton swab

Coverslip

Test tube

Balloon

Fruit or leaves

Methylene blue (second period)

Cultures (as assigned)

Baker's yeast

Rhodotorula rubra

Candida albicans

Saccharomyces cerevisiae

Techniques Required

Compound light microscopy, Exercise 1

Colony morphology, Exercise 9

Wet-mount technique, Exercise 2

Plate streaking, Exercise 11

Fermentation tests, Exercise 14

Procedure

Yeasts

1. Gently suspend a pinch of baker's yeast in a small amount of lukewarm water in a test tube, creating a milky solution.
2. Each pair of students will use one of the yeast cultures and the suspension of baker's yeast.
   a. Divide one Sabouraud agar plate in half. Streak one-half with a known yeast culture and the other half with the baker's yeast suspension.
   b. Inoculate each organism into a glucose fermentation tube and a sucrose fermentation tube.
3. Incubate all media at 35°C until growth is seen.
4. Make a wet mount of each culture by using a small drop of methylene blue. Record your observations.
5. After the yeasts have grown, record your results. Examine cultures of the yeasts you did not culture, and record pertinent results. (See Color Plate IV.1.)

Yeast Isolation

1. Cut the fruit or leaves into small pieces. Place them in the bottle of glucose–yeast extract broth. Cover the mouth of the bottle with a balloon. Incubate the bottle at room temperature until growth has occurred. Record the appearance of the broth after incubation. Was gas produced? ______________
2. Divide a Sabouraud agar plate in half. Each partner inoculates half of the medium, following either procedure a or procedure b.
   a. Swab the surface of your tongue with a sterile swab. Inoculate one-half of the agar surface with the swab. Discard the swab in the disinfectant. Why will few bacteria grow on this medium? ______________________________
   b. Using a sterile inoculating loop, streak one-half of the agar surface with a loopful of broth from the bottle just prepared in step 1. Replace the balloon.
3. Incubate the plate, inverted, at room temperature until growth has occurred. Prepare wet mounts with methylene blue from different-appearing colonies. Record your results.

Exercise 34

Fungi: Yeasts

LABORATORY REPORT

NAME ______________________

DATE ______________________

LAB SECTION ______________________

Purpose ______________________

____

____

Data

Yeasts

Fermentation tubes:

Organism	Glucose			Sucrose		
	Acid	Gas	Fermentation	Acid	Gas	Fermentation
Rhodotorula rubra						
Candida albicans						
Saccharomyces cerevisiae						
Baker's yeast						

Sabouraud agar plates:

Organism	Draw a Typical Colony	Wet Mount
Rhodotorula rubra	Color: ______________	______ ×
Candida albicans	Color: ______________	______ ×

Organism	Draw a Typical Colony	Wet Mount
Saccharomyces cerevisiae	Color: ____________	______ ×
Baker's yeast	Color: ____________	______ ×

Yeast Isolation

Plants used: ______________________________

Describe the appearance of the glucose broth after ________ days' incubation. ______________________________

____

Was gas produced? ______________

Sabouraud agar:

Area Sampled	Colony Appearance	Color	Size	Wet Mount
Tongue:				
Bottle:				

Any bacteria seen? ______________________________

If so, which colonies? ______________________________

Questions

1. Could you identify the genus of baker's yeast? ______________________

____

2. Did you culture yeast from your mouth? ______ From the plants? ______ How do you know? ______

____

3. What was the purpose of the balloon on the glucose–yeast extract broth bottle? ______________________

____

4. Define the term *yeast*. ______________________

____

____

____

____

5. Compare and contrast yeast and bacteria regarding their appearance both on solid media and under the microscope.

____

____

____

6. Why are yeast colonies larger than bacterial colonies? ______________________

____

____

____

____

Critical Thinking

A 45-year-old HIV-positive woman using continuous intravenous antibiotic infusion to manage a kidney infection developed a 39°C fever. Cultures of blood, the needle tip, and the insertion site revealed large ovoid cells that reproduced by budding.

1. What is your preliminary identification of the infecting organism?

2. How might the antibiotic treatment have affected this patient?

3. What do the findings suggest about the portal of entry for the infection?

Exercise 35

Fungi: Molds

And what is weed? A plant whose virtues have not been discovered.

RALPH WALDO EMERSON

Objectives

After completing this exercise, you should be able to:

1. Characterize and classify fungi.
2. Compare and contrast fungi and bacteria.
3. Identify common saprophytic molds.
4. Explain dimorphism.

Background

The multicellular filamentous fungi are called **molds.** Because of the wide diversity in mold morphology, morphology is very useful in identifying these fungi.

A macroscopic mold colony is called a **thallus** and is composed of a mass of strands called **mycelia.** Each strand is a **hypha,** with the **vegetative hyphae** growing in or on the surface of the growth medium. Aerial hyphae, called **reproductive hyphae,** originate from the vegetative hyphae and produce a variety of asexual reproductive **spores** (Figure 35.1). The hyphal strand of most molds is composed of individual cells separated by a cross wall, or **septum.** These hyphae are called **septate hyphae.** A few fungi, including *Rhizopus*, shown in Figure 35.1a, have hyphae that lack septa and are a continuous mass of cytoplasm with multiple nuclei. These are called **coenocytic hyphae.**

In a laboratory, fungi are identified by the appearance of their colony (color, size, and so on), hyphal organization (septate or coenocytic), and the structure and organization of reproductive spores. Because of the importance of colony appearance and organization, culture techniques and microscopic examination of fungi are very important.

Until recently, fungi were classified on the basis of their sexual spores. Fungi that do not produce sexual spores were put in a "holding" phylum called

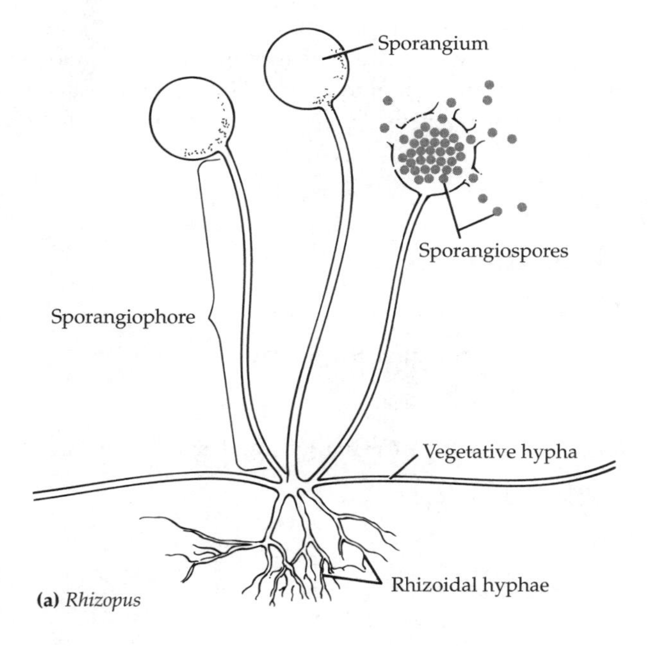

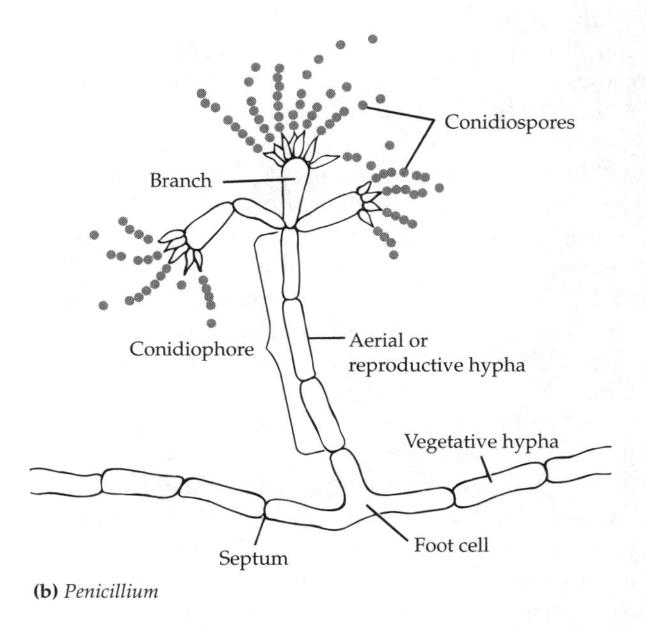

Figure 35.1

Asexual spores are produced by aerial hyphae. **(a)** Sporangiospores are formed within a sporangium. **(b)** Conidiospores are formed in chains. One possible arrangement is shown here.

Table 35.1

Characteristics of Common Saprophytic Fungi

Phylum	Growth Characteristics	Asexual Reproduction	Sexual Reproduction
Fungi Zygomycota ("conjugation fungi")	Coenocytic hyphae	Sporangiospores	Zygospores
Ascomycota ("sac fungi")	Septate hyphae; yeastlike	Conidiospores, budding	Ascospores
Basodiomycota ("club fungi")	Septate hyphae; includes fleshy fungi (mushrooms)	Fragmentation	Basidiospores
Fungus-like algae Oomycota ("water molds")	Coenocytic hyphae	Zoospores	Oospores

deuteromycota. Now fungi are classified into phyla based on similarities in their rRNA. This eliminated the need for deuteromycota and revealed that water molds (Oomycota) are actually more closely related to diatoms and kelp than to fungi.

The fungus-like algae in the phylum **Oomycota** are included here because they look like fungi to the unaided eye and their growth requirements are similar to fungi. The oomycota are usually found in aquatic habitats and form sexual spores called oospores, formed by the fusion of two cells, and motile asexual spores called **zoospores.** One oomycote, *Phytophthora infestans*, causes late potato blight, which resulted in the Irish famine in the mid-19th century.

Members of the phylum **Zygomycota** are saprophytic molds that have coenocytic hyphae. A **saprophyte** obtains its nutrients from dead organic matter; in healthy animals and plants, they do not usually cause disease. A common example is *Rhizopus* (bread mold) (Color Plate IV.2). The asexual spores are formed inside a **sporangium,** or spore sac, and are called **sporangiospores** (Figure 35.1a). Sexual spores called **zygospores** are formed by the fusion of two cells.

The **Ascomycota** include molds with septate hyphae and some yeasts. They are called sac fungi because their sexual spores, called **ascospores,** are produced in a sac, or **ascus.** The saprophytic molds usually produce **conidiospores** asexually (Figure 35.1b). The arrangements of the conidiospores are used to identify these fungi. Common examples are *Penicillium* (Color Plate IV.3) and *Aspergillus* (Color Plate IV.4).

The **Basidiomycota** include the fleshy fungi, or mushrooms, and have sexual spores called **basidiospores** produced by the club-shaped cap called a **basidium.**

A classification scheme for saprophytic fungi is shown in Table 35.1. A majority of the fungi classified in Table 35.1 are saprophytes, but each phylum contains a few genera that are **pathogens,** causing disease in plants and animals.

Fungi, especially molds, are important clinically and industrially. Spores in the air are also the most common source of contamination in the laboratory. Sabouraud agar is a selective medium that is commonly used to isolate fungi.

Some pathogenic fungi exhibit **dimorphism**—that is, they have two growth forms. In pathogenic fungi, dimorphism is usually temperature dependent. The fungus is a yeastlike at 37°C and a moldlike at 25°C.

In this exercise, we will examine different molds and demonstrate dimorphism. The dimorphism of *Mucor* is not temperature dependent; it will be left for you to decide what physical condition affects the growth of *Mucor*.

Materials

Mold Culture

Petri plates containing Sabouraud agar (2)

Melted Sabouraud agar

Vaspar (one-half petroleum jelly and one-half paraffin)

Sterile Petri dish

Coverslips

Bent glass rod

Pasteur pipette

Dimorphic Gradient

Sabouraud agar, 5 ml, melted at 48°C

5-ml plastic microbeakers or small paper cups (2)

Razor blade (second period)

Tape

Cultures (as assigned)

Rhizopus stolonifer

Aspergillus niger

Penicillium chrysogenum

Mucor indicus

Prepared Slides

Zygospores

Ascospores

Techniques Required

Compound light microscopy, Exercise 1

Aseptic techniques, Exercise 10

Dissecting microscope, Appendix E

Sabouraud agar, Exercise 34

Procedure

First Period

1. Contaminate one Sabouraud agar plate in any manner you desire. Expose it to the air (outside, hall, lab, or wherever) for 15 to 30 minutes, or touch it. Incubate the plate, inverted, at room temperature for 5 to 7 days.
2. Inoculate the other Sabouraud agar plate with the mold culture assigned to you. Make one line in the center of the plate. Why don't you streak it? ____

   ____

   Incubate the plate, inverted, at room temperature for 5 to 7 days. (See Color Plates IV.2, IV.3, and IV.4.)
3. Set up a slide culture of your assigned mold (Figure 35.2).
   a. Clean a slide and dry it.
   b. Place a drop of Sabouraud agar on the slide, flatten it to the size of a dime, and let it solidify.
   c. Using a sterile hot loop, carefully scrape off half of the circle of agar, leaving a smooth edge.
   d. Gently shake the mold culture to resuspend it, and inoculate the straight edge of the agar with a loopful of mold.
   e. Place a coverslip on the inoculated agar.

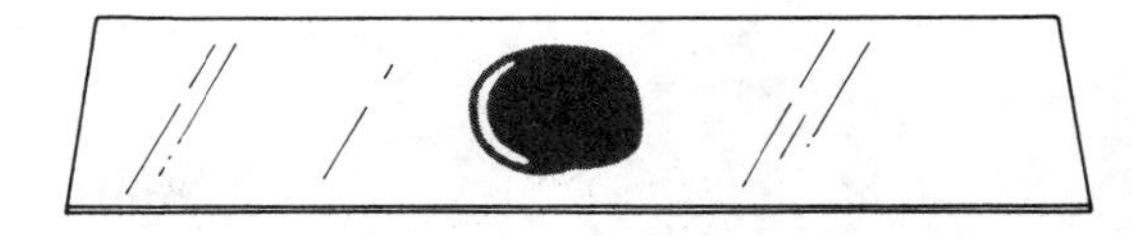

(a) Place agar medium on the slide.

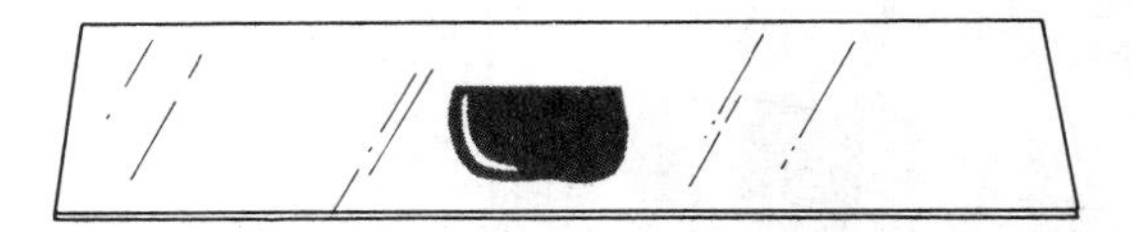

(b) Cut a straight edge on one side of the solidified agar.

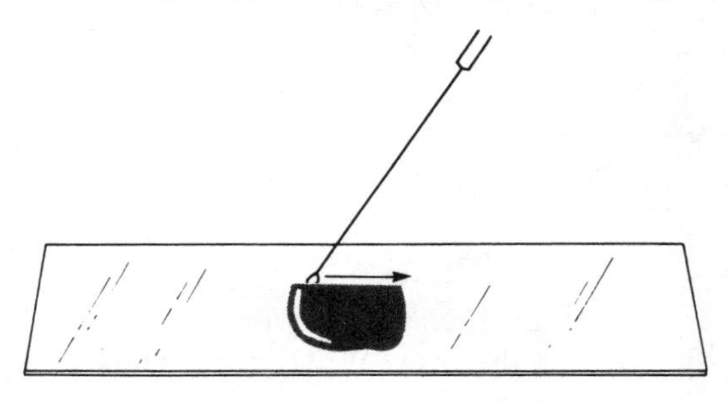

(c) Inoculate the mold onto the straight edge.

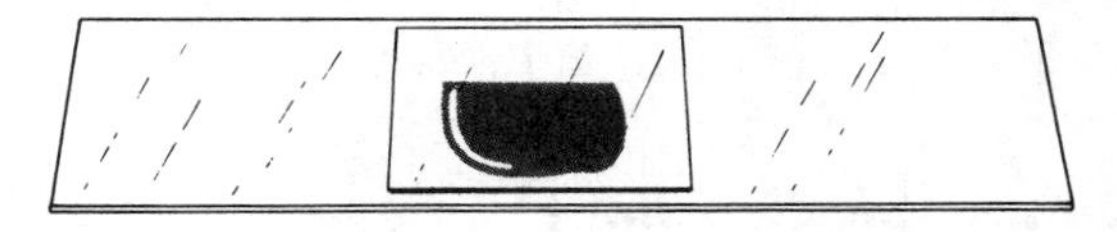

(d) Place a coverslip on the agar.

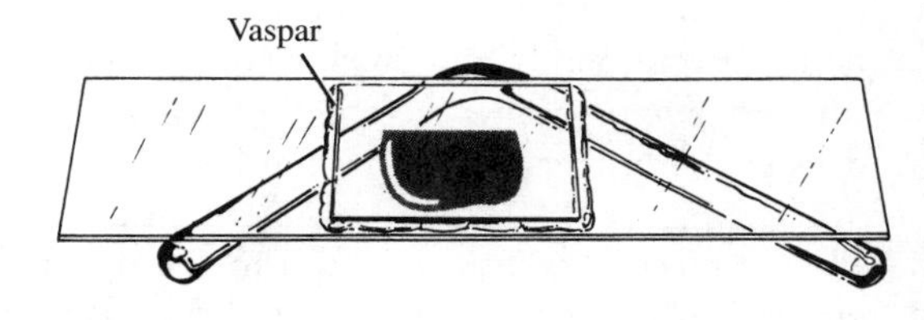

(e) Seal the three uninoculated edges with vaspar, and place the slide on a glass rod in a Petri plate.

Figure 35.2
Preparation of a slide culture.

   f. Using a Pasteur pipette, add melted vaspar to seal three edges. Do not seal the inoculated edge.
   g. Put a piece of wet paper towel in the bottom of a Petri dish. Why? ____

   ____

   Place the slide culture on the glass rod on top of the towel and close the Petri dish. Incubate

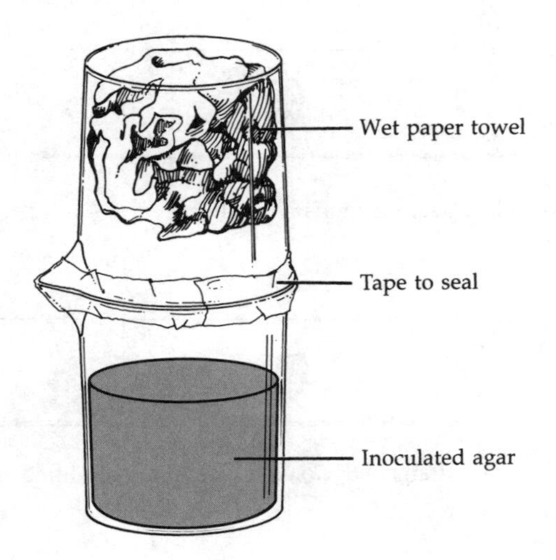

Figure 35.3
Tape the two beakers together as shown.

it at room temperature for 2 to 5 days. Do *not* invert the dish. What is the purpose of the glass rod? ______________________________

____

4. Observe the prepared slides showing sexual spores. Carefully diagram each of the spore formations in the spaces provided in the Laboratory Report.

*Dimorphic Gradient**

1. Flick the *Mucor* culture tube to resuspend the culture. Inoculate the melted agar with 2 or 3 loopfuls of the fungal culture. Mix the tube by rolling it between your hands, and quickly pour the contents into an empty beaker before they harden. Why doesn't the beaker need to be sterile? ______________

____
2. Place a piece of wet paper towel in the remaining empty beaker, and invert it over the beaker containing the agar (Figure 35.3). Tape the edges.
3. Incubate the taped beakers at room temperature until growth occurs (5 to 7 days).

Second Period

1. Examine plate cultures of *each* mold (without a microscope), and describe their color and appearance. Then examine them with a dissecting microscope. Look at both the top and the underside.

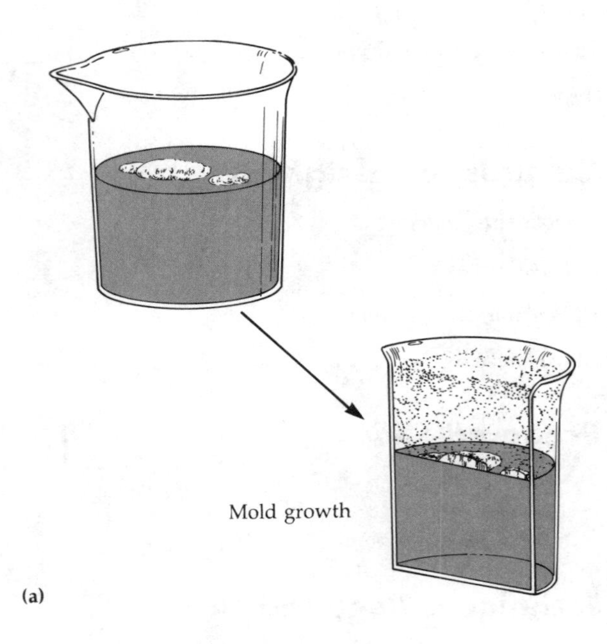

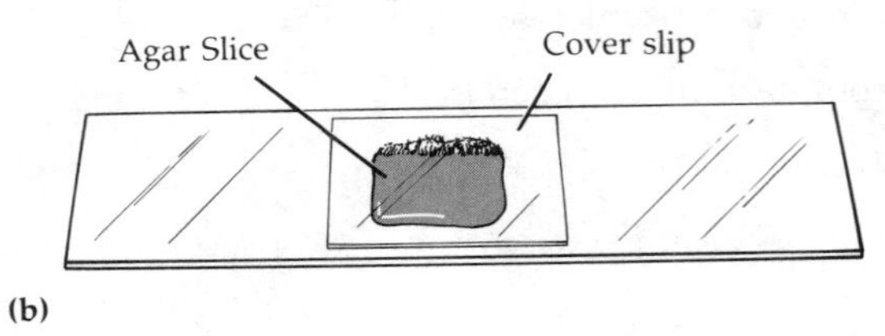

Figure 35.4
Observing fungal growth microscopically. **(a)** Cut the bottom beaker in half with a razor blade, and cut a vertical slice of agar. **(b)** Place the thin slice of agar on a slide, and cover it with a coverslip.

2. Examine your contaminated Sabouraud plate and describe the results.
3. Examine slide cultures of each mold using a dissecting microscope. Record your observations.
4. To examine your dimorphic gradient, remove the tape and cut the beaker in half vertically with a razor blade (Figure 35.4a).
5. Cut a thin (approximately 1 mm) vertical slice of the inoculated agar with a razor blade. Carefully place the slice of agar on a slide and cover it with a coverslip (Figure 35.4b). Discard the razor blade by placing it in the sharps container.
6. Observe the slide under low and high power. Scan the agar from the bottom of the slice to the top. Many focal planes are present.
7. Discard your beaker in the biohazard container and your slide in the autoclave container.

*Adapted from S. Bertnicki-Garcia. "The Dimorphic Gradient of *Mucor rouxii:* A Laboratory Exercise." *ASM News* 38(9):486–488, 1972. Permission granted by the American Society for Microbiology.

Exercise 35

LABORATORY REPORT

Fungi: Molds

NAME ______________________

DATE ______________________

LAB SECTION ______________________

Purpose

____

____

____

Observations

Plate Cultures

Colony Appearance	Organisms			
	Rhizopus stolonifer	*Aspergillus niger*	*Penicillium chrysogenum*	Unknown from Contaminated Plate
Macroscopic Hyphae color				
Spore color				
Underside color				
Microscopic Diagram ___ ×				

Slide Cultures

Colony Appearance	Organisms		
	Rhizopus stolonifer	*Aspergillus niger*	*Penicillium chrysogenum*
Macroscopic Hyphae color			
Spore color			
Underside color			
Microscopic Diagram ___ ×			

Prepared Slides

Sexual spores

Phylum:	Zygomycetes	Ascomycetes
Genus:	____________	____________
Total magnification:	___×	___×

Dimorphic Gradient

Fungus: ______________________________

Top of agar (draw the structures seen):

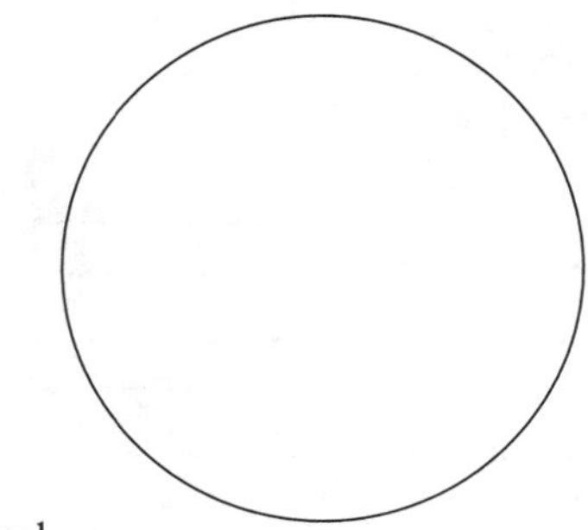
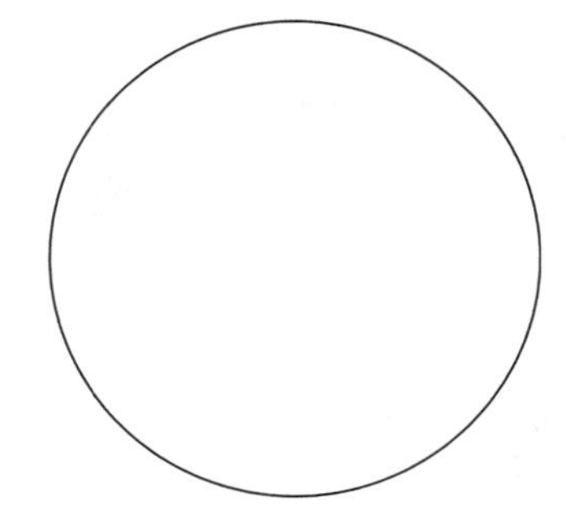
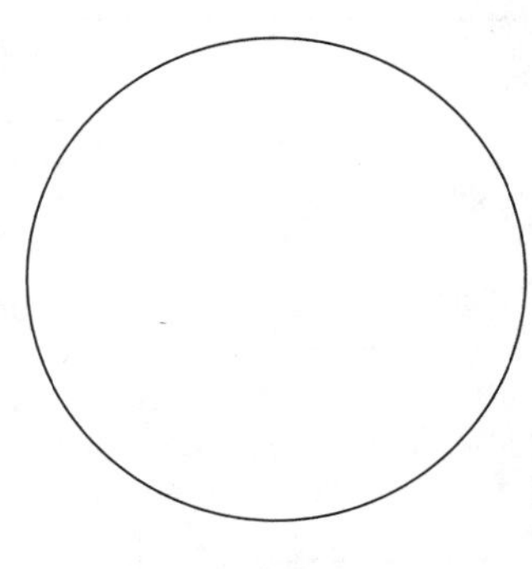

Total magnification: ___× ___× ___×

Middle of agar:

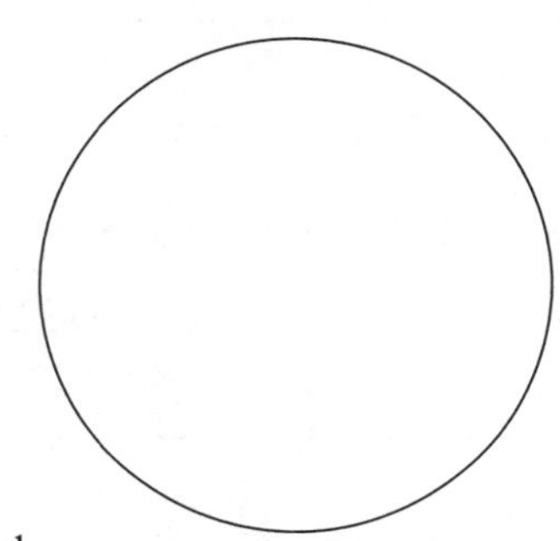
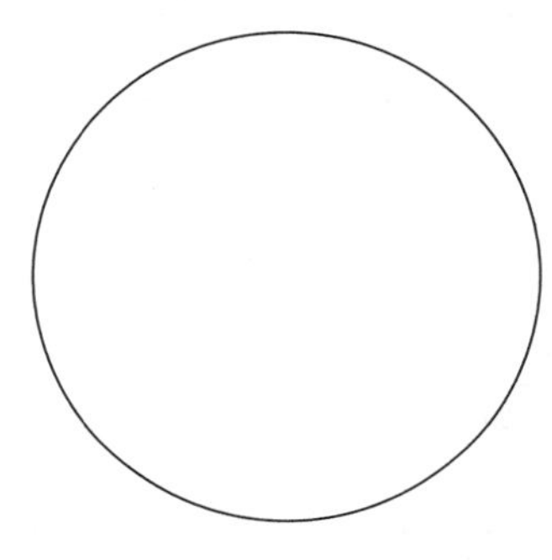
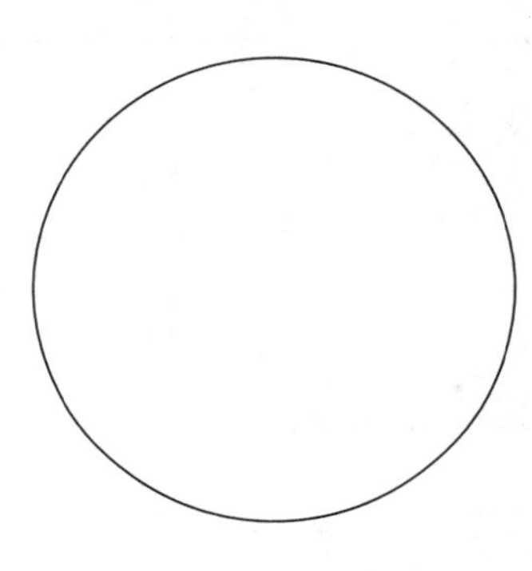

Total magnification: ___× ___× ___×

Bottom of agar:

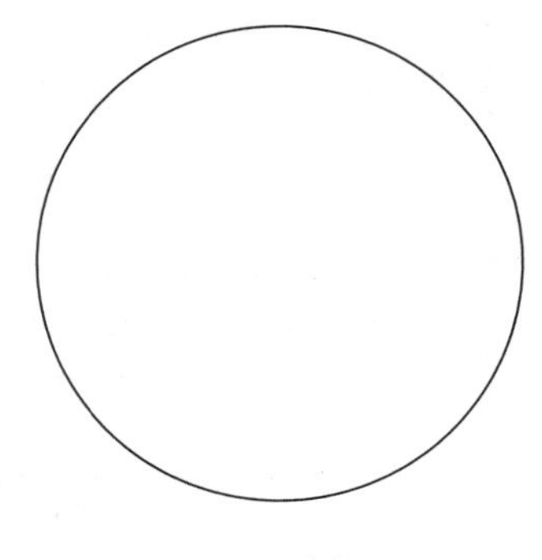
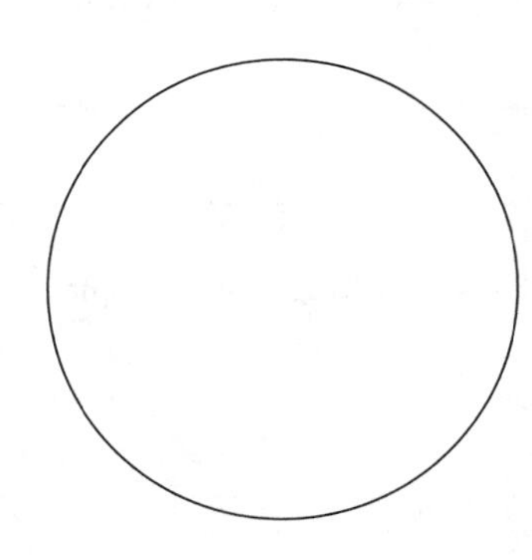

Total magnification: ___× ___× ___×

Coenocytic or septate: ______________________________

Questions

1. Fill in the following table.

Organism	Asexual Spores	Sexual Spores	Phylum
Rhizopus			
Aspergillus			
Penicillium			

2. How do mold spores differ from bacterial endospores? ______

3. What determines the form of *Mucor* seen in the dimorphic gradient? ______

4. Why do media used to culture fungi contain sugars? ______

5. Why are antibiotics frequently added to Sabouraud agar for isolation of fungi from clinical samples?

6. How can fungi cause respiratory-tract infections? ______

7. Why weren't pathogenic fungi used in this exercise? ______

Critical Thinking

An 82-year-old man was hospitalized with pneumonia. He was treated with penicillin and discharged after 7 days when his symptoms subsided. Two weeks later, the man was readmitted to the hospital for pneumonia. No bacteria or viruses were cultured from a biopsy specimen. Broad-spectrum antibiotic therapy was administered. The patient developed pneumonia again after returning home from an 8-day hospital stay. A bronchoscopy showed septate hyphae in his lung tissue. An investigation of his home showed that he did not keep pets, but some sparrows had nested outside the living room window; he was not a gardener or outdoorsman. *Aspergillus* was cultured from his humidifier.

1. What is your preliminary identification of the pathogen?

2. Why was the diagnosis delayed?

3. Why did his infection recur?

Exercise 36

Phototrophs: Algae and Cyanobacteria

Objectives

After completing this exercise, you should be able to:

1. List criteria used to classify algae.
2. List general requirements for the growth of phototrophic organisms.
3. Compare and contrast algae with fungi and bacteria.

Background

Most freshwater phototrophs belong to the groups listed in Table 36.1. Of primary interest to microbiologists are the **cyanobacteria.** Cyanobacteria have prokaryotic cells and belong to the Domain Bacteria.

Algae is the common name for photosynthetic eukaryotic organisms that lack true roots, stems, and leaves. Algae may be found in the ocean and in freshwater and on moist tree bark and soil. (See Color Plate IV.5.) Algae may be unicellular, colonial, filamentous, or multicellular. They exhibit a wide range of shapes: from the giant brown algae or kelp and delicate marine red algae to spherical green-algal colonies. Algae are identified according to pigments, storage products, the chemical composition of their cell walls, and flagella.

While the growth of phototrophs is essential in providing oxygen and food for other organisms, some filamentous algae, such as *Spirogyra,* are a nuisance to humans because they clog filters in water systems (see Color Plate IV.6). And others, such as *Alexandrium,* produce toxins that are harmful to vertebrates. Phototrophs can be used to determine the quality of water. Polluted waters containing excessive nutrients from sewage or other sources have more cyanobacteria and fewer diatoms than clean waters do. Additionally, the *number* of algal cells indicates water quality. More than 1000 algal cells per milliliter indicates that excessive nutrients are present.

Table 36.1

Some Characteristics of Major Groups of Phototrophs Found in Freshwater

	Bacteria	Algae		
Characteristics	Cyanobacteria	Euglenoids	Diatoms	Green Algae
Color	Blue-green	Green	Yellow-brown	Green
Cell wall	Bacteria-like	Lacking	Readily visible with regular markings	Visible
Cell type	Prokaryote	Eukaryote	Eukaryote	Eukaryote
Flagella	Absent	Present	Absent	Present in some
Cell arrangement	Unicellular or filamentous	Unicellular	Unicellular or colonial	Unicellular, colonial, or filamentous
Nutrition	Autotrophic	Facultatively heterotrophic	Autotrophic	Autotrophic
Produce O_2	Yes	Yes	Yes	Yes

Materials

Pond water samples:

A. Incubated in the light for 4 weeks.
B. Incubated in the dark for 4 weeks.
C. With nitrates and phosphates added; incubated in the light for 4 weeks.
D. With copper sulfate added; incubated in the light for 4 weeks.

Techniques Required

Compound light microscopy, Exercise 1

Hanging-drop procedure, Exercise 2

Procedure

1. Prepare a hanging-drop slide from a sample of pond water A. Take your drop from the bottom of the container. Why? ______________________________
____
2. Examine the slide using the low and high-dry objectives. Identify the algae present in the pond water. Refer to Color Plate XIII for identification. Draw those algae that you cannot identify. Record the relative amounts of each type of alga from 4+ (most abundant) to + (one representative seen).
3. Repeat the observation and data collection for the remaining pond water samples.

Exercise 36

LABORATORY REPORT

Phototrophs: Algae and Cyanobacteria

NAME ____________________

DATE ____________________

LAB SECTION ____________________

Purpose ____________________

Data

Name of Alga or Cyanobacterium	Relative Abundance in Pond Water			
	Incubated in Light	Incubated in Dark	Incubated with NO_3^- and PO_4^{3-}	Incubated with $CuSO_4$
Drawings of other algae seen				
Total number of species				
Total number of organisms				

Which sample is the control? ____________________

Conclusions

Compare the number and various groups of algae observed in each environment with the control.

Dark (sample B) vs. light (sample A)

In which sample did you expect more algae? ______

Why? ______

______

______

Did your findings agree with expected results? If not, briefly explain why. ______

______

______

Nitrate and phosphate (sample C) vs. no nitrate and phosphate (sample A)

In which sample did you expect more algae? ______

Why? ______

______

______

Did your findings agree with expected results? If not, briefly explain why. ______

______

______

Copper (sample D) vs. no copper (sample A)

In which sample did you expect more algae? ______

Why? ______

______

______

Did your findings agree with expected results? If not, briefly explain why. ______

______

______

Questions

1. Why can algae and cyanobacteria be considered indicators of productivity as well as of pollution?

2. How can algae be responsible for the production of more oxygen than land plants?

3. Why aren't algae included in the Kingdom Plantae?

4. Describe one way in which algae and fungi differ.

   How are they similar?

Critical Thinking

1. Outbreaks of cyanobacterial intoxication associated with lakes and ponds are reported annually. What would cause an increased number of cases in summer months? Why aren't cyanobacterial intoxications associated with swimming pools?

2. Cyanobacteria were once called "blue-green algae." What characteristics would lead to the name "blue-green algae"? What caused biologists to reclassify them as cyanobacteria?

Exercise 37

Protozoa

Objectives

After completing this exercise, you should be able to:

1. List three characteristics of protozoa.
2. Explain how protozoa are classified.

Background

Protozoa are unicellular eukaryotic organisms. Many protozoa live in soil and water, and some are normal microbiota in animals. A few species of protozoa are parasites.

Protozoa are heterotrophs, and most are aerobic. They feed on other microorganisms and on small particulate matter. Protozoa lack cell walls; in some, the outer covering is a thick, elastic membrane called a **pellicle.** Cells with a pellicle require specialized structures to take in food. The **contractile vacuole** may be visible in some specimens (Figure 37.1). This organelle fills with freshwater and then contracts to eliminate excess freshwater from the cell, allowing the organism to live in low-solute environments. Would you expect more contractile vacuoles in freshwater or marine protozoa? ______________________________

In this exercise, we will examine live, free-living protozoa, as well as prepared slides of three parasitic protozoa. The **Rhizopoda** phylum consists of protozoa that move by using pseudopods. The **amoebas** (Figure 37.1a) move by extending lobelike projections of cytoplasm called **pseudopods.** As pseudopods flow from one end of the cell, the rest of the cell flows toward the pseudopods.

The **Euglenozoa** (Figure 37.1b) have one or more flagella. Although many euglenozoa are heterotrophs, the organism used in this exercise is a facultative heterotroph. It grows photosynthetically in the presence of light and heterotrophically in the dark.

Members of the phylum **Ciliophora** (Figure 37.1c) have many cilia extending from the cell. In some ciliates, the cilia occur in rows over the entire surface of the cell. In ciliates that live attached to solid surfaces, the cilia occur only around the oral groove. Why only around the oral groove? ______________________
Food is taken into the **oral groove** through the cytostome (mouth) and into the cytopharynx, where a **food vacuole** forms.

Archaezoa generally have multiple flagella and they lack mitochondria. Many archaezoa live in the digestive tracts of animals. The absence of mitochondria is

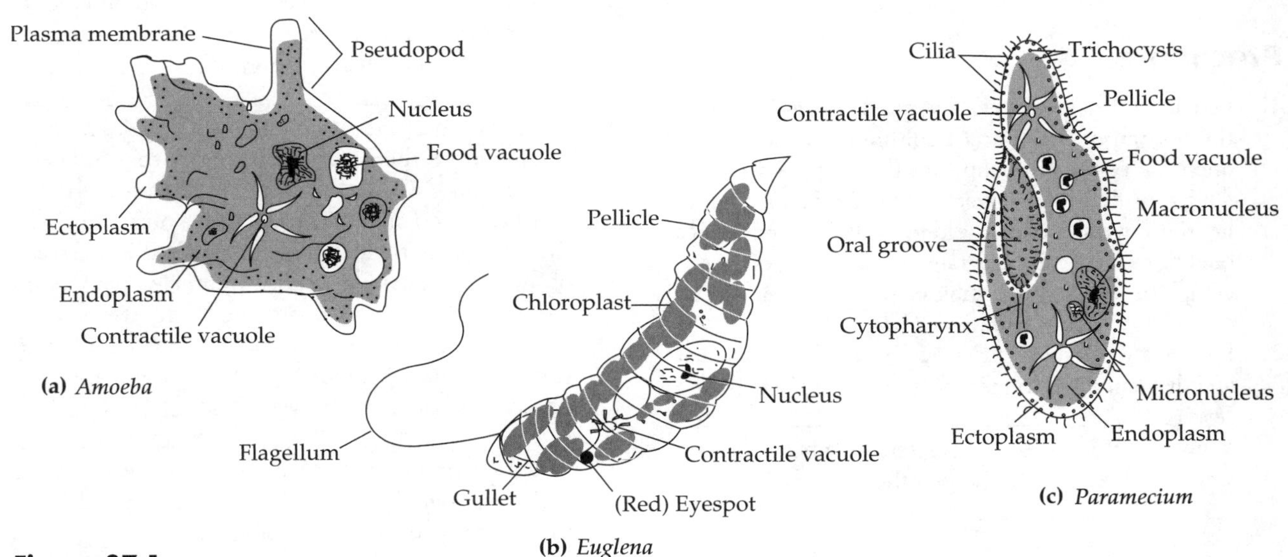

Figure 37.1

Protozoa. **(a)** An amoeba moves by extending pseudopods. **(b)** A euglena has a whiplike flagellum. **(c)** A paramecium has cilia over its surface.

probably not a disadvantage in this anoxic (without oxygen) environment. **Apicomplexa** are nonmotile, obligate intracellular parasites. They have complex life cycles that ensure their transmission to new hosts.

Materials

Methylcellulose, 1.5%

Acetic acid, 5%

Pasteur pipettes

Coverslips

Cultures

Amoeba

Paramecium

Paramecium feeding on Congo red-yeast suspension

Euglena

Prepared Slides

Giardia

Plasmodium

Trypanosoma

Techniques Required

Compound light microscopy, Exercise 1

Wet-mount technique, Exercise 2

Procedure

1. Prepare a wet mount of *Amoeba*. Place a drop from the bottom of the *Amoeba* culture on a slide. Place one edge of the coverslip into the drop, and let the fluid run along the coverslip (Figure 37.2). Gently lay the coverslip over the drop. Observe the amoeboid movement, and diagram it. Which region of the cytoplasm has more granules: the ectoplasm or the endoplasm? (Refer to Figure 37.1a.) ____________

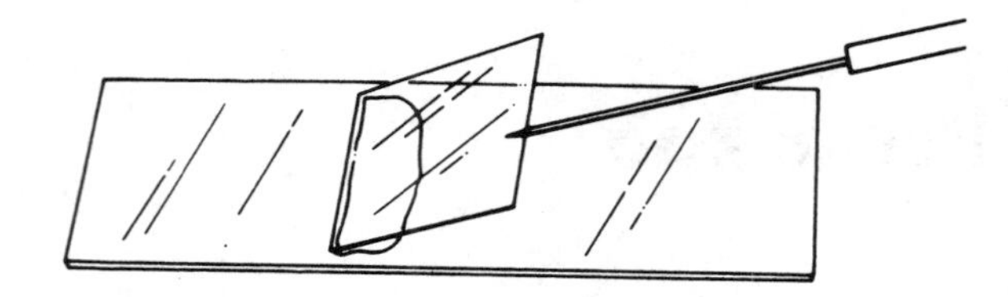

Figure 37.2
Gently lower the coverslip.

2. Prepare a wet mount of *Euglena*. Follow one individual and diagram its movement. Can you see *Euglena's* red "eyespot"? ____________

   Why do you suppose it is present in photosynthetic strains and not in nonphotosynthetic strains? ______

   ____

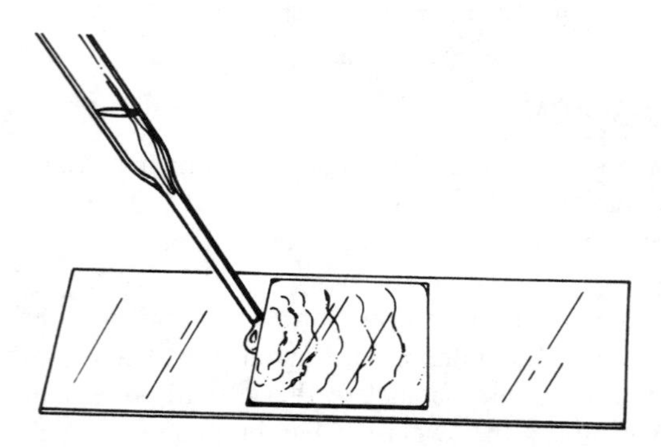

Figure 37.3
Add a drop of acetic acid at one edge of the coverslip. Allow it to diffuse into the wet mount.

   Allow a drop of acetic acid to seep under the coverslip (Figure 37.3). How does *Euglena* respond?

   ____

3. Prepare a wet mount of *Paramecium* and observe its movement. (See Color Plate IV.6.) Why do you suppose it rolls and *Amoeba* does not? ____________

   ____

4. Place a drop of methylcellulose on a slide. Make a wet mount of the *Paramecium* culture that has been feeding on a Congo red-yeast suspension in this mixture. The *Paramecium* will move more slowly in the viscous methylcellulose. Observe the ingestion of the red-stained yeast cells by *Paramecium*. Congo red is a pH indicator. As the contents of food vacuoles are digested, the indicator will turn blue (pH 3). What metabolic products would produce acidic conditions in the vacuoles? ____________

   ____

   Count the number of red and blue food vacuoles in a *Paramecium*. Sketch a *Paramecium* and identify the locations of the food vacuoles.

Exercise 37

Protozoa

LABORATORY REPORT

NAME ______________________

DATE ______________________

LAB SECTION ______________________

Purpose ______________________

____

____

Observations

Use a series of diagrams to illustrate the following:

The movement of *Amoeba* across the field of vision. ____ ×

The movement of *Euglena* and its flagellum. ____ ×

The movement of *Paramecium* and its cilia. ____ ×

The ingestion of food, formation of a food vacuole, and movement of food vacuoles in *Paramecium*. Note any color changes in the food vacuoles. ____ ×

Prepared Slides

Sketch a field with each parasite.

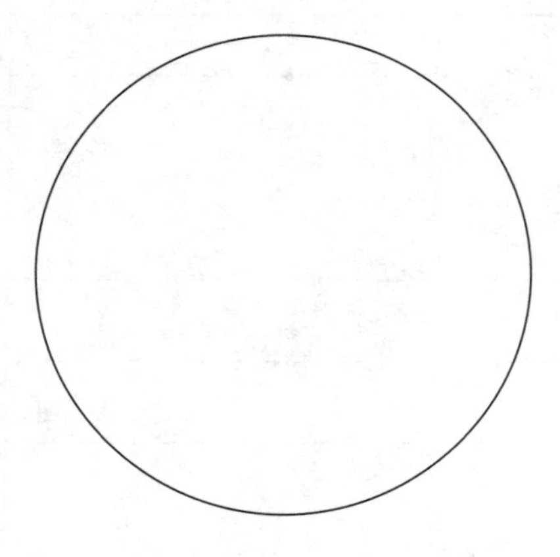

Trypanosoma ____ ×

Label the flagellum.

Disease caused: ______________

Phylum: ______________

Giardia ____ ×

Label the flagella.

Disease caused: ______________

Phylum: ______________

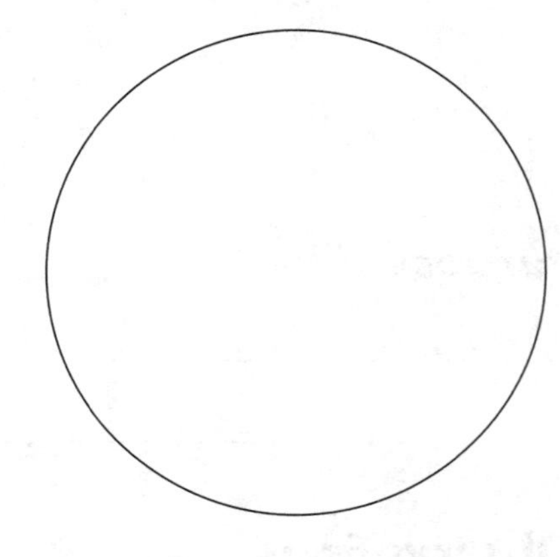

Plasmodium ____ ×

Label the nucleus and cytoplasm.

Disease caused: ______________

Phylum: ______________

Questions

1. How did *Euglena* respond to the acetic acid? ______________

2. Which of the live organisms observed in this exercise would you bet on in a race? ______________

3. Describe the arrangement of cilia on *Paramecium*. ______________

4. *Trypanosoma* and *Plasmodium* are both found in blood. How do they differ in their locations relative to red blood cells? ______________

5. What advantage does *Trypanosoma*'s shape provide? ______________

6. Match the characteristics (a–g) to the phyla listed. Name one genus from each phylum.

Phylum	Characteristics	Example Genus
Archaezoa		
Apicomplexa		
Dinoflagellates		
Ciliophora		
Euglenozoa		

Characteristics
a. Apical enzymes
b. Cilia
c. Flagella
d. Have cellulose
e. No mitochondria
f. Photosynthetic
g. Pseudopods

Critical Thinking

1. Why is *Euglena* often used to study algae and protozoa?

2. Why are *Giardia* and *Trypanosoma* not classified into the same phylum? Which two genera in this exercise are most closely related?

3. *Pneumocystis* was classified as a protozoan since its discovery in 1908. However, rRNA sequencing now shows that it is a fungus. Why might it be important to have an accurate classification of this organism?

4. Laboratory eyewashes should be flushed once a month to remove *Acanthamoeba* accumulations from the pipes. Why is this removal necessary for eyewashes but not other water outlets?

Part Nine Viruses

EXERCISES

Viruses are fundamentally different from the bacteria studied in previous exercises, as well as from all other organisms. Wendell M. Stanley, who was awarded the Nobel Prize in 1946 for his work on tobacco mosaic virus, described the virus as

> *one of the great riddles of biology. We do not know whether it is alive or dead, because it seems to occupy a place midway between the inert chemical molecule and the living organism.**

Viruses are submicroscopic, filterable, infectious agents (see the illustration on the next page). Viruses are too small to be seen with a light microscope and can be seen only with an electron microscope. Filtration is frequently used to separate viruses from other microorganisms. A suspension containing viruses and bacteria is filtered through a membrane filter with a small pore size (0.45 μm) that retains the bacteria but allows viruses to pass through.

Viruses are *obligate intracellular parasites* with a simple structure and organization. Viruses contain DNA or RNA enclosed in a protein coat. Viruses exhibit quite strict specificity for host cells. Once inside the host cell, the virus uses the host cell's synthetic machinery and material to cause synthesis of virus particles that can infect new cells.

Viruses are widely distributed in nature and have been isolated from virtually every eukaryotic and prokaryotic organism. Based on their host specificity, viruses are divided into general categories, such as **bacterial viruses (bacteriophages), plant viruses,** and **animal viruses.** A virus's host cells must be grown in order to culture viruses in a laboratory.

In Exercise 38, we will isolate bacteriophages and estimate the number of phage particles. In Exercise 39, we will isolate plant viruses in plant leaves.

---

*Quoted in H. A. Lechevalier and M. Solotorovsky. *Three Centuries of Microbiology*. New York: Dover Publications, 1974, p. 25.

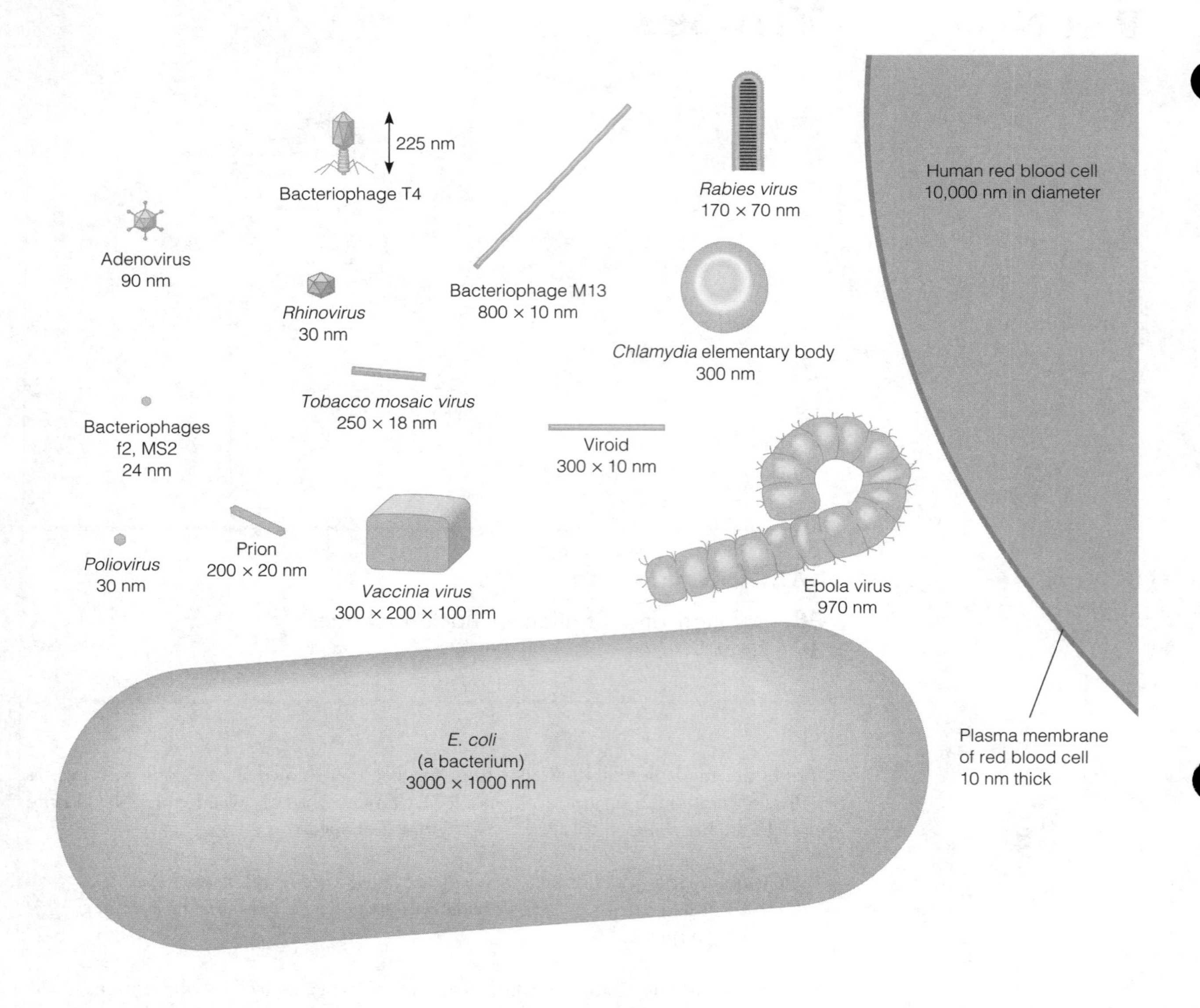

Comparative sizes of several viruses, bacteria, and a human red blood cell.

Exercise 38

Isolation and Titration of Bacteriophages

Objectives

After completing this exercise, you should be able to:

1. Isolate a bacteriophage from a natural environment.
2. Describe the cultivation of bacteriophages.
3. Determine the titer of a bacteriophage sample using the broth-clearing and plaque-forming methods.

Background

Bacteriophages, a term coined around 1917 by Félix d'Hérelle meaning "bacteria eater," parasitize most, if not all, bacteria in a very specific manner. Some bacteriophages, or **phages,** such as T-even bacteriophages, have a **complex structure** (Figure 38.1a). The protein coat consists of a polyhedral head and a helical tail, to which other structures are attached. The head contains the nucleic acid. To initiate an infection, the bacteriophage **adsorbs** onto the surface of a bacterial cell by means of its tail fibers and base plate. The bacteriophage injects its nucleic acid into the bacterium during **penetration** (Figure 38.1b). The tail sheath contracts, driving the tail core through the cell wall and injecting the nucleic acid into the bacterium.

Bacteriophages can be grown in liquid or solid cultures of bacteria. The use of solid media makes location of the bacteriophage possible by the **plaque-forming method.** Host bacteria and bacteriophages are mixed together in melted agar, which is then poured into a Petri plate containing hardened nutrient agar. Each bacteriophage that infects a bacterium multiplies, releasing several hundred new viruses. The new viruses infect other bacteria, and more new viruses are produced. All the bacteria in the area surrounding the original virus are destroyed, leaving a clear area, or **plaque,** against a confluent "lawn" of bacteria. The lawn of bacteria is produced by the growth of uninfected bacterial cells.

In this exercise, we will isolate a bacteriophage from host cells in a natural environment (i.e., in sewage or houseflies). Because the numbers of phages in a natural source are low, the desired host bacteria and additional nutrients are added as an **enrichment procedure.** After incubation, the bacteriophage can be isolated by centrifugation of the enrichment media and membrane filtration. **Filtration** has been used for removing microbes from liquids for purposes of sterilization since 1884, when Pasteur's associate, Charles Chamberland, made the first out of porcelain. Today most filtration of viruses is done using nitrocellulose or polyvinyl membrane filters with pore sizes (usually 0.45 μm) that physically exclude bacteria from the filtrate.

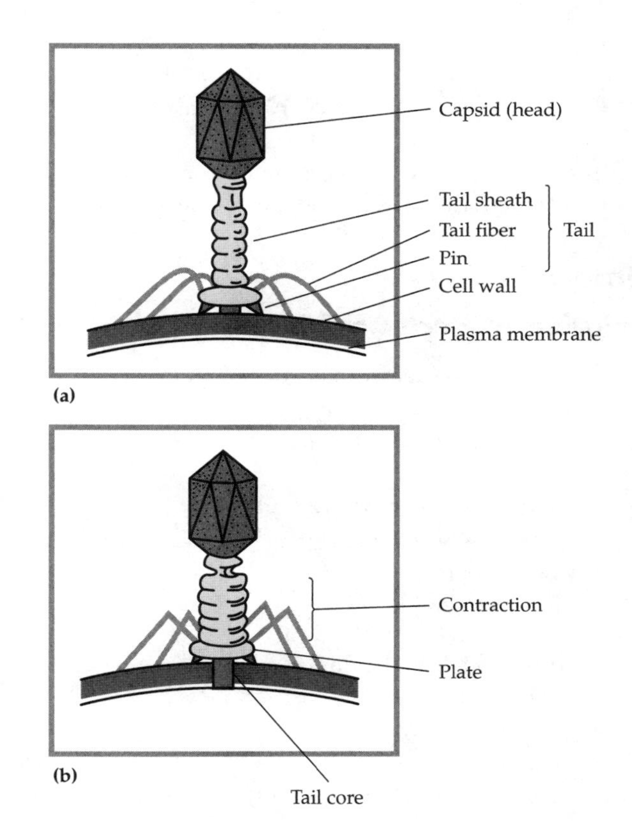

Figure 38.1

T-even bacteriophage. **(a)** Diagram of a bacteriophage, showing its component parts and adsorption onto a host cell. **(b)** Penetration of a host cell by the bacteriophage.

You will measure the viral activity in your sample by performing sequential dilutions of the viral preparation and assaying for the presence of viruses. In the **broth-clearing assay,** the **endpoint** is the highest dilution (smallest amount of virus) producing lysis of bacteria and clearing of the broth. The **titer,** or concentration, that results in a recognizable effect is the reciprocal of the endpoint. In the plaque-forming method, the titer is determined by counting plaques. Each plaque theoretically corresponds to a single infective virus in the initial suspension. Some plaques may arise from more than one virus particle, and some virus particles may not be infectious. Therefore, the titer is determined by counting the number of **plaque-forming units (pfu).** The titer, plaque-forming units per milliliter, is determined by counting the number of plaques and dividing by the amount plated times the dilution. For example, 32 plaques with 0.1 ml plated of a 1:10^3 dilution is equal to

$$\frac{32}{0.1 \times 10^{-3}} = 3.2 \times 10^5 \text{ pfu/ml}$$

In this exercise, we will determine the viral activity by a plaque-forming assay and a broth-clearing assay.

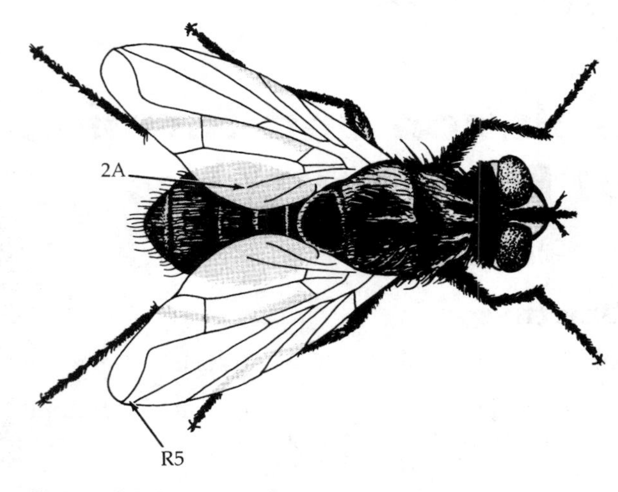

Figure 38.2
Houseflies may harbor bacteriophages. To identify a housefly, observe that the 2A vein doesn't reach the wing margin, and the R5 cell narrows.

Materials

Isolation of Bacteriophage

Use flies, sewage, or bacteriophages to culture.

Flies as source:
- Houseflies (fresh) (20 to 25); bring your own (Figure 38.2)
- Trypticase soy broth (20 ml)
- Mortar and pestle

Sewage as source:
- Raw sewage (45 ml)
- 10× nutrient broth (5 ml)
- 50-ml graduated cylinder
- Funnel

Bacteriophage as source:
- Bacteriophage
- Trypticase soy broth (20 ml)

Sterile 125-ml Erlenmeyer flask

Sterile 5-ml pipette (1)

Centrifuge tubes

Screw-capped tube

Sterile membrane filter (0.45 μm)

Sterile membrane filter apparatus

Safety goggles

Gloves

Titration of Bacteriophage

Tubes containing 9 ml of Trypticase soy broth (6)

Petri plates containing nutrient agar (6)

Tubes containing 3 ml of melted soft Trypticase soy agar (0.7% agar) (6)

Sterile 1-ml pipettes (7)

Cultures

Escherichia coli broth

Bacteriophage culture or bacteriophage from isolation procedure

Techniques Required

Pour plate technique, Exercise 11

Pipetting, Appendix A

Serial dilution technique, Appendix B

Membrane filtration, Appendix F

Procedure

Isolation of Bacteriophage

 **Wear gloves and goggles when working with raw sewage. Handle flies carefully.**

1. Follow enrichment procedure a, b, or c.
   a. Wear gloves. Add the flies to half of the Trypticase soy broth in the mortar, and grind with the pestle to a fine pulp. Transfer this mixture to a sterile flask, and add the remainder of the broth and 2 ml of *E. coli*. Place the mortar and pestle in the To Be Autoclaved basket and the gloves in the biohazard container.

   b. Wear gloves and safety glasses. Add 45 ml of sewage to 5 ml of 10× nutrient broth in a sterile flask. Add 5 ml of *E. coli* broth. Mix gently. Place gloves in the biohazard container. Why is the broth 10 times more concentrated than normal? ______________________________

   c. Add 0.5 ml of bacteriophage to 20 ml of Trypticase soy broth in a sterile flask. Add 2 ml of *E. coli* broth. Mix gently.

2. Incubate the enrichment for 24 hours at 35°C.
3. Decant 10 ml of the enrichment into a centrifuge tube. Place the tube in a centrifuge.

Balance the centrifuge with a similar tube containing 10 ml of water.

   Centrifuge at 2500 rpm for 10 minutes to remove most bacteria and solid materials.

4. Filter the supernatant through a membrane filter (Appendix F). Decant the clear liquid into a screw-capped tube. How does filtration separate viruses from bacteria? ______________________________

   ____

   Store at 5°C until the next lab period.

Titration of Bacteriophage (Figure 38.3)

Read carefully before proceeding.

1. Label the plates and the broth tubes "1" through "6."
2. Aseptically add 1 ml of phage suspension (from step 4, above) to tube 1. Mix by carefully aspirating up and down three times with the pipette. Using a different pipette, transfer 1 ml to the second tube, mix well, and then put 1 ml into the third tube. Why is the pipette changed? ________ Continue until the fifth tube. After mixing this tube, discard 1 ml into a container of disinfectant. What dilution exists in each tube?

	Tube 1	Tube 2	Tube 3
Dilutions:	______	______	______

	Tube 4	Tube 5	Tube 6
Dilutions:	______	______	______

   What is the purpose of tube 6? ______________

3. With a pipette, add 0.1 ml of *E. coli* to the soft agar tubes and place them back in the water bath. Keep the water bath at 43°C to 45°C at all times. Why?

   ____

4. With the remaining pipette, start with broth tube 6 and aseptically transfer 0.1 ml from tube 6 to the soft agar tube. Mix by swirling, and quickly pour the inoculated soft agar evenly over the surface of Petri plate 6. Then, using the same pipette, transfer 0.1 ml from tube 5 to a soft agar tube, mix, and pour over plate 5. Continue until you have completed tube 1. Why can the procedure be done with one pipette? ______________________

   ____

5. After completing step 4, add two loopfuls of *E. coli* to each of the remaining broth tubes and mix. Incubate the tubes at 35°C. Observe them in a few hours. In the broth tubes, record the highest dilution that was clear as the endpoint. The titer is the reciprocal of the endpoint.
6. Incubate the plates at 35°C until plaques develop. The plaques may be visible within hours. (See Color Plate VII.1.) Plates can be stored at 5°C.
7. Select a plate with between 25 and 250 plaques. Count the number of plaques, and calculate the number of plaque-forming units (pfu) per milliliter. Record your results.

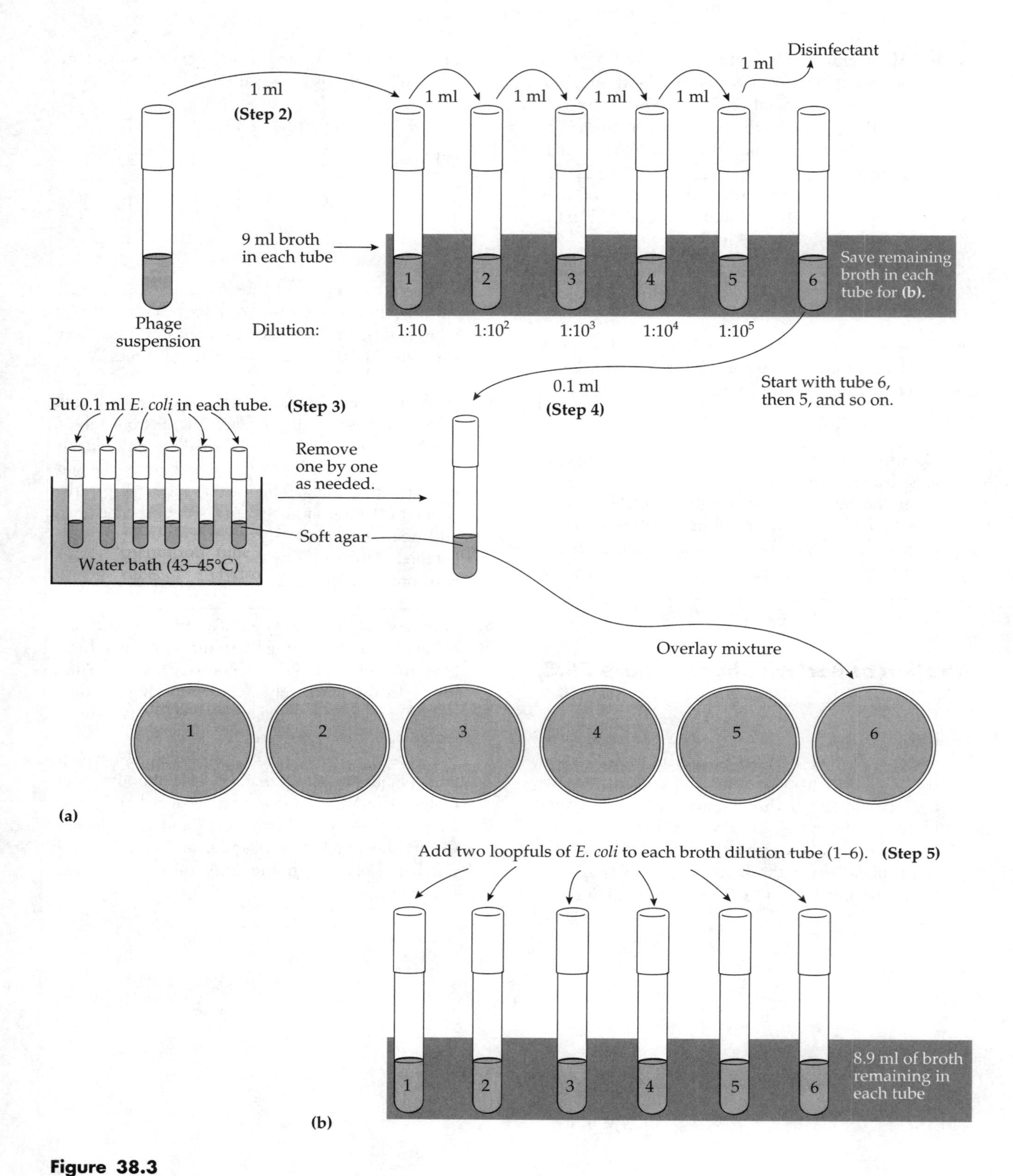

Figure 38.3

Procedure for titration of bacteriophages. **(a)** Plaque-forming assay. **(b)** Broth-clearing assay.

Exercise 38

LABORATORY REPORT

Isolation and Titration of Bacteriophages

NAME ____________________

DATE ____________________

LAB SECTION ____________________

Purpose ____________________

____

____

Data

Plaque-Forming Assay

Choose one plate with 25 to 250 plaques.

Draw what you observed.

Number of plaques = ____________

Dilution used for that plate = ____________

Broth-Clearing Assay

Indicate whether each tube was turbid or clear.

Incubated ________ hours.

Tube	Turbid or Clear	Dilution
1		
2		
3		
4		
5		
6		

Conclusions

Plaque-Forming Assay

pfu/ml = ____________

Show your calculations.

Broth-Clearing Assay

What was the endpoint? ______________________________

What was the titer? ______________________________

Questions

1. Are any contaminating bacteria present? How can you tell? ______________________________

2. Why did you add *Escherichia coli* to sewage, which is full of bacteria? ______________________________

3. Are all the plaques the same size? Briefly explain why or why not. ______________________________

4. What does the endpoint represent in a broth-clearing assay? ______________________________

5. Which of these assays is more accurate? Briefly explain. ______________________________

Critical Thinking

1. How would you develop a pure culture of a phage?

2. If there were no plaques on your plates, offer an explanation. How would you explain turbidity in all of the tubes in a broth-clearing assay?

3. How would you isolate a bacteriophage for a species of *Bacillus*?

4. Titration of a phage on a pure bacterial lawn resulted in some cloudy plaques. Briefly explain this result.

Exercise 39

Plant Viruses

Smoking was definitely dangerous to your health in seventeenth-century Russia. Czar Michael Federovitch executed anyone on whom tobacco was found. But Czar Alexei Mikhailovitch was easier on smokers; he merely tortured them until they told who their suppliers were.

ANONYMOUS

Objectives

After completing this exercise, you should be able to:

1. Isolate a plant virus.
2. Describe the cultivation of a plant virus.
3. Determine the host range of a plant virus.

Background

Plant viruses are important economically in that they are the second leading cause of plant disease. (Fungi are the leading cause of plant disease.) In order for most plant viruses to infect a plant, the virus must enter through an abrasion in the leaf or stem. Insects called **vectors** carry viruses from one plant to another. Most viruses cause either a **localized infection,** in which the leaf will have necrotic lesions (brown plaques), or a **systemic infection,** in which the infection will run throughout the entire plant. One of the most studied viruses is tobacco mosaic virus (TMV), which was the first virus to be purified and crystallized.

In this experiment, we will attempt to isolate TMV from various tobacco products and determine its host range.

Materials

Tobacco plant

Tomato plant

Bean plant (*Chenopodium*)

Tobacco products of various types (bring some, if possible)

Mortar and pestle

Fine sand or Carborundum

Wash bottle of water

Paper labels or labeling tape

Cotton

Sterile water

Techniques Required

None

Procedure

1. Select a plant, and label the pot with your name and lab section.
2. Place labels around the petioles (see Figure 39.1) of several leaves, with names corresponding to the various tobacco products available. Label one petiole "C" for control.
3. *Wash your hands carefully before and after each inoculation.*
4. Spray the control leaf with a small amount of water, and dust the leaf with Carborundum or sand.

 **Be careful not to inhale Carborundum dust. It is a lung irritant.**

Gently rub the leaf with a cotton ball dampened in sterile water. Wash the leaf with a wash bottle of water. What is the purpose of this leaf? ____________

____

5. Select a tobacco product, place a "pinch" of it into a mortar, and add sterile water. Grind the mixture into a slurry with a pestle. Spray one leaf with water, dust it with Carborundum or sand, and gently rub it with a cotton ball soaked with the tobacco slurry. Wash the leaf with water. Record the name of the tobacco source.
6. Using the same tobacco product, repeat steps 1 through 5 on a different species of plant. Repeat steps 5 and 6 with another tobacco product, cleaning the mortar and pestle with soap and water between products.

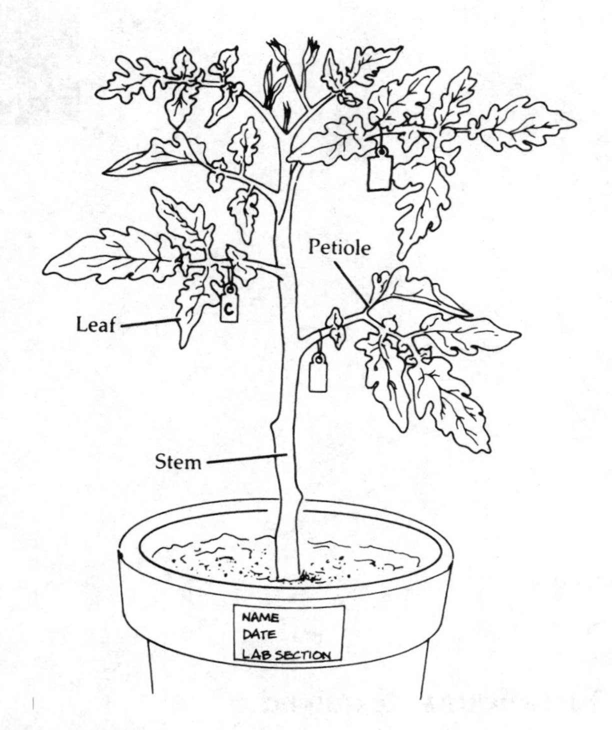

Figure 39.1

Label a plant. Place labels close to the leaves that will be treated, as shown.

7. Replace the plant in the rack.
8. Check the plant at each laboratory period for up to 3 to 4 weeks for brown plaques (localized infection) or wilting (systemic infection). Water your plants as needed. Discard plants in the To Be Autoclaved area. (See Color Plate VII.2.)

Exercise 39

LABORATORY REPORT

Plant Viruses

NAME ______________________

DATE ______________________

LAB SECTION ______________________

Purpose

____

____

____

Data

Obtain data from your classmates for the other plant species and tobacco products.

Identify Plant Species and Tobacco Product Used		Appearance of Leaves							
		Lab Period							
		1	2	3	4	5	6	7	8
1.	Control								
	Test								
2.	Control								
	Test								
3.	Control								
	Test								
4.	Control								
	Test								
5.	Control								
	Test								
6.	Control								
	Test								

Draw a plant showing locations of the labeled leaves and any signs of infection.

Conclusions

What can you conclude about the host range of TMV? ______

______

______

______

Questions

1. Did systemic disease occur? How can you tell? ______

   ______

2. If the control leaf was damaged, what happened? ______

   ______

3. Why is the leaf rubbed with sand or Carborundum? ______

   ______

Critical Thinking

1. Design an experiment to prove that damage to the test leaves was caused by a viral infection.

2. A reservoir of infection is a continual source of that infection. What is a likely reservoir for TMV?

3. How could TMV be used to make inexpensive therapeutic proteins, such as insulin, or to make vaccines?

Part Ten

Interaction of Microbe and Host

EXERCISES

In 1883, after detecting microorganisms in the environment, Robert Koch asked:

> *What significance do these findings have? Can it be stated that air, water, and soil contain a certain number of microorganisms and yet are without significance to health?**

Koch felt that the causes, or **etiologic agents,** of infectious diseases could be found in the study of microorganisms. He proved that anthrax and tuberculosis were caused by specific bacteria, and his work provided the framework for the study of the etiology of any infectious disease.

Today we refer to Koch's protocols for identifying etiologic agents as **Koch's postulates** (see the illustration on the next page; also see Exercise 41). The study of how and when diseases occur is called **epidemiology** (Exercise 40).

*Quoted in R. N. Coetsch, ed. *Microbiology: Historical Contributions from 1776 to 1908*. New Brunswick, NJ: Rutgers University Press, 1960, p. 130.

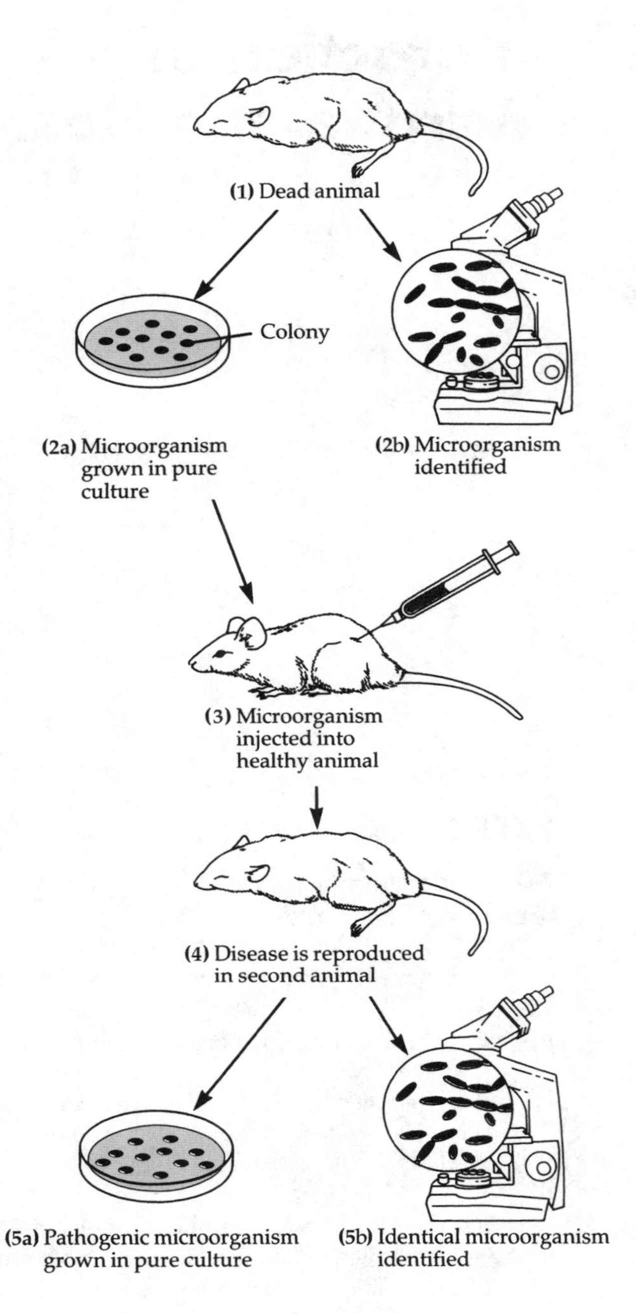

Koch's postulates. **(1)** Microorganisms from a diseased or dead animal are **(2a)** grown in pure culture and **(2b)** identified, if possible. **(3)** A sample of the pure culture is inoculated into a healthy animal, and **(4)** the disease is reproduced in the animal. **(5a)** Pathogenic microorganisms are again grown in pure culture and **(5b)** identified, confirming that they are identical with the original pathogen.

Exercise 40

Epidemiology

Objectives

After completing this exercise, you should be able to:

1. Define the following terms: epidemiology, epidemic, reservoir, and carrier.
2. Describe three methods of transmission.
3. Define index case and case definition.
4. Determine the source of a simulated epidemic.

Background

In every infectious disease, the disease-producing microorganism, the **pathogen,** must come in contact with the **host,** the organism that harbors the pathogen. **Communicable diseases** can be spread either directly or indirectly from one host to another. Some microorganisms cause disease only if the body is weakened or if a predisposing event such as a wound allows them to enter the body. Such diseases are called **noncommunicable diseases**—that is, they cannot be transmitted from one host to another. The science that deals with when and where diseases occur and how they are transmitted in the human population is called **epidemiology. Endemic diseases** such as pneumonia are constantly present in the population. When many people in a given area acquire the disease in a relatively short period of time, it is referred to as an **epidemic disease.** The first reported patient in a disease outbreak is the **index case.** One of the first steps in analyzing a disease outbreak is to make a **case definition,** which should include the typical symptoms of patients so you know who should be included.

Diseases can be transmitted by **direct contact** between hosts. **Droplet infection,** when microorganisms are carried on liquid drops from a cough or sneeze, is a method of direct contact. Diseases can also be transmitted by contact with contaminated inanimate objects, or **fomites.** Drinking glasses, bedding, and towels are examples of fomites that can be contaminated with pathogens from feces, sputum, or pus.

Some diseases are transmitted from one host to another by vectors. **Vectors** are insects and other arthropods that carry pathogens. In **mechanical transmission,** insects carry a pathogen on their feet and may transfer the pathogen to a person's food. For example, houseflies may transmit typhoid fever from the feces of an infected person to food. Transmission of a disease by an arthropod's bite is called **biological transmission.** An arthropod ingests a pathogen while biting an infected host. The pathogen can multiply or mature in the arthropod and then be transferred to a healthy person in the arthropod's feces or saliva.

The continual source of an infection is called the **reservoir.** Humans who harbor pathogens but who do not exhibit any signs of disease are called **carriers.**

An **epidemiologist** compiles data on the incidence of a disease and its method of transmission and tries to locate the source of infection in order to decrease the incidence. The time course of an epidemic is shown by graphing the number of cases and their date of onset. This **epidemic curve** gives a visual display of the outbreak's magnitude and time trend. An epidemic curve provides a great deal of information. First, you will usually be able to tell where you are in the course of the epidemic and possibly be able to project its future course. Second, if you have identified the disease and know its usual incubation period, you may be able to estimate a probable time period of exposure and can then develop a questionnaire focusing on that time period. Finally, you may be able to draw inferences about the epidemic pattern, for example, whether it is an outbreak resulting from a common-source exposure, from a person-to-person spread, or both.

In this exercise, an epidemic will be simulated. Although you will be in the "epidemic," you will be the epidemiologist who, by deductive reasoning and with luck, determines the source of the epidemic.

Materials

Petri plate containing nutrient agar

Latex or vinyl glove, or small plastic sandwich bag

One unknown swab per student (one swab has *Serratia marcescens*, *Rhodotorula rubra*, or *Micrococcus roseus* on it. Your instructor will tell you which organism.)

Techniques Required

Colony morphology, Exercise 9

Procedure (Figure 40.1)

1. Divide the Petri plate into five sectors, labeled "1" through "5."
2. Record the number of your swab in your Lab Report.

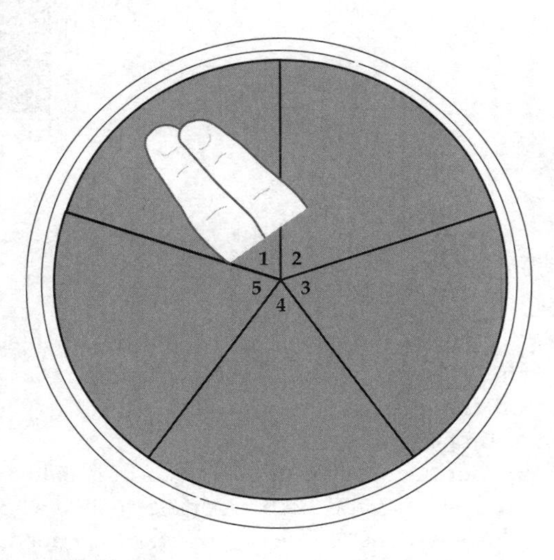

Figure 40.1

After touching the palm of one classmate with your gloved fingers, immediately touch sector 1 of the plate. Record the person's name and swab number. After the finger–palm touch with a second classmate, touch sector 2. Continue until you have inoculated all five sectors.

Carefully read steps 3 through 9 before proceeding.

3. Put a glove on your left hand. Carefully unwrap your swab without touching the cotton. Holding the swab with your right hand, rub it on the palm of your left hand (i.e., on the glove).
4. Discard the swab in a container of disinfectant.
5. Using your gloved left hand, touch the gloved palm of a classmate when the instructor gives the signal. Your gloved fingers touch the other's gloved palm and vice versa.
6. After the finger–palm touch, touch your fingers to the first sector of the nutrient agar. Record the person's name and swab number.
7. Repeat the finger–palm touch with a different classmate. Then touch your fingers to the second sector of the nutrient agar. Record the person's name and swab number.
8. Repeat steps 5 and 6, doing the finger–palm touch with another classmate and then touching the third sector of the nutrient agar.
9. Repeat steps 5 and 6, doing the finger–palm touch with two other classmates. Remember to touch your fingers to the corresponding sector of the nutrient agar after each handshake. Keep good records.
10. Discard the glove in the To Be Autoclaved basket.
11. Incubate the plate, inverted, at room temperature.
12. At the next lab period, record your results. Data can be keyed into a computer database. Using your group's data, deductively try to determine who had the contaminated fomite. (See Color Plates V.4 and XIV.1.)

Exercise 40

Epidemiology

Laboratory Report

NAME ____________________

DATE ____________________

LAB SECTION ____________________

Purpose

____

Data

Your swab # __________ Which organism was on the contaminated swab? ____________________

What is the case definition for this classroom "epidemic"? ____________________

Sector	Student's Name	Swab Number	Appearance of Colonies on Nutrient Agar
1.			
2.			
3.			
4.			
5.			

Attach your group's or the class's data.

Conclusions

1. Who was the index case (the contaminated swab)? ____________________
2. What number was the contaminated swab? __________ Explain how you arrived at your conclusion.

   ____
3. Diagram the path of the epidemic from the index case to all "infected" class members.

Questions

1. Could you be the "infected" individual and not have growth on your plate? Explain. ______

2. Do all people who contact an infected individual acquire the disease? ______

3. How can an epidemic stop without medical intervention (e.g., quaratine, chemotherapy, vaccines)?

4. Are any organisms other than the culture assigned for this experiment growing on the plates? How can you tell?

5. What was the method of transmission of the "disease" in this experiment? ______

Critical Thinking

1. Assume you work in the Infectious Disease Branch of the Centers for Disease Control and Prevention. You are notified of the following two incidents. (1) On June 20, cruise ship X reported that 84 of 2,318 passengers had reported to the infirmary with norovirus gastroenteritis during a 7-day vacation cruise. According to federal regulations, when the incidence of acute gastroenteritis among passengers and crew exceeds 3%, an outbreak is defined and requires a formal investigation. Is this an outbreak? ________________ (2) A nursing home reported 125 cases of norovirus gastroenteritis among residents and staff during the week of June 23rd.

   Data collected from the cruise ship and nursing home are shown in the following table. Use these data to answer the questions.

	Cruise Ship Data		Nursing Home Data	
	Date	Number of Cases	Date	Number of Cases
Cruise one	6/9	2	6/23	1
	6/10	4	6/24	8
	6/11	5	6/25	12
	6/12	3	6/26	12
	6/13	3	6/27	50
	6/14	2	6/28	32
	6/15	1	6/29	10
Cruise two	6/16	2	6/30	8
	6/17	10		
	6/18	13		
	6/19	13		
	6/20	84		
	6/21	46		
	6/22	20		
Cruise three	6/23	10		
	6/24	23		
	6/25	39		
	6/26	41		
	6/27	18		
	6/28	17		
	6/29	9		

   a. On a separate page, graph the epidemic curve for these data.

   b. How is norovirus transmitted?

c. What happened on cruise 2?

d. What would you do before cruise 4?

e. Three nursing home residents were passengers on cruise 2. What can you conclude?

2. During a 3-month period, acute hepatitis B virus (HBV) infection was diagnosed in nine residents of a nursing home. Serological testing of all residents revealed that nine people had acute HBV infection, two had chronic infection, five were immune, and 58 were susceptible. Medical charts of residents were reviewed for history of medications and use of ancillary medical services. Infection-control practices at the nursing home were assessed through interviews with personnel and direct observations of nursing procedures. A summary of the medical charts is shown below.

	Case Patients*	Susceptible Residents†
Received insulin injections	11	58
Patients having capillary blood taken by fingersticks	11	6
≥60 fingersticks/month	7	0
<60 fingersticks/month	4	6
Average number of venous blood draws/month	23	6
Patients having both capillary and venous blood drawn	11	39
Visited by dentist	1	5
Visited by podiatrist	10	52
Recieved blood transfusion	0	0

*A case patient has hepatitis B.
† Susceptible residents live in the nursing home but do not have hepatitis B.

a. What is the usual method of transmission for hepatitis B? ____________________

b. What is the probable source of infection in hospitals? ____________________

c. How was *this* infection transmitted? ____________________

____

____

3. A health department received a report from hospital A that 15 patients had been admitted on October 12 with unexplained pneumonia. On October 21, hospital B, located 15 miles from hospital A, reported a higher-than-normal pneumonia census for the first 2 weeks of October. *Legionella pneumophila* was eventually identified in 23 patients, 21 were hospitalized, and two died. To identify potential exposures associated with *L. pneumophila*, a questionnaire was developed, and a case-control study was initiated on November 2 to identify the source of infection. Three healthy controls were selected for each confirmed case; controls were matched by age, gender, and underlying medical conditions. Of the 15 case patients for whom a history was available, 14 had visited a large home improvement center 2 weeks before onset of illness. Results of the questionnaire are shown below.

	Case Patients	Healthy Controls
Number of patients	15	45
Visited home improvement center	14	12
Average time at center (min)	79	29
Looked at whirlpool spa X	13	9
Looked at whirlpool spa Y	13	1
Visited greenhouse sprinkler system display	10	10
Visited decorative fish pond	14	12
Used drinking fountain	13	10
Used urinals	10	4
Used restroom hot water	6	4
Used restroom cold water	8	8

a. What is the most likely source of this outbreak of legionellosis? ____________________

b. How would you prove this was the source?

c. How was this disease transmitted?

d. Provide an explanation of the infected patient who did not go to the home improvement center.

4. A salmonellosis outbreak occurred after a brunch in Livonia, New York. All fresh food items were delivered to the cooking area the morning of May 22. The menu consisted of omelets, bacon, fruit salad, and pastries.

Omelets consisted of grade A eggs from a Maryland farm. They were beaten and mixed with milk, salt, pepper, and diced onions. The egg batter was fried in small batches and served from warming trays heated by alcohol burners.

Bacon was purchased from a meat wholesaler. The bacon was fried on a large griddle and served from warming trays.

Fruit salad was made from fresh apples, bananas, and oranges; they were cut up and mixed with commercially canned peaches and pears.

Muffins and other breads were purchased from a bakery and served with butter.

	Foods Eaten*					Time of		Onset of Symptoms	
Case	1	2	3	4	Beverages†	Meal	Symptoms‡	Day	Hr
1	x	x	x	x	C, J	1100	D, V, N, A	Sun	2400
2		x	x	x	M, J	1200			
3	x	x	x	x	C	1200	N, A	Mon	0200
4	x	x	x	x	C	1200	D, N, A	Mon	0600
5				x	T	1100			
6		x		x	C, J	1200			
7			x	x	C, J	1300			
8				x	C, J	1300			
9	x	x	x	x	C	1100	D, V, N, A	Mon	1100
10				x	T	1200			
11				x	M	1300			
12	x	x	x	x	M	1300	D, N, A	Mon	1400
13				x	C	1200			
14		x	x	x	C, J	1500			
15	x	x		x	T	1500	D, N, V	Mon	2400
16				x	M, J	1100			
17		x	x	x	C	1100			
18		x		x	C	1200	N	Sun	1300
19	x	x			T	1200	D, V, N, A	Mon	1700
20			x	x	C, J	1200			
21	x	x		x	M	1300	D, N, A	Mon	0200
22				x	M	1400	A	Tues	0800
23				x	M, J	1400			
24	x	x	x	x	C	1100	D, A	Mon	1500
25	x	x		x	C	1100	V, N, A	Mon	1100
26			x	x	T	1200			
27				x	T	1100			
28	x	x	x	x	C	1200	D, A	Mon	0200
29		x		x	M	1300			
30					C, J	1400			
31		x	x		C	1100			
32	x	x	x	x		1200	D, A	Mon	1200

*Foods Eaten: 1, omelet; 2, bacon; 3, fruit salad; 4, pastry.
†Beverages: C, coffee; T, tea; M, milk; J, orange juice.
‡Symptoms: D, diarrhea; N, nausea; V, vomiting; A, abdominal cramps.

a. What was the source of this disease? ____________________

b. How could this outbreak have been prevented? ____________________

Exercise 41

Koch's Postulates

Objectives

After completing this exercise, you should be able to:

1. Define the following terms: etiologic agent, pathogenicity, and virulence.
2. List and explain Koch's postulates.

Background

The **etiologic agent** is the cause of an infectious disease. Microorganisms are the etiologic agents of a wide variety of infectious diseases in all forms of life. Microbes that cause diseases are called pathogens, and the process of disease initiation and progress is called **pathogenesis.** The interaction between the microbe and host is complex. Whether or not a disease occurs depends on the host's vulnerability or **susceptibility** to the pathogen and on the virulence of the pathogen. **Virulence** is the degree of pathogenicity. Factors influencing virulence include the number of pathogens, the portal of entry into the host, and toxin production.

The actual cause of many diseases is hard to determine. Although many microorganisms can be isolated from a diseased tissue, their presence does not prove that any or all of them caused the disease. A microbe may be a secondary invader or part of the normal microbiota or transient microbiota of that area. While working with anthrax and tuberculosis, Robert Koch established four criteria, now called **Koch's postulates,** to help identify a particular organism as the causative agent for a particular disease. (See the figure on p. 290.) Koch's postulates are the following:

1. The same organism must be present in every case of the disease.
2. The organism must be isolated from the diseased tissue and grown in pure culture in the laboratory.
3. The organism from the pure culture must cause the disease when inoculated into healthy, susceptible laboratory animals.
4. The organism must again be isolated—this time from the inoculated animals—and must be shown to be the same pathogen as the original organism.

These criteria are used by most investigators, but they cannot be applied to all infectious diseases. For example, viruses cannot be cultured on artificial media, and they are not readily observable in a host. Moreover, many viruses cause diseases only in humans and not in laboratory animals.

In this exercise, we will demonstrate Koch's postulates with one of two different bacteria: *Bacillus thuringiensis* isolated from milky spore disease of tomato hornworm larvae, or *Pectobacterium carotovorum* isolated from soft rot of carrots.

Materials

Bacillus: *Milky Spore Disease*

Petri plate containing nutrient agar

Petri plate containing starch agar

Second period

Scalpel and scissors

Tube containing sterile nutrient broth

Tomato hornworm larvae (2)

Sterile Pasteur pipettes (2)

Gram-staining reagents

Demonstration (Second Period)

Dissected control larvae

Pectobacterium: *Carrot Soft Rot*

Petri plate containing nutrient agar

Carrot

Scalpel or razor blade

Potato peeler

Forceps

Alcohol

Disinfectant (bleach)

Sterile water

Sterile Petri plate with filter paper

Cultures

Bacillus: *Milky Spore Disease*

Bacillus thuringiensis

Pectobacterium: *Carrot Soft Rot*

Pectobacterium carotovorum

Techniques Required

Compound light microscopy, Exercise 1

Gram staining, Exercise 5

Plate streaking, Exercise 11
Starch hydrolysis, Exercise 13

Procedure

Bacillus: *Milky Spore Disease*

1. Divide each Petri plate into three sectors; label them "A," "B," and "C." Streak a loopful of nutrient broth onto sector A and a loopful of *Bacillus* onto sector B of each plate. Incubate the plates, inverted, at room temperature until growth occurs, and then store them in the refrigerator.
2. Make a smear of *Bacillus* and heat-fix it; store it in your drawer.
3. Carefully examine your larvae. Include records of their size, weight, and color.
4. Infect one larva by adding a few drops of *B. thuringiensis* to its food supply. As the infected larva begins to show symptoms of the disease, and as the disease progresses, record the symptoms.
5. Feed the remaining larva in the same manner with nutrient broth. What is the purpose of this step?

 ____

 ____

6. The inoculated larva should die within 1 week. Once it is dead, record its size, weight, and external appearance; place it in a wet towel; and keep it in the refrigerator until you have time to necropsy the larva. *If your larva does not die,* provide some reasons for its survival.
7. Observe the intestine of the dissected control larva.
8. Work with the larva on a disinfectant-saturated towel. Disinfect your instruments by putting them in alcohol and burning off the alcohol immediately before and after use. See Color Plate VII.4.

 **Keep the flame away from the beaker of alcohol.**

 Slit the ventral surface open from anus to mouth; then make horizontal cuts to expose the intestine (Figure 41.1). Examine and record your findings. Transfer a loopful of the hemolymph (blood) to sector C on your plates stored previously in the refrigerator (step 1), and incubate them at room temperature for 48 hours. Make a smear of hemolymph on the same slide you used in step 2. Discard the larva as instructed.

 **These insects are agricultural pests and cannot be released.**

9. Prepare a Gram stain of the smears and examine the growth on the plates. Growth on the plates can also be Gram stained. Flood the starch plate with Gram's iodine. Record your results.

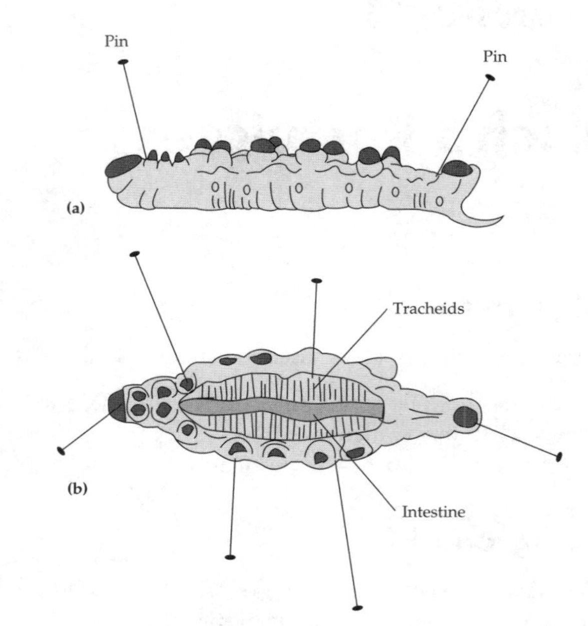

Figure 41.1
Hornworm larva necropsy. **(a)** Dissection. **(b)** Internal anatomy.

Pectobacterium: *Carrot Soft Rot**

1. Wash the carrot well, and peel it to eliminate the outer surface; then dry it, wash it with disinfectant, and rinse it with sterile water. Dip your scalpel in alcohol, burn off the alcohol, and cut the carrot into four cross-sectional slices 5- to 8-mm thick.
2. Put the four slices on filter paper in the bottom of a Petri plate.
3. Inoculate the center of three slices with a loopful of the *Pectobacterium* culture. Why not all four?

 ____

 Saturate the filter paper with sterile water. Incubate the plate right-side up at room temperature until soft rot appears. (See Color Plate VII.3.) More sterile water may be added if the disease process is slow.
4. Divide the nutrient agar plate in half. Inoculate one-half of the nutrient agar with the *Pectobacterium* broth; incubate the plate, inverted, for 48 hours at room temperature; and then refrigerate it. Make a smear of the *Pectobacterium*. Heat-fix the smear and store it in your drawer.
5. Streak an inoculum from the diseased carrot on the remaining half of the Petri plate. Incubate the plate, inverted, at room temperature for 48 hours.
6. Make a smear from the diseased carrot. Perform a Gram stain on both smears and record your observations.
7. Observe the nutrient agar plate and record your results. Prepare a Gram stain from the nutrient agar cultures if time permits.

*Adapted from R. S. Hogue. "Demonstration of Koch's Postulates." *American Biology Teacher* 33:174–175, 1971.

Exercise 41 **Laboratory Report**

Koch's Postulates

NAME ______________________

DATE ______________________

LAB SECTION ______________________

Purpose ______________________

____

____

Bacillus: Milky Spore Disease

Gram Stains

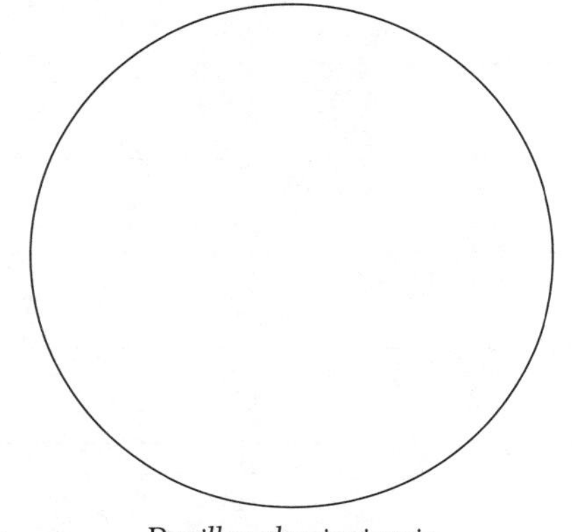

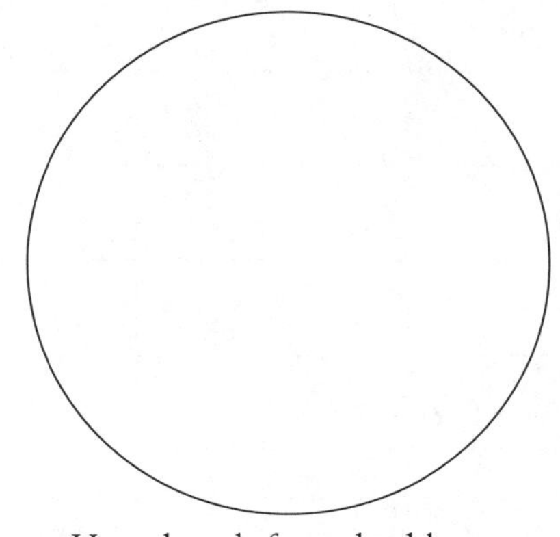

Specimen: *Bacillus thuringiensis* | Hemolymph from dead larva

Morphology: ______________________ Morphology: ______________________

Gram reaction: ______________________ Gram reaction: ______________________

Cultures

Describe the growth on your plates.

Inoculum	Nutrient Agar	Gram Stain	Starch Agar	Starch Hydrolysis
Broth				
Bacillus culture				
Hemolymph				

Describe the larvae by filling in the following table.

	Larva 1	Larva 2
Initial appearance Date, time		
Length		
Weight		
Color		
Activity		
Which larva is infected?		
After death of infected larva Date, time		
Length		
Weight		
Color		
Appearance of intestine		
Appearance of hemolymph		

Conclusions

Questions

1. What evidence would indicate that the microbe obtained from the dead larva was *B. thuringiensis*?

2. How would you determine the ID_{50} for *B. thuringiensis?*

3. What parts of Koch's postulates did this experiment fulfill?

4. What is the normal method of transmission of *Bacillus thuringiensis?*

5. *Bacillus thuringiensis* endospores in an inert matrix are sold in garden supply stores. Describe the properties of this insecticide that make it better than chemical insecticides.

Pectobacterium: Carrot Soft Rot

Gram Stains

Specimen: *Pectobacterium carotovorum*

Morphology:

Gram reaction:

Soft Rot

Morphology:

Gram reaction:

Cultures

Describe the growth on your plates.

Inoculum	Colony Description	Gram Stain
Pectobacterium culture		
Soft rot		

Describe the appearance of the carrots.

Conclusions

Questions

1. Why is the disease called soft rot?

2. Why will the addition of water speed the disease process?

3. What parts of Koch's postulates did this experiment fulfill?

4. What is the normal method of transmission of *Pectobacterium carotovorum*?

Critical Thinking

1. Eighty-one patients became ill and 14 died in an outbreak at St. Elizabeth's Hospital, Washington, D.C., in 1965. Epidemiologic evidence suggested a link between infection and windblown dust from excavations on hospital grounds. In 1968, 144 cases of self-limited illness occurred in employees at the health department building in Pontiac, Michigan. Investigations at that time demonstrated that the etiologic agent was present in the condenser of a malfunctioning air-conditioning system. In 1974, at least 20 persons attending a convention at the Bellevue Stratford Hotel in Philadelphia developed pneumonia, and two died. In 1976, 182 people became ill and 29 died at another convention at the Bellevue Stratford Hotel. What would you need to do to identify the causative agent of these outbreaks?

2. Near Four Corners, New Mexico, on May 14, two people living in the same household died within 5 days of each other. Their illnesses were characterized by flulike symptoms with rapid respiratory failure. The third case was the death of an 8-year-old girl in Mississippi. Over the next 120 days, 53 more people became ill near Four Corners, and 26 died. One man became ill 2 weeks after returning home from a visit to New Mexico. On July 9, a previously healthy woman living in eastern Texas died following acute respiratory distress. The woman had not traveled outside of Texas. What evidence do you have that this is an infectious disease?

Part Eleven Immunology

EXERCISES

In this part, we will examine our defenses against invasion by microorganisms. In 1883, Elie Metchnikoff began his investigations into the body's defenses with this entry in his diary:

> *These wandering cells in the body of the larva of a starfish, these cells eat food, they gobble up carmine granules—but they must eat up microbes too! Of course—the wandering cells are what protect the starfish from microbes! Our wandering cells, the white cells of blood—they must be what protects us from invading germs.**

These white blood cells (see the photograph on the next page) and certain chemicals are considered agents of **innate** or **nonspecific immunity** because they will combat any microorganism that invades the body. Factors involved in innate immunity are the topic of Exercise 42.

The role of specific defenses—or **adaptive immunity**—was demonstrated when Emil von Behring developed what he called diphtheria "antitoxin" and, in 1894, successfully treated humans with it. Adaptive immunity involves the production by the body of specific proteins called **antibodies,** which are directed against distinct microorganisms. If microorganism A invades the body, antibodies are produced against A; if microorganism B invades the body, antibodies are produced against B.

*Quoted in P. DeKruif. *Microbe Hunters*. New York: Harcourt, Brace, & World, 1953, p. 195.

The antibodies produced against microorganism B will not react with microorganism A. The chemical substance that induces antibody production is called an **antigen.** In the examples just mentioned, microorganisms A and B possess antigens on their surfaces.

Adaptive and innate immunity involve blood. **Blood** consists of a fluid called **plasma** and formed elements: cells and cell fragments. The cells that are of immunologic importance are the white blood cells. **Serum** is the fluid portion that remains after blood has clotted. We will study the techniques and mechanisms of **serology,** or antigen–antibody reactions in vitro, in Exercises 43 through 45.

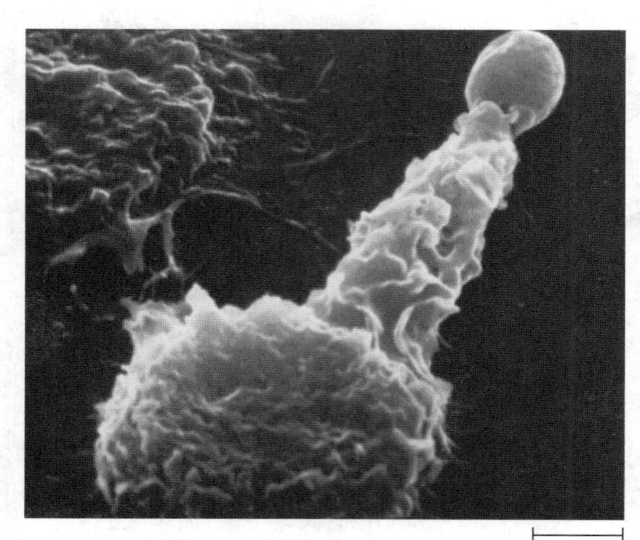

Early stage of phagocytosis by an alveolar macrophage (a type of macrophage found in the alveoli of the lungs). The "wrinkled" macrophage has contacted and adhered to a smooth, roughly spherical yeast cell by means of a pseudopod.

Exercise 42

Innate Immunity

Objectives

After completing this exercise, you should be able to:

1. Differentiate between adaptive and innate immunity.
2. List and discuss the functions of at least two chemicals involved in innate immunity.
3. Use a spectrophotometer.
4. Determine the effects of normal serum bactericidins.

Background

Our ability to ward off diseases through our body's defenses is called **immunity.** Immunity can be divided into two kinds: adaptive (specific) and innate (nonspecific). **Adaptive immunity** is the defense against a specific microorganism. **Innate immunity** refers to all our defenses that protect us from invasion by any microorganism. Physical barriers, such as the skin and mucous membranes, are part of innate immunity. Inflammation and phagocytosis play major roles in innate immunity, as do chemicals produced by the body. As early as 1888, Metchnikoff noted that bacteria "show degenerative changes" when inoculated into samples of mammalian blood.

Lysozyme and complement are examples of chemicals involved in innate immunity. These proteins are collectively called **bactericidins. Lysozyme** is an enzyme found in body fluids that is capable of breaking down the cell walls of gram-positive bacteria and a few gram-negative bacteria. **Complement** is a group of proteins found in normal serum that are involved in enhancing phagocytosis and lysis of bacteria. Complement can be activated by bacterial cell wall polysaccharides and antigen–antibody reactions.

In this exercise, we will examine chemicals involved in innate immunity.

Materials

Lysozyme Activity

Lysozyme buffer, 4.5 ml

Egg-white lysozyme, 1:10^5 dilution, 2.5 ml

Spectrophotometer tubes (2)

1-ml pipette

5-ml pipette (5)

Petri dish

Spectrophotometer

Normal Serum Bactericidins

Nutrient agar, melted and cooled to 45°C

Sterile Petri dishes (9)

Sterile 1-ml pipettes (10)

Sterile 99-ml dilution blanks (3)

Sterile serological tubes (9)

Serological test-tube rack

Sterile 0.85% saline solution

Normal animal serum

Animal serum heated to 56°C for 30 minutes

Cultures (as assigned)

Micrococcus luteus suspension, 5 ml

Escherichia coli

Staphylococcus aureus

Techniques Required

Pour plate procedure, Exercise 11

Pipetting, Appendix A

Serial dilution technique, Appendix B

Spectrophotometry, Appendix C

Graphing, Appendix D

Procedure

 **Work only with your own tears or saliva.**

Lysozyme Activity

1. Collect tears or saliva in a Petri dish, as explained by your instructor.
2. Prepare a 1:10 dilution of tears or saliva by adding 0.5 ml of tears or saliva to a 4.5-ml lysozyme buffer. What is the purpose of the buffer? ______________

____
3. Mix 2.5 ml of the tear or saliva preparation with a 2.5 ml of *Micrococcus luteus* suspension in a spectrophotometer tube. Carefully pipette the contents of the tube up and down three times to mix.
4. Record the absorbance at 540 nm at 30 seconds, 60 seconds, 120 seconds, 180 seconds, 240 seconds, 5 minutes, and at 5-minute intervals for the next 15 minutes (Appendix C).
5. Repeat steps 3 and 4, using egg-white lysozyme.
6. Plot your data in the Laboratory Report.

 **Place the Petri plate, dilution tube, and spectrophotometer tube in disinfectant.**

Normal Serum Bactericidins (Figure 42.1)

Work in groups as assigned.

1. Aseptically prepare $1:10^2$, $1:10^4$, and $1:10^6$ dilutions of the culture (*S. aureus* or *E. coli*) assigned to you (Figure 42.1a).
2. Label three sterile serological tubes "6A," "6B," and "6C." Aseptically transfer 0.1 ml of culture from the highest dilution (1:_______) to each tube (Figure 42.1b).
3. Repeat step 2 with the other dilutions, using the same pipette. Label each tube appropriately. How many tubes do you have? ______________
4. To each A tube, add 0.4 ml of unheated normal serum; to each B tube, add 0.4 ml of serum heated to 56°C for 30 minutes; to each C tube, add 0.4 ml of saline (Figure 42.1c, d, and e). What is the purpose of the C tubes? ______________
5. Mix all tubes and incubate them at 35°C for 1 hour (Figure 42.1f).
6. Label nine sterile Petri dishes in the same manner as the tubes: "$1:10^2$ A," "$1:10^2$ B," "$1:10^2$ C," "$1:10^4$ A," "$1:10^4$ B," and so on.
7. Remove 0.1 ml from the A tube of the highest dilution, and place it in the appropriate Petri dish. Using the same pipette, transfer 0.1 ml from the $1:10^4$ A tube to the corresponding Petri dish. Repeat this procedure with the $1:10^2$ A tube (Figure 42.1g). Why are the samples transferred from the highest dilution first? ______________
8. Using another pipette, repeat step 7 with the B tubes. With your remaining pipette, repeat step 7 with the C tubes.
9. Pour melted, cooled nutrient agar into each dish to a depth of approximately 5 mm. Gently swirl to mix the contents, and allow the agar to solidify (Figure 42.1h).
10. Incubate the plates for 24 to 48 hours at 35°C. Record the number of colonies on each plate. Plates with fewer than 25 colonies should be reported as TFTC (too few to count) and those with more than 250 colonies as TNTC (too numerous to count).
11. Calculate the number of colony-forming units (cfu) per milliliter for the control. Choose one plate with between 25 and 250 colonies:

$$\frac{\text{Number of colonies}}{0.1 \text{ ml} \times \text{dilution}} = \text{cfu/ml}$$

    Calculate the number of colony-forming units per milliliter for the unheated and heated sera.
12. Calculate the percent of increase or decrease in bacterial numbers caused by exposure to heated and unheated serum as compared to the control.

$$\frac{\text{Control} - \text{Experiment}}{\text{Control}} \times 100 = \%$$

13. Compare your data with those of a group using the other bacterial species.

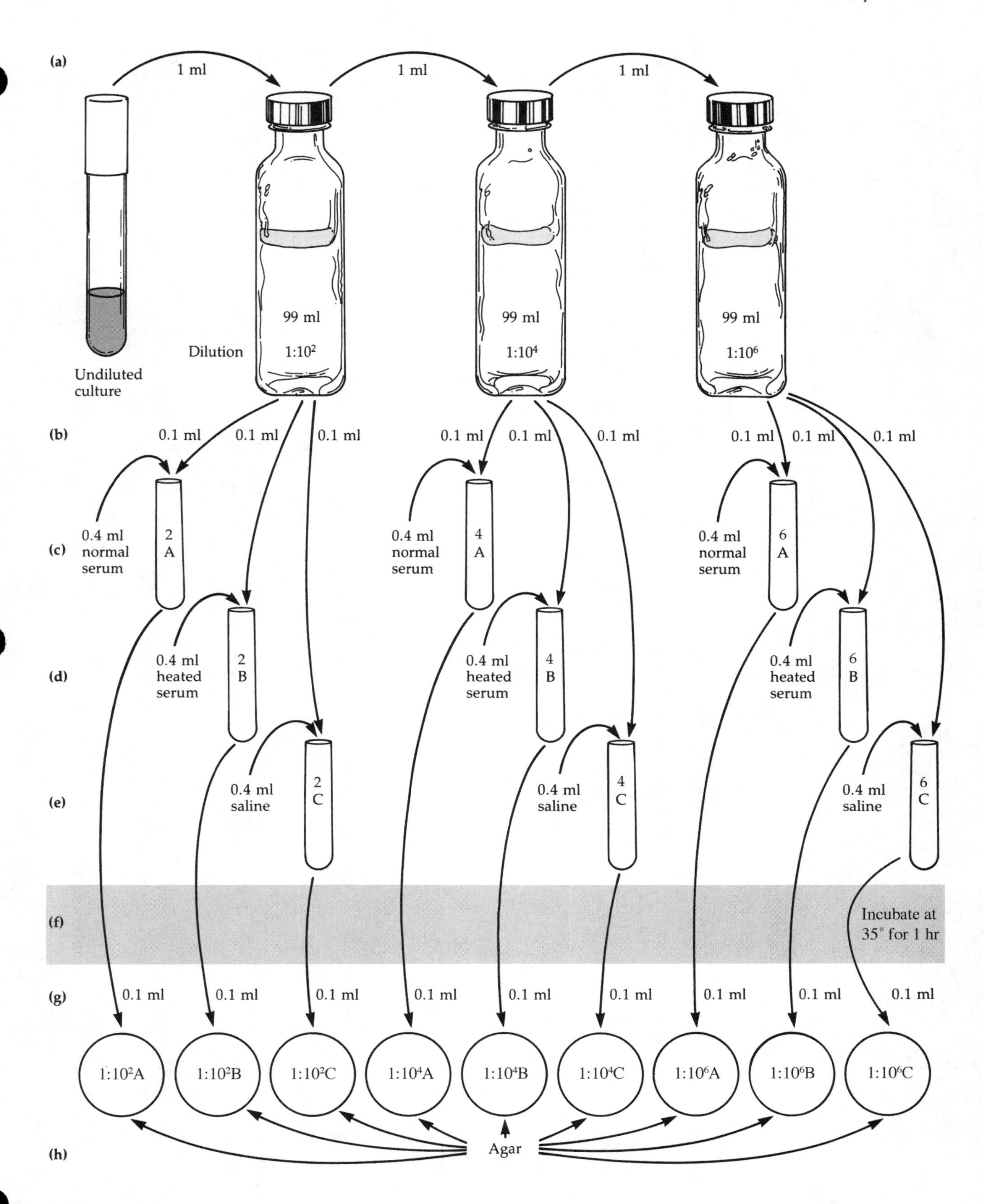

Figure 42.1

Determining the bactericidal activity of serum.

Exercise 42

Innate Immunity

LABORATORY REPORT

NAME ______________________

DATE ______________________

LAB SECTION ______________________

Purpose ______________________

Data

Lysozyme Activity

	Absorbance		
Time	Saliva	Tears	Egg-White Lysozymes
30 sec			
60 sec			
120 sec			
180 sec			
240 sec			
5 min			
10 min			
15 min			
20 min			

Plot your data for lysozyme activity using a computer application or on the graph paper on the next page. Absorbance is marked on the Y-axis and time on the X-axis. Make one line for tears or saliva and another line for egg-white lysozyme.

Normal Serum Bactericidins

	Number of Colonies			cfu/ml
	1:10^2	1:10^4	1:10^6	
Unheated serum				
Heated serum				
Control				

Your calculations:

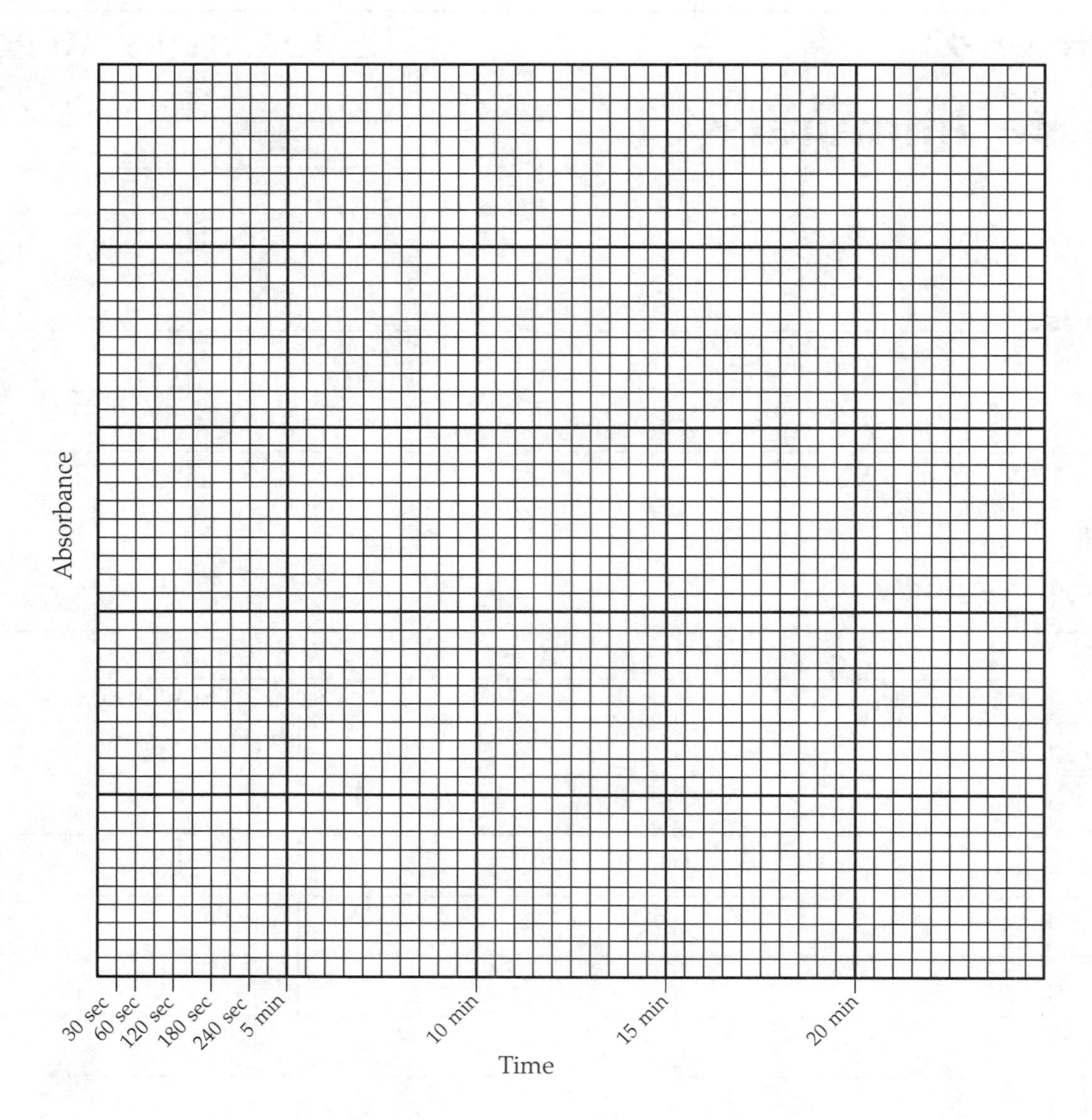

Conclusions

What effect did lysozyme have on the bacterial suspension? ______________________________

____

	% Increase or Decrease	
	S. aureus	*E. coli*
Unheated serum		
Heated serum		

Your calculations:

Questions

1. Why are tears and saliva potential biohazards? ____________________

____

2. Was lysozyme present in your tears or saliva? __________ How can you tell? __________

____

3. Did you test tears or saliva? __________ Compare your data with someone who tested the other. Which secretion has more lysozyme activity? __________ How can you tell? __________

____

____

____

4. Does normal serum have bactericidal properties? Explain. ____________________

____

____

Critical Thinking

1. What other factors contribute to the innate immunity of your eyes? Of your mouth?

2. What can you conclude regarding the effect of heat on the bactericidal properties of normal serum?

3. Design an experiment to determine whether plant fluids have bactericidal properties.

Exercise 43

Blood Group Determination: Slide Agglutination

Objectives

After completing this exercise, you should be able to:

1. Compare and contrast the terms agglutination and hemagglutination.
2. Determine ABO and Rh blood types.
3. Determine possible compatible transfusions.

Background

Agglutination reactions occur between high-molecular-weight, particulate antigens and antibodies. Because many antigens are on cells, agglutination reactions lead to the clumping, or **agglutination,** of cells. When the cells involved are red blood cells, the reaction is called **hemagglutination.**

Hemagglutination reactions are used in the typing of blood. The presence or absence of two very similar carbohydrate antigens (designated A and B) located on the surface of red blood cells is determined using specific antisera (Figure 43.1). Hemagglutination occurs when anti-A antiserum is mixed with type A red blood cells. When anti-A antiserum is mixed with type B red blood cells, no hemagglutination occurs. People with type AB blood possess both A and B antigens on their red blood cells, and those with type O blood lack A and B antigens (see Figure 43.1).

Many other blood antigen series exist on human red blood cells. Another surface antigen on red blood cells is designated the Rh factor. The **Rh factor** is a complex of many antigens. The Rh factor that is used routinely in blood typing is the **Rh_0 antigen,** or **D antigen.** Individuals are Rh-positive when D antigen is present. The presence of the Rh factor is determined by a hemagglutination reaction between anti-D antiserum and red blood cells with D antigen on their surfaces.

A person possesses antibodies to the alternate antigen. Thus, people of blood type A will have antibodies to the B antigen in their sera (see Figure 43.1). Rh-negative individuals do not naturally have anti-D antibodies in their sera. Anti-D antibodies are produced when red blood cells with D antigen are introduced into Rh-negative individuals.

The ABO and Rh systems place restrictions on how blood may be transfused from one person to another. An incompatible transfusion results when the antigens of the donor red blood cells react with the antibodies in the recipient's serum or induce the formation of antibodies.

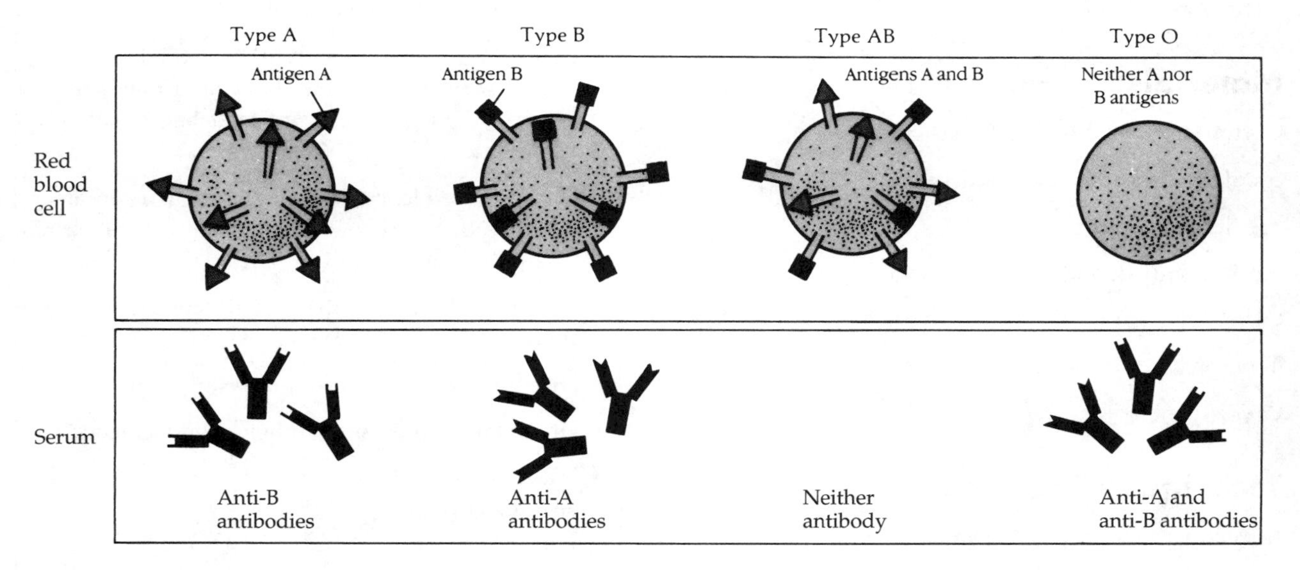

Figure 43.1

Relationship of antigens and antibodies involved in the ABO blood group system.

Table 43.1
The ABO and Rh Blood Group Systems

	Blood Group					
Characteristic	A	B	AB	O	Rh^+	Rh^-
Antigen present on the red blood cells	A	B	Both A and B	Neither A nor B	D	No D
Antibody normally present in the serum	Anti-B	Anti-A	Neither anti-A nor anti-B	Both anti-A and anti-B	No anti-D	No anti-D*
Serum causes agglutination of red blood cells of these types	B, AB	A, AB	None	A, B, AB	Neither Rh^+ nor Rh^-	Neither Rh^+ nor Rh^-*
Percent occurrence in a mixed White population	41	10	4	45	85	15
Percent occurrence in a mixed Black population	27	20	7	46	90	10
Percent occurrence in a mixed Asian population	28	27	5	40	98	2

*Anti-D antibodies are not naturally present in the serum of Rh^- people. Anti-D antibodies can be produced upon exposure to the D antigen through blood transfusions or pregnancy.
Source: Adapted from G. J. Tortora, B. R. Funke, and C. L. Case. *Microbiology: An Introduction*, 9th ed. San Francisco, CA: Benjamin Cummings, 2007.

A summary of the major characteristics of the ABO and Rh blood groups is presented in Table 43.1. Blood group antigens are genetically determined, and thus they are inherited. Differences in groups of people are created by geographical isolation during the development of populations.

Safety Precautions for Working with Blood in the Laboratory

- Do not perform this lab experiment if you are sick.
- Work only with **your own** blood* or wear gloves.
- **Disinfect** your finger with 70% alcohol. Use any finger except your thumb. Unwrap a **new, sterile** lancet and pierce the disinfected finger with the sterile lancet.
- Place the used lancet **in disinfectant** in a container designated by the instructor or a biohazard "sharps" container.
- Stop the bleeding with a sterile cotton ball and **apply an adhesive bandage.**
- **Dispose of** the used cotton **in disinfectant.**
- **Discard** used slides and toothpicks **in disinfectant.**
- **Wash** any spilled blood from your work area **with disinfectant.**

*Blood obtained from blood banks is tested for hepatitis B virus and HIV. However, no test method can offer complete assurance that laboratory specimens do not contain these viruses.

Materials

Cotton moistened with 70% ethyl alcohol

Sterile cotton balls and bandage

Sterile lancet

Anti-A, anti-B, and anti-D antisera

Glass slides (2)

Toothpicks

Grease pencil

Techniques Required

Dissecting microscope, Appendix E

Carefully read the Safety Precautions for Working with Blood in the Laboratory.

Procedure

1. With a grease pencil, draw two circles on a clean glass slide, and label one "A" and the other "B." Draw a circle on the second slide and label it "D."
2. Disinfect a finger with cotton saturated in alcohol (Figure 43.2). Pierce the disinfected finger with a sterile lancet.
3. Let a drop of blood fall into each circle. Stop the bleeding with a sterile cotton ball, and apply an adhesive bandage.
4. To the A circle, add one drop of anti-A antiserum. Add one drop of anti-B antiserum to the B circle and one drop of anti-D antiserum to the D circle.
5. Mix each suspension with toothpicks. Use a different toothpick for each antiserum. Why? ________

   ____
6. Observe for agglutination, and determine your blood type. (See Color Plate IX.1.) A dissecting microscope may help you see hemagglutination.

Discard the slides, toothpicks, cotton balls, and lancet in the disinfectant.

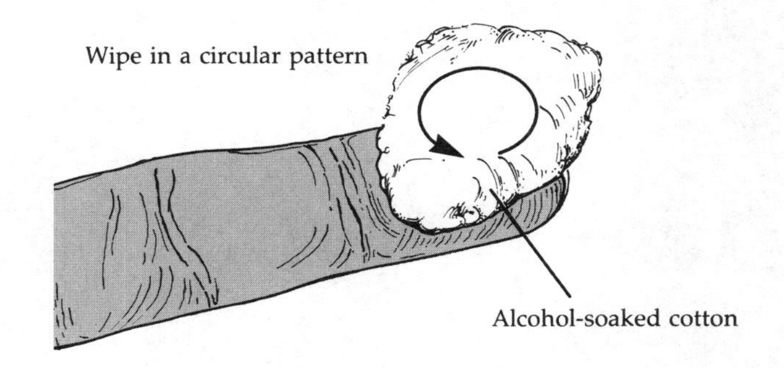

Figure 43.2
To disinfect skin, rub it with 70% ethyl alcohol.

Exercise 43

LABORATORY REPORT

Blood Group Determination: Slide Agglutination

NAME ____________________

DATE ____________________

LAB SECTION ____________________

Purpose ____________________

____

____

Data

Antiserum	Hemagglutination
Anti-A	
Anti-B	
Anti-D	

Blood Type:

ABO ____________________

Rh ____________________

Data Analysis

On the blackboard, tabulate the blood types for the class or laboratory section. Calculate the percentage of each blood type, and compare with the percent distribution given in Table 43.1.

Blood Type	Number of Students	Percent
A		
B		
AB		
O		
Rh^+		

Questions

1. What is your blood type? ____________________

2. What is the blood type of the person on your left? ____________________

   Is your blood potentially compatible? __________ Explain briefly. ____________________

   ____

3. What is the blood type of the person on your right? ______

Is your blood potentially compatible? ______ Explain briefly. ______

______

Critical Thinking

1. What is hemolytic disease of the newborn? How does RhoGAM prevent it?

2. How can blood type O be considered the "universal donor"?

3. Individuals may have antibodies to the A and B blood antigens not found on their red blood cells. What is the origin of these antibodies in someone who has never had a transfusion?

4. A woman with type A blood can have a healthy baby with type B blood. Why doesn't the baby develop hemolytic disease of the newborn?

5. A woman with type A, Rh-negative blood can have a healthy baby with type B Rh-positive blood and usually will not develop antibodies to the Rh antigen. Why?

Exercise 44

Agglutination Reactions: Microtiter Agglutination

Objectives

After completing this exercise, you should be able to:

1. Define agglutination and titer.
2. Determine the titer of antibodies by the agglutination method.
3. Provide two applications for agglutination reactions.

Background

The surfaces of bacterial cells contain antigens that can be used in agglutination reactions. **Agglutination reactions** occur between *particulate* antigens—such as cell walls, flagella, or capsules *bound* to cells—and antibodies. **Agglutination,** or the clumping of bacteria by antibodies, is a useful laboratory diagnostic technique.

In a slide agglutination test, an unknown bacterium is suspended in a saline solution on a slide and mixed with a drop of known antiserum. This test is done with different antisera on separate slides. The bacteria will agglutinate when mixed with antibodies produced against the same species and strain. A positive test can identify the bacterium.

An **agglutination titration** can estimate the concentration (*titer*) of antibody in serum. This test is used to determine whether a particular organism is causing the patient's symptoms. If an increase in titer is shown in successive daily tests, the patient probably has an infection caused by the organism used in the test.

In the titration, bacteria are mixed with dilutions of serum (e.g., 1:10, 1:20, 1:40, and so on). The *endpoint* is the greatest dilution of serum showing an agglutination reaction. The *reciprocal* of this dilution is the *titer*. For example, in Figure 44.1a, the titer is 20.

In this exercise, we will perform a test to diagnose typhoid fever, the **Widal test.** Typhoid fever is caused by *Salmonella enterica* Typhi. However, the Widal test also detects nontyphoidal *Salmonella,* thus allowing us to use *S. enterica* Typhimurium in this exercise.

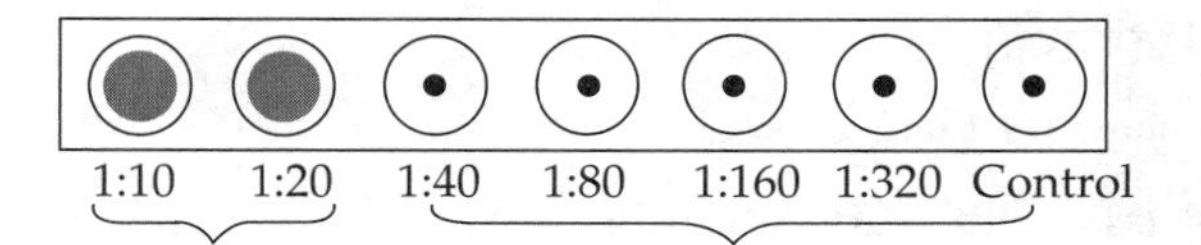

(a) Top view of wells

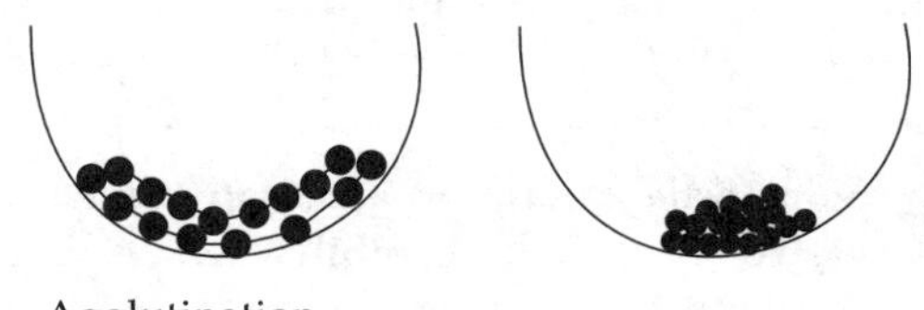

Agglutination

(b) Enlarged side view of wells

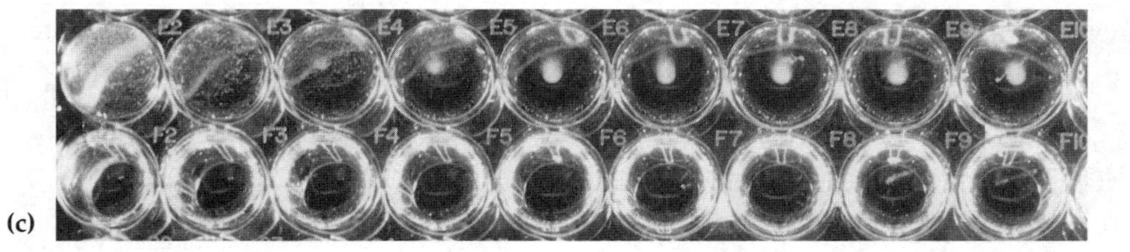

(c)

Figure 44.1

(a) Each well in this microtiter plate contains, from left to right, only half the concentration of serum in the preceding well. Each well contains the same concentration of bacterial cells. **(b)** In a positive reaction, sufficient antibodies are present in the serum to link the antigens together, forming an antibody–antigen mat that sinks to the bottom of the well. In a negative reaction, insufficient antibodies are present to cause the linking of antigens. **(c)** Appearance of wells after refrigeration for 24 hours. The first two wells in the top row show agglutination. The endpoint is 1:20.

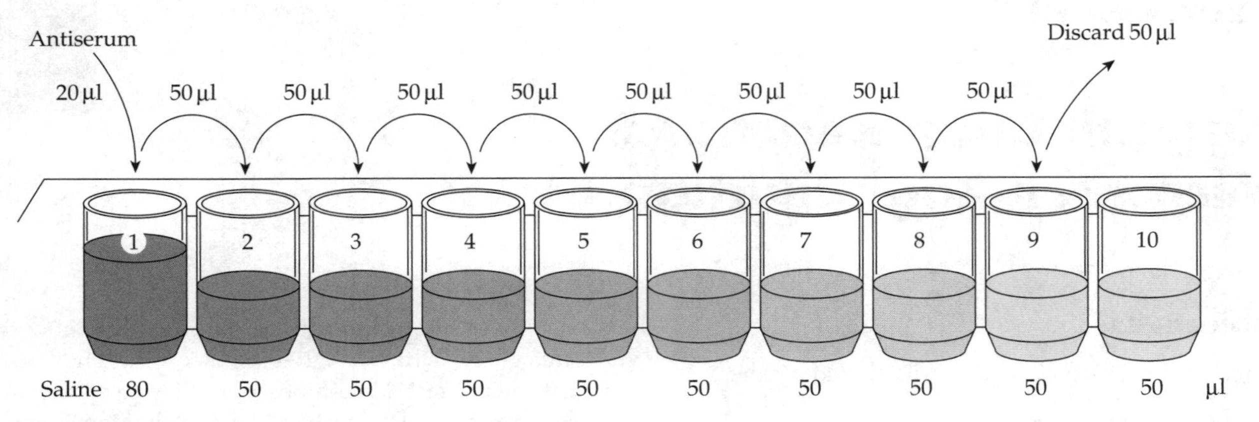

Figure 44.2

Dilution of antiserum. Twenty µl of patient's serum is mixed with 80 µl of saline in well 1. Serial dilutions are made by transferring 50 µl from one well to the next, through well 9. Discard 50 µl from well 9.

Materials

0.85% saline solution

"Patient's" serum

Microtiter plate and lid

Micropipettes, 20 to 80 µl

Micropipette tips (13)

Culture

Salmonella enterica Typhimurium (heated for 30 minutes at 56°C in a water bath)

***Salmonella enterica* is a causative agent of gastroenteritis. Be careful when handling this pathogenic bacterial species.**

Techniques Required

Pipetting, Appendix A

Serial dilution technique, Appendix B

Dissecting microscope, Appendix E

Procedure

1. Label 10 wells in the microtiter plate "1" through "10" (Figure 44.2).
2. Add 80 µl of saline to the first well and 50 µl to each of the remaining wells.
3. Add 20 µl of patient's serum to the first well. Mix up and down three times and, with a new pipette tip, transfer 50 µl to the second well. Change the pipette tip, mix, and transfer 50 µl to the third well, and so on. Continue until you have reached the ninth well. Discard 50 µl from that well, as shown in Figure 44.2. The patient's serum has now been diluted. The dilutions are as follows:

Well 1	1:5
Well 2	1:10
Well 3	1:20
Well 4	1: _______
Well 5	1: _______
Well 6	1: _______
Well 7	1: _______
Well 8	1: _______
Well 9	1: _______
Well 10	no serum

   Fill in the missing dilutions.
4. Carefully add 50 µl of the antigen to each well. What is the antigen? ____________________
   How does the addition of 50 µl of antigen affect the dilutions? ____________________
5. Cover the plate with the microtiter plate lid.
6. Shake the plate carefully in a horizontal direction to mix the contents in each well. Place the plate in a 35°C incubator for 60 minutes.
7. Refrigerate the plate until the next laboratory period.
8. Observe the bottom of each well for agglutination. A dissecting microscope may help you to see agglutination. Which well serves as the control?

   ____

   ____

   What should occur in this well? ____________________

   ____

   Determine the endpoint and the titer.

Exercise 44 **LABORATORY REPORT**

Agglutination Reactions: Microtiter Agglutination

NAME ______________________

DATE ______________________

LAB SECTION ______________________

Purpose ______________________

____

____

Data

Well #	Final Dilution	Agglutination
1		
2		
3		
4		
5		
6		
7		
8		
9		
10		

Diagram the appearance of the bottom of a positive and a negative well:

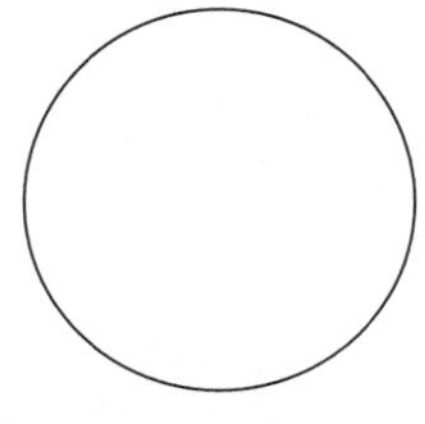

Positive

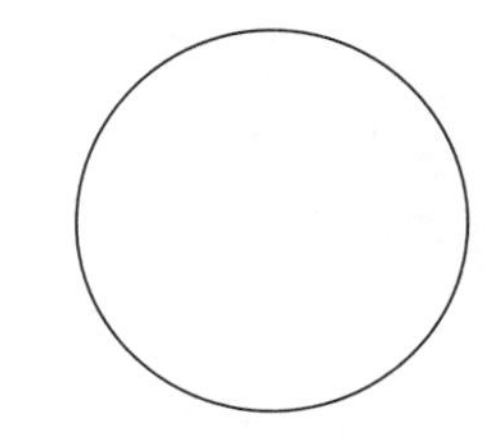

Negative

Questions

1. How did you determine the presence of agglutination? ______________________________

____

2. What was the endpoint? ______________________________

3. What was the antibody titer? ______________________________

4. What was the antiserum used in this experiment? ______________________________

Critical Thinking

1. How would agglutination reactions help in locating the source of an epidemic?

2. The following antibody titers were obtained for three patients:

Patient	Antibody Titer		
	Day 1	Day 5	Day 12
A	128	128	128
B	128	256	512
C	0	0	0

What can you conclude about each of these patients?

3. An agglutination test was performed on a patient's blood. The titer was 512, and upon further testing, the antibody involved was identified as IgM. Can any conclusions be drawn?

Exercise 45

ELISA Technique

A scientist is like a learned child. At some point, though, children have to put away their things and stop their playing. But a scientist can stay that way for a lifetime.

Scientific discovery is a bit like digging for gold. The pioneering work gets done by individualists, frequently following lonesome trails, commonly a little eccentric, and often failing in their quests. Every now and then an inspired or lucky strike turns up a vein rich and promising enough to draw a crowd.

GEORGE WALD

Objectives

After completing this exercise, you should be able to:

1. Explain how ELISA tests can be used clinically to detect antibodies or antigens.
2. Differentiate between direct and indirect ELISA tests.
3. Determine the titer of antibodies by the ELISA method.

Background

Diseases can be diagnosed by using known antibodies to detect the presence of an antigen or by detecting specific antibodies in a patient. These types of tests are called **immunoassays.** Immunoassays are based on detectable interactions between antigens and antibodies such as precipitation, agglutination, or complement fixation. Increased sensitivity in detecting antigens can be achieved by labeling antibodies with substances such as radioactive chemicals (e.g., iodine-125), fluorescent compounds, magnetic beads, or enzymes. **ELISA (enzyme-linked immunosorbent assay)** or **EIA (enzyme immunoassay)** is the most widely used immunoassay in labs today. ELISAs use enzyme labels because of their stability, reproducibility, safety, ease of detection, and relatively low cost. The enzymes horseradish peroxidase and alkaline phosphatase are most often used. When the appropriate substrate is added, the enzyme reacts with the substrate to make a colored product, which can be detected visually or by a spectrophotometer. The amount of product produced is directly proportional to the amount of enzyme—and, therefore, antibody—present, allowing the technique to be quantitative as well as qualitative.

In the *direct* ELISA technique, enzyme-labeled antibodies are used to identify an antigen. In the direct ELISA technique, a known antibody is adsorbed to the wells in a microtiter plate. The unknown microorganism from a patient is added to the wells. If the antibody in the well is specific for the microorganism, the microbe will be bound to the antibody. A second antibody specific for the antigen is then added. This antibody is linked to an enzyme. The reaction between the antigen and antibody is made visible by addition of the substrate for the enzyme (Figure 45.1a).

Indirect ELISAs are used to detect a specific antibody in a patient's serum. In the indirect technique, the known antigen is attached to the wells of a microtiter plate. Dilutions of the suspected antibody are added and, after washing, enzyme-labeled anti-antibody is added. A color change after adding the substrate indicates that the antibody was present (Figure 45.1b). The ELISA test for HIV antibodies is one example of the indirect technique.

In this exercise, we will perform an ELISA test to diagnose salmonellosis.

***Salmonella enterica* is a causative agent of gastroenteritis. Be careful when handling this pathogenic bacterial species.**

Materials

Coating buffer

Washing buffer

Blocking buffer

Patient's serum

Alkaline phosphatase–labeled anti-antibodies

BCIP/NBT substrate

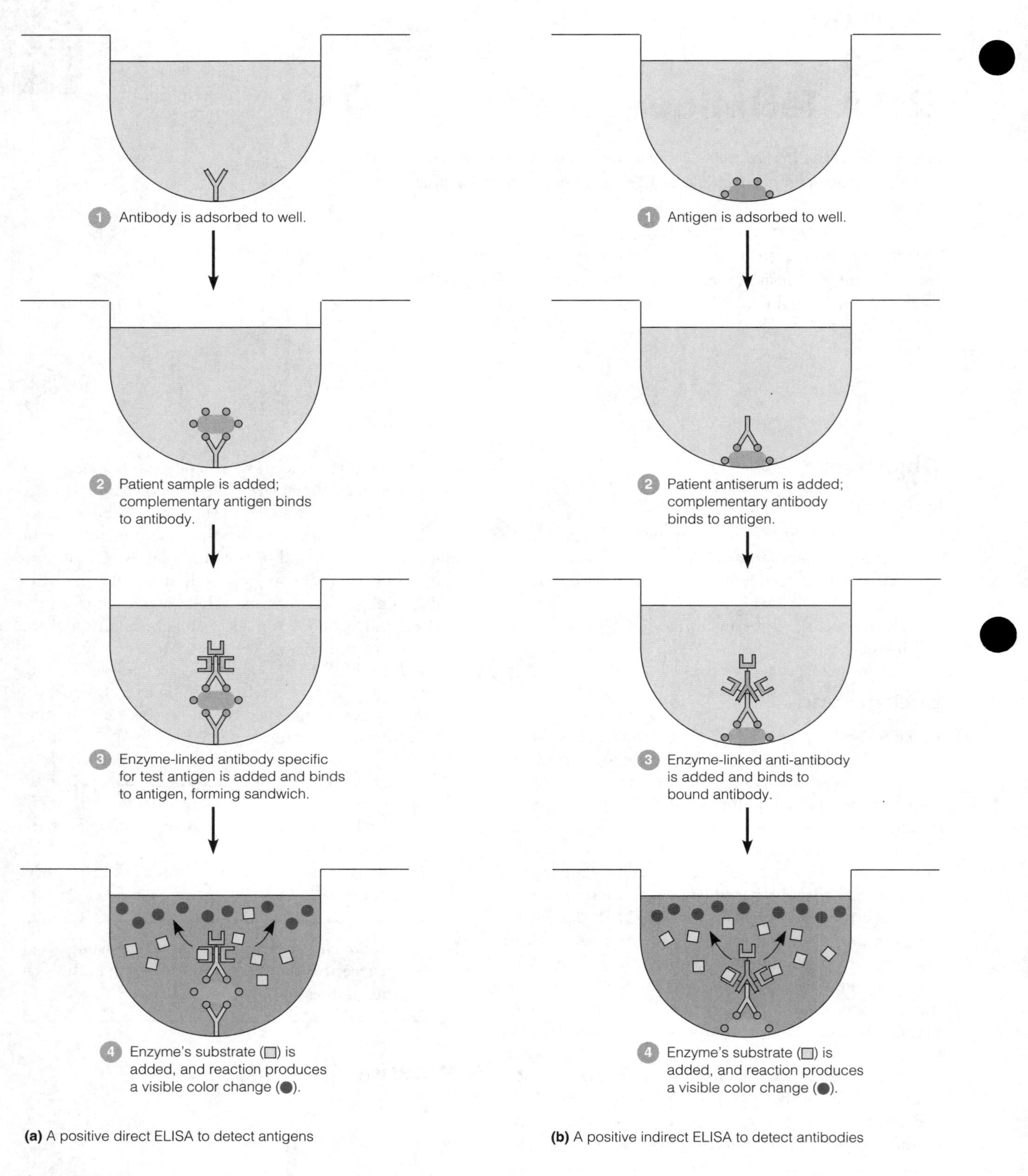

Figure 45.1

The ELISA method. The components are usually contained in small wells of a microtiter plate.

Flat-bottom microtiter plate

Clear plastic tape

Micropipettes, 50 to 100 μl

Micropipette tips

Culture

Salmonella enterica Typhimurium (heated for 30 minutes at 56°C in a water bath)

Techniques Required

Pipetting, Appendix A

Serial dilution technique, Appendix B

Microtiter dilutions, Exercise 44

Procedure

1. Add 100 μl of coating buffer to each well of one row (wells 1–12) of the microtiter plate.
2. Add 100 μl of *S. enterica* Typhimurium to each well.
3. Seal the wells with a strip of plastic tape, and refrigerate the plate at 5°C for 1 to 7 days.
4. Remove your plate from the refrigerator and carefully remove the tape. Shake the inverted plate with a quick shake to remove the liquid into disinfectant.
5. Fill the wells with washing buffer and shake to remove. Wash two more times.
6. Add 100 μl of blocking buffer. Leave for 30 to 90 minutes as directed by your instructor.
7. Perform dilutions of the patient's serum by placing 100 μl in the first well. Mix up and down three times and, with a new pipette tip, transfer 100 μl to the second well. Mix up and down three times, change pipette tips, and transfer 100 μl to the third well, and so on. Continue until you have reached the 11th well. Discard 100 μl from that well. The patient's serum has now been diluted from 1:2 to 1:____ in the 11th well. What is in well 12? ________________________________

   What is the purpose of well 12? ________________________________
   ____
8. Incubate the plate at 35°C for 60 minutes.
9. Shake the inverted plate with a quick shake to remove the contents. Wash three times with washing buffer as described in step 5.
10. Add 100 μl of alkaline phosphatase–labeled anti-antibody to each well (1–12). Seal the wells with tape and incubate the plate at 35°C for 45 minutes. Plates can be sealed and stored at 5°C until the next lab period. What is the antigen? ________________________________
11. Remove the tape carefully, shake out the contents, and wash the wells three times with washing buffer.
12. Add 100 μl of the alkaline phosphatase substrate (BCIP/NBT) to each well in the row.
13. Leave at room temperature for 10 to 30 minutes until color develops; well 12 will be colorless.
14. Record the results. The highest dilution with a blue color is the endpoint. The titer is the reciprocal of the dilution of the endpoint.

Exercise 45

ELISA Technique

LABORATORY REPORT

NAME ______________________

DATE ______________________

LAB SECTION ______________________

Purpose

____

Data

Well #	Final Dilution	Color
1		
2		
3		
4		
5		
6		
7		
8		
9		
10		
11		
12		

Questions

1. What was the endpoint? ______________________________

2. What was the antibody titer? ______________________________

3. Was the test performed in this exercise a direct or an indirect ELISA test? ______________________________

4. Why was the control well colorless? ______________________________

____

____

Critical Thinking

1. Compare and contrast direct and indirect ELISA techniques.

2. Describe how you would determine if a bacterium isolated from a patient was *Salmonella* using the ELISA technique with a known anti-*Salmonella* antiserum.

3. Why would the ELISA technique be valuable for identification of viruses?

4. ELISAs are commonly used in clinical laboratories. What advantage does the ELISA technique have over fluorescent antibody or radioactive antibody techniques?

Part Twelve

Microorganisms and Disease

EXERCISES

Many microorganisms grow abundantly both inside and on the surface of the normal adult body. The microorganisms that establish more or less permanent residence without producing diseases are known as **normal microbiota**. Microorganisms that may be present for a few days or months are called **transient microbiota.** At one time, bacteria and fungi were thought to be plants and thus the term *flora* was used.

A microorganism that causes disease is called a **pathogen.** The pathogens that cause disease consist of many different organisms. Robert Koch speculated in one of his early publications:

> *On this . . . I take my stand, and, till the cultivation of bacteria from spore to spore shows that I am wrong, I shall look on pathogenic bacteria as consisting of different species . . .**

In a clinical laboratory, samples from diseased tissue are cultured, and pathogens must be distinguished from normal and transient microbiota.

---

*Quoted in T. D. Brock, ed. *Milestones in Microbiology*. Washington, DC: American Society of Microbiology, 1961, p. 99.

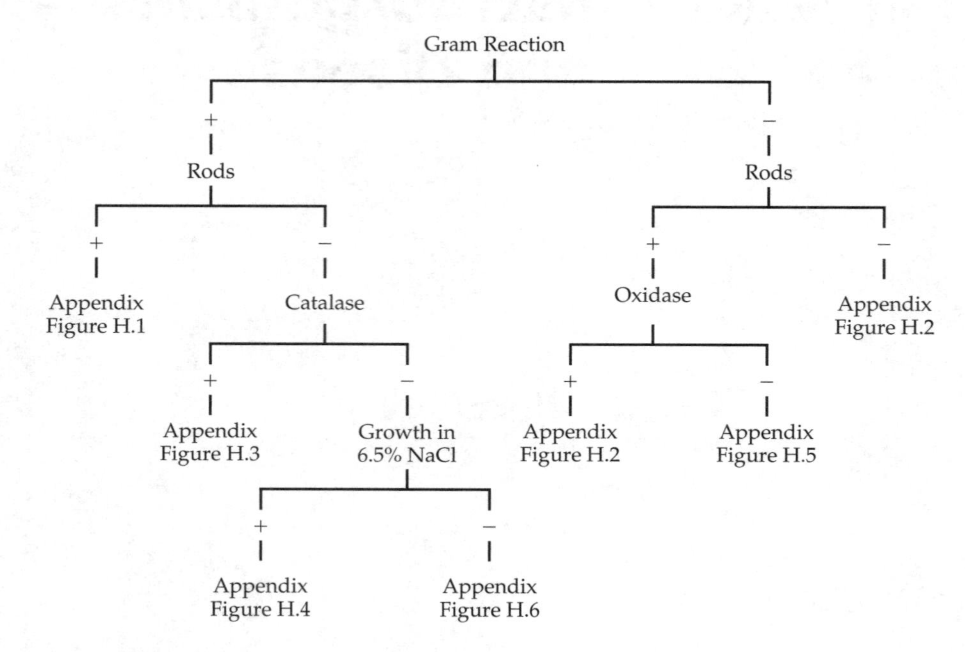

A dichotomous key to Appendix H. There is no single dichotomous key. This is one example of a key to Appendix H. In this key, the first test is a Gram stain. If the bacterium is gram-positive, you move to the left side of the diagram, ignoring all of the gram-negative bacteria. If the bacterium is a gram-positive coccus, you need to do a catalase test. If the bacterium is catalase-positive, go to Appendix H, Figure H.3 to complete the identification.

Identification of pathogens is imperative to initiating proper treatment and to tracing the source of the infection. Morphological characteristics (Part 2 of this lab manual), differential stains (Part 2), and biochemical testing (Part 3) provide information used to identify microorganisms in a clinical laboratory.

Dichotomous keys are essential tools for sorting through the information obtained from these laboratory tests in an organized manner (see the figure). In a dichotomous key, identification is based on successive questions, and each question has two possible answers. (*Dicho-* means "cut in two.") After answering one question, the microbiologist is directed to another question until an organism is identified. For example, a dichotomous key for bacteria would begin with an easily determined characteristic, such as gram reaction, and move on to the ability to ferment a sugar. There is no single key for all bacteria, and there is no one correct way to make a dichotomous key. Keys are made for specific groups of bacteria as they suit the needs of the microbiologist. Several dichotomous keys to help you identify bacteria in Part 12 are in Appendix H.

In Exercises 46 through 50, we will examine the variety of bacteria associated with the human body. In Exercise 51, we will identify bacteria in a simulated clinical sample.

Exercise 46

EXERCISE 46

Bacteria of the Skin

Objectives

After completing this exercise, you should be able to:

1. Isolate and identify bacteria from the human skin.
2. Provide an example of normal skin microbiota.
3. List characteristics used to identify the staphylococci.
4. Explain why many bacteria are unable to grow on human skin.

Background

The skin is generally an inhospitable environment for most microorganisms. The dry layers of keratin-containing cells that make up the epidermis (the outermost layer of the skin) are not easily colonized by most microbes. Sebum, secreted by oil glands, inhibits bacterial growth, and salts in perspiration create a hypertonic environment. Perspiration and sebum are nutritive for certain microorganisms, however, which establishes them as part of the normal microbiota of the skin.

Normal microbiota of the skin tend to be resistant to drying and to relatively high salt concentrations. More bacteria are found in moist areas, such as the axilla (armpit) and the sides of the nose, than on the dry surfaces of arms or legs. Transient microbiota are present on hands and arms in contact with the environment.

Propionibacterium live in hair follicles on sebum from oil glands. The propionic acid they produce maintains the pH of the skin between 3 and 5, which suppresses the growth of other bacteria. Most bacteria on the skin are gram-positive and salt-tolerant.

Staphylococcus aureus is part of the normal microbiota of the skin and is also considered a pathogen. *S. aureus*, which produces **coagulase,** an enzyme that coagulates (clots) the fibrin in blood, is pathogenic. A test for the presence of coagulase is used to distinguish *S. aureus* from other species of *Staphylococcus*.

Although many different bacterial genera live on human skin, in this exercise we will attempt to isolate and identify a catalase-positive, gram-positive coccus.

Materials

Petri plate containing mannitol salt agar

Sterile cotton swab

Sterile saline

Second Period

Petri plate containing mannitol salt agar

3% hydrogen peroxide

Gram-staining reagents

Fermentation tubes

Coagulase plasma (as needed)

Toothpick

Techniques Required

Gram staining, Exercise 5

Plate streaking, Exercise 11

Selective media, Exercise 12

Fermentation tests, Exercise 14

Catalase test, Exercise 17

Procedure

1. Wet the swab with saline, and push the swab against the wall of the test tube to express excess saline (Figure 46.1a). Swab any surface of your skin. The sides of the nose, axilla, an elbow, or a pus-filled sore are possible areas.
2. Swab one-third of the plate with the swab.

 **Discard the swab in disinfectant.**

   Using a sterile loop, streak back and forth into the swabbed area a few times, and then streak away from the inoculum (Figure 46.1b), to cover about one-third of the agar. Sterilize your loop and spread the bacteria over the rest of the agar.
3. Incubate the plate, inverted, at 35°C for 24 to 48 hours.
4. Examine the colonies. Record the appearance of the colonies and any mannitol fermentation (yellow halos). (See Color Plate V.1.) Perform a Gram stain of the colonies and test for catalase production. Perform the catalase test by making a

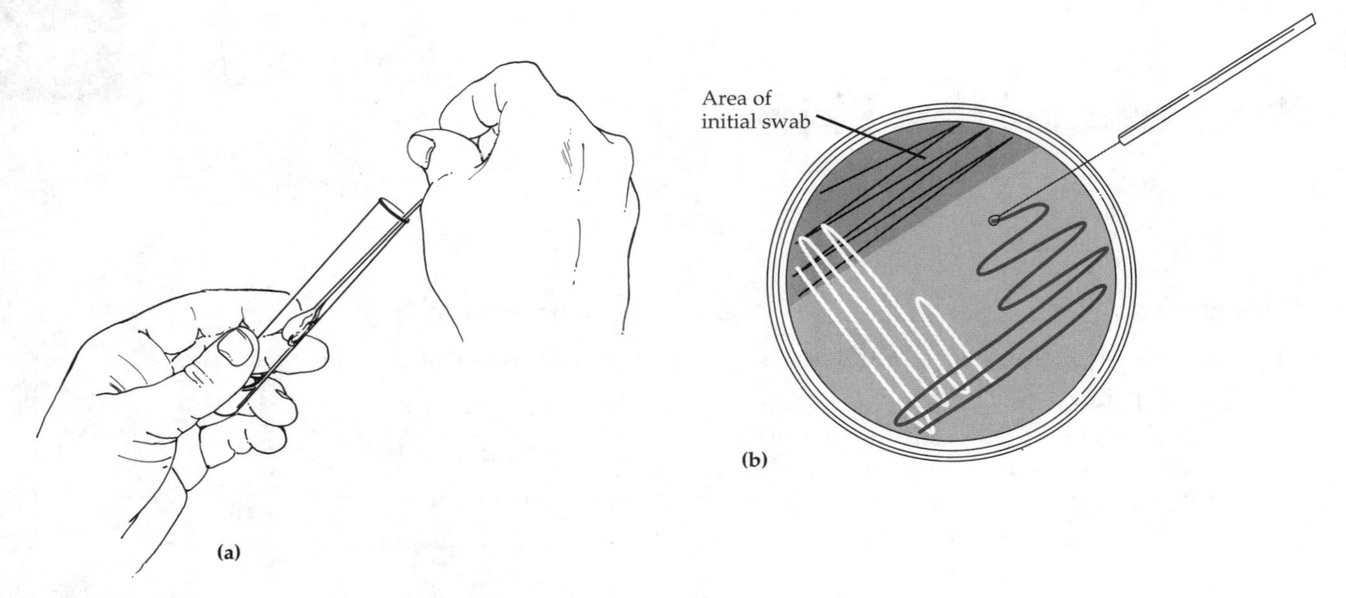

Figure 46.1

Taking a sample from the skin. **(a)** Aseptically moisten a sterile cotton swab in saline. **(b)** After swabbing half the plate, use a sterile loop to streak the inoculum over the agar.

suspension of the desired colony on a slide with a toothpick and adding a drop of H_2O_2. Discard the toothpick in the biohazard container.

5. Subculture a catalase-positive, gram-positive coccus on another mannitol salt plate. Do not attempt to subculture a colony that has not been tested for catalase. Why? ______________________
6. Using the key in Appendix H, Figure H.3, proceed to identify your isolate.
   a. To test for coagulase, place a loopful of rehydrated coagulase plasma on a clean slide. Add a loopful of water and make a heavy suspension of the bacteria to be tested. Observe for clumping of the bacterial cells (clumping = coagulase-positive; no clumping = coagulase-negative). (See Color Plate VIII.3.)
   b. Inoculate the appropriate fermentation tubes. Incubate the fermentation tubes at 35°C for 24 to 48 hours.

Exercise 46

LABORATORY REPORT

Bacteria of the Skin

NAME ______________________

DATE ______________________

LAB SECTION ______________________

Purpose

____

____

____

Data

Source of inoculum: ______________________

First Mannitol Salt Plate

	Colony		
	1	2	3
Colony description			
Pigment			
Mannitol fermentation			

Which colony was isolated? ______________________

Catalase reaction: ______________________

Gram stain

Reaction: ______________________

Morphology: ______________________

Arrangement: ______________________

Additional Biochemical Tests

____

____

____

Conclusion

What organism did you identify? ______

Questions

1. Why is mannitol salt agar used as a selective medium for normal skin microbiota? ______

2. After you have observed a gram-positive coccus, what is the additional information you need before performing a coagulase test? ______

3. List three identifying characteristics of *Staphylococcus aureus*. ______

4. List three factors that protect the skin from infection. ______

Critical Thinking

1. What is coagulase? How is it related to pathogenicity?

2. Assume that you isolated *S. aureus* from your skin. How would you determine whether it was penicillin-resistant?

3. An 8-year-old male went to the pediatrician complaining of pain in his right hip. He had no previous injury to the hip or leg. His temperature was 38°C. A bone scan revealed damage to the head of the femur. Bacterial cultures of blood and fluid from the hip joint were positive for a gram-positive, catalase-positive, coagulase-positive cocci that ferment mannitol. Use Appendix H to identify the species of this bacterium.

Exercise 47

EXERCISE 47

Bacteria of the Respiratory Tract

Objectives

After completing this exercise, you should be able to:

1. List representative normal microbiota of the respiratory tract.
2. Differentiate the pathogenic streptococci based on biochemical testing.

Background

The respiratory tract can be divided into two systems: the upper and lower respiratory tracts. The **upper respiratory tract** consists of the nose and throat, and the **lower respiratory tract** consists of the larynx, trachea, bronchial tubes, and alveoli. The lower respiratory tract is normally sterile because of the efficient functioning of the ciliary escalator. The upper respiratory tract is in contact with the air we breathe—air contaminated with microorganisms.

The throat is a moist, warm environment, allowing many bacteria to establish residence. Species of many different genera—such as *Staphylococcus*, *Streptococcus*, *Neisseria*, and *Haemophilus*—can be found living as normal microbiota in the throat. Despite the presence of potentially pathogenic bacteria in the upper respiratory tract, the rate of infection is minimized by microbial **antagonism.** Certain microorganisms of the normal microbiota suppress the growth of other microorganisms through competition for nutrients and production of inhibitory substances.

Streptococcal species are the predominant organisms in throat cultures, and some species are the major cause of bacterial sore throats (acute pharyngitis). Streptococci are identified by biochemical characteristics, including hemolytic reactions, and antigenic characteristics (Lancefield's system). Hemolytic reactions are based on hemolysins that are produced by streptococci while growing on blood-enriched agar. Blood agar is usually made with defibrinated sheep blood (5.0%), sodium chloride (0.5%) to minimize spontaneous hemolysis, and nutrient agar. Three patterns of hemolysis can occur on blood agar:

1. **Beta-hemolysis:** Complete hemolysis, giving a clear zone with a clean edge around the colony. (See Color Plate VIII.5.)
2. **Alpha-hemolysis:** Green, cloudy zone around the colony. Partial destruction of red blood cells due to bacteria-produced hydrogen peroxide. (See Color Plate VIII.4.)
3. **Gamma-hemolysis:** No hemolysis, and no change in the blood agar around the colony.

Streptococci that are alpha-hemolytic and gamma-hemolytic are usually normal microbiota, whereas beta-hemolytic streptococci are frequently pathogens.

The streptococci can be antigenically classified into Lancefield groups A through O by antigens in their cell walls. Over 90% of streptococcal infections are caused by beta-hemolytic group A streptococci. These bacteria are assigned to the species *S. pyogenes*. *S. pyogenes* is sensitive to the antibiotic bacitracin; other streptococci are resistant to bacitracin (see Color Plate VII.5). Optochin sensitivity is used to distinguish pathogenic *S. pneumoniae* from other alpha-hemolytic streptococci.

Materials

Throat Culture

Petri plate containing blood agar

Sterile cotton swab

Gram-staining reagents

Toothpicks

Hydrogen peroxide, 3% (second period)

Streptococcus

Petri plate containing blood agar

Sterile cotton swabs (2)

Forceps and alcohol

Optochin disk

Bacitracin disk

10% bile salts (second period)

Tubes containing 2 ml of nutrient broth (2) (second period)

Cultures

Streptococcus pyogenes

Streptococcus pneumoniae

Techniques Required

Gram staining, Exercise 5

Plate streaking, Exercise 11

Catalase test, Exercise 17

Kirby-Bauer technique, Exercise 25

Procedure

Throat Culture

 **Work only with swabs collected from your own throat.**

1. Swab your throat with a sterile cotton swab. The area to be swabbed is between the "golden arches" (glossopalatine arches), as shown in Figure 47.1. Do not hit the tongue.
2. After obtaining an inoculum from the throat, swab one-half of a blood agar plate. Streak the remainder of the plate with a sterile loop (Figure 46.1b).

 **Discard the swab in disinfectant.**

3. Incubate the plate, inverted, at 35°C for 24 hours. Observe the plate for hemolysis. Transfer some colonies to a slide with a toothpick, and perform a catalase test. Discard the toothpick in the biohazard container. Why can't the catalase test be done on blood agar? ____

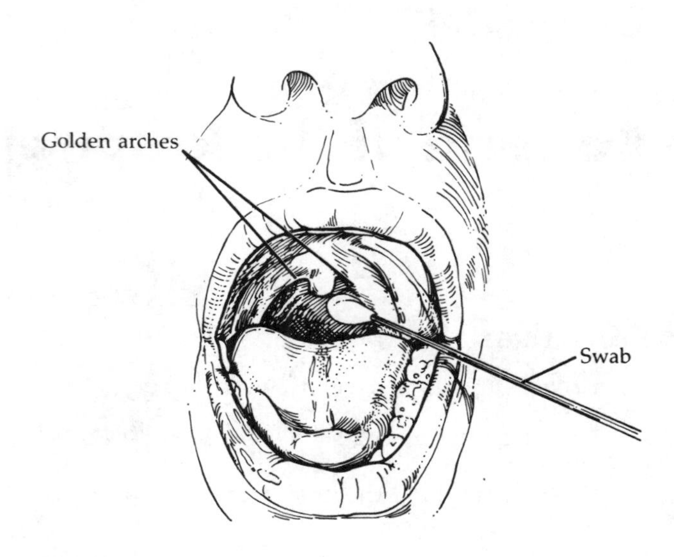

Figure 47.1
Swab your throat with a sterile swab.

Streptococcus

1. Inoculate each half of a blood agar plate, one side with *S. pyogenes* and the other half with *S. pneumoniae*. Use swabs to obtain confluent growth.

 **Be careful—these bacteria are pathogens.**

2. Dip forceps in alcohol and burn off the alcohol.

 **Keep the forceps tip pointed down while burning. Keep the beaker of alcohol away from the flame.**

   Using the forceps, place and press a bacitracin disk and an optochin disk on each half. Space the disks so that zones of inhibition may be observed.
3. Incubate the plates, inverted, at 35°C for 24 hours. Observe for hemolysis and inhibition of growth by bacitracin and optochin.
4. Prepare Gram stains from smears of each organism. Do the organisms differ microscopically? ________________
5. Using a sterile loop, prepare a suspension of each organism in a tube of nutrient broth.
6. Add a few drops of 10% bile salts to each tube. Observe the tubes after 15 minutes for lysis of the cells. Which culture is "bile soluble"? ________

Exercise 47

LABORATORY REPORT

Bacteria of the Respiratory Tract

NAME ______________________

DATE ______________________

LAB SECTION ______________________

Purpose ______________________

____

____

Data

Throat Culture

	Colony			
	1	2	3	4
Appearance of colonies on blood agar				
Hemolysis				
Catalase reaction				

Streptococcus

	Organism	
	S. pyogenes	*S. pneumoniae*
Gram stain Gram reaction		
Morphology		
Arrangement		
Blood agar plate Appearance of colonies		
Hemolysis		

	Organism	
	S. pyogenes	*S. pneumoniae*
Inhibition by Optochin		
Bacitracin		
Bile solubility		

Questions

1. Is blood agar selective or differential? ______________ Briefly explain. ______________________

____

2. Why does addition of a bacitracin disk to a blood agar plate make a rapid identification technique?

____

____

3. Is the Gram stain of significant importance in the identification of the organisms studied in this exercise?

Explain. ____

____

____

4. You have isolated a gram-positive coccus from a throat culture that you cannot identify as staphylococci or streptococci. A test for one enzyme can be used to distinguish *quickly* between these bacteria. What is the enzyme?

____

Critical Thinking

1. A 45-year-old male was admitted to the hospital with chest and back pain. On examination, his chest was dull to percussion. A chest X ray shows lower left lung infiltrates. A sputum culture reveals alpha-hemolytic, gram-positive, catalase-negative cocci that were inhibited by optochin. The bacteria produce acid from lactose. Use Appendix H to identify the species of this bacterium.

2. Assume that you isolated non–acid-fast, gram-positive, catalase-positive rods from your throat. Does this represent a disease state? Briefly explain.

Exercise 48

EXERCISE 48

Bacteria of the Mouth

Objectives

After completing this exercise, you should be able to:

1. List characteristics of streptococci found in the mouth.
2. Describe the formation of dental caries.
3. Explain the relationship between sucrose and caries.

Background

The mouth contains millions of bacteria in each milliliter of saliva. Some of these bacteria are transient microbiota carried on food. Some species of *Streptococcus* are part of the normal microbiota of the mouth. An identification scheme for streptococci found in the mouth is shown in Appendix H, Figure H.4.

Streptococcus mutans, *S. salivarius*, and *S. sanguinis* produce sticky polysaccharides specifically from sucrose. Bacterial exoenzymes hydrolyze sucrose into its component monosaccharides, glucose and fructose. The energy released in the hydrolysis is used by *S. mutans* and *S. sanguinis* for polymerization of the glucose to form **dextran.** Fructose released from the sucrose is fermented to produce lactic acid. Enzymes of *S. salivarius* have similar specificity for sucrose, but the fructose is polymerized into **levan,** and the liberated glucose is fermented.

The dextran capsule enables the bacteria to adhere to surfaces in the mouth. *S. salivarius* colonizes the surface of the tongue, and *S. sanguinis* and *S. mutans*, the teeth. Masses of bacterial cells, dextran, and debris adhering to the teeth constitute dental **plaque.** Production of lactic acid by bacteria in the plaque initiates dental caries by eroding tooth enamel. Streptococci and other bacteria, such as *Lactobacillus* species, are able to grow on the exposed dentin and tooth pulp.

Other carbohydrates, such as glucose or starch, may be fermented by bacteria, but they are not converted to dextran and hence do not promote plaque formation. "Sugarless" candies contain mannitol or sorbitol, which cannot be converted to dextran, although they may be fermented by such normal microbiota as *S. mutans*. Acid production increases the extent of dental caries.

The number of *S. mutans* present in stimulated saliva has been correlated with the potential for the formation of caries. In this exercise, we will observe and count *S. mutans* and other polysaccharide-producing streptococci in a stimulated saliva sample. The media used contain sucrose to promote capsule formation, and mitis-salivarius-bacitracin (MSB) agar inhibits the growth of most oral bacteria, except *S. mutans*. Oral streptococci require an elevated CO_2 atmosphere. This can be achieved with a disposable plastic bag for one or two culture plates. In this technique, the bag is coated with antifogging chemicals, and a wet sodium bicarbonate tablet is added to generate CO_2 (Figure 48.1a).

Materials

Rodac plate containing MSB agar

Petri plate containing sucrose blood agar

Chewing gum

Tongue blade

50-ml beaker

Sterile 1-μl calibrated loop

CO_2 jar or CO_2 envelope

3% hydrogen peroxide (second period)

Techniques Required

Catalase test, Exercise 17

Dissecting microscopy, Appendix E

Procedure*

Work only with your own saliva, and discard all contaminated materials in the To Be Autoclaved area.

Getting Ready

1. Label an MSB plate and a sucrose blood agar plate.
2. Chew a piece of chewing gum until the flavor is gone. Do not swallow your saliva.

*Adapted from C. Hoover. "Indigenous Flora of Saliva and Supragingival Plaque." Unpublished exercise for dental students. San Francisco: University of California, Department of Stomatology, 1991.

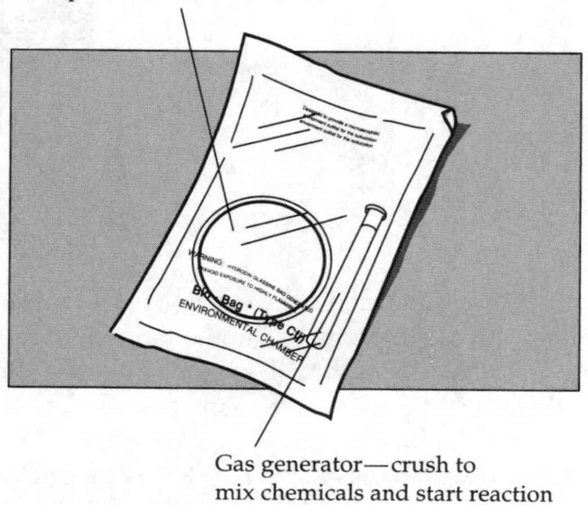

(a) CO_2 pouch. Your instructor will demonstrate how to crush the reagent packet to start the reaction.

(b) Candle jar. Light the candle and screw the lid onto the jar. Do not let the flame touch the Petri plates. The flame will produce CO_2 and will burn until the air in the jar has a lowered oxygen concentration.

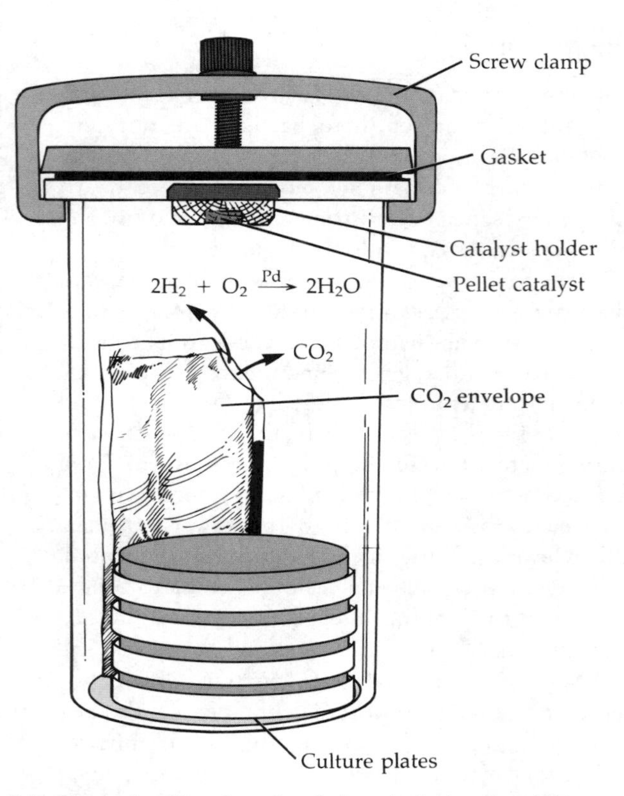

(c) Brewer jar. Place inoculated plates in the jar. Cut off the corner of the GasPak along the dotted line and place the envelope in the jar. Break the CO_2 indicator ampule. Shake well and place the CO_2 indicator on top of the plates. Add 10 ml of water through the cut corner of the envelope. Close the jar promptly after the envelope is activated.

Figure 48.1

CO_2 incubation techniques.

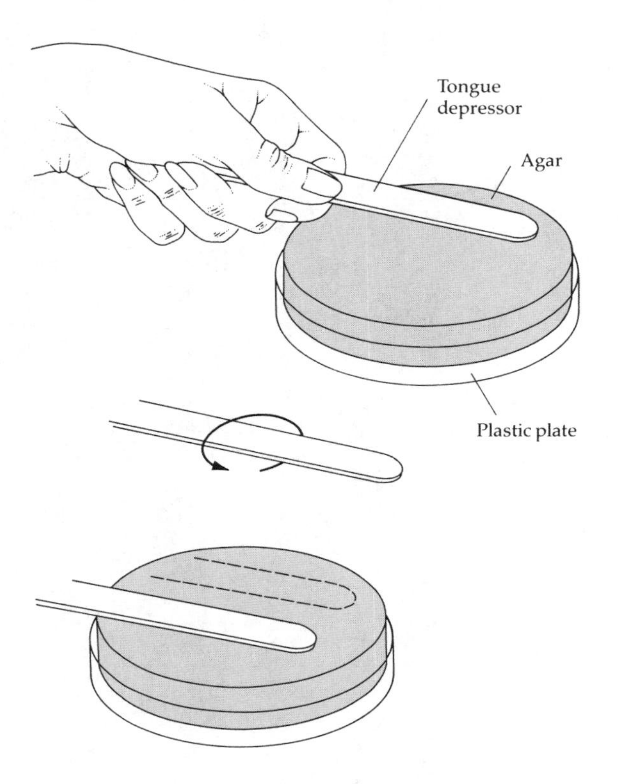

Figure 48.2
To inoculate the Rodac plate, press one side of the tongue blade on the agar. Turn the blade over, and press the other side on the remaining half of the agar.

MSB Agar

1. Rotate a sterile tongue blade in your mouth 10 times so both sides of the blade are thoroughly inoculated with saliva. Remove the tongue blade through closed lips to remove excess saliva. Don't swallow yet.
2. Press one side of the tongue blade onto one-half of the surface of the MSB agar. Then press the other side of the blade onto the other side of the agar, as shown in Figure 48.2. Discard the tongue blade in disinfectant.
3. Use one of the CO_2-incubation techniques shown in Figure 48.1.
4. Incubate the plate, inverted, in the CO_2 jar for 72 to 96 hours at 35°C.
5. Examine the MSB plate with a dissecting microscope (Appendix E) for the presence of *S. mutans*. *S. mutans* colonies are light blue to black, raised, and rough; their surface resembles etched glass or "burnt" sugar. Other bacteria occasionally grow on MSB; therefore, careful observation of colony morphology is necessary to accurately estimate the number of *S. mutans* present.

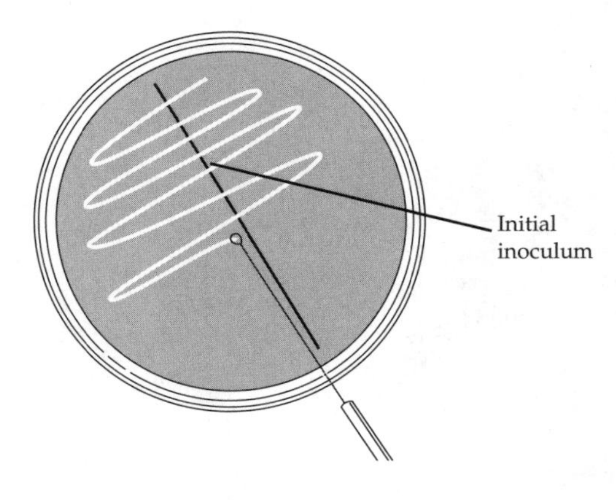

Figure 48.3
Using a calibrated loop, streak one line across the agar; then streak back and forth across that line.

Sucrose Blood Agar

1. Collect your saliva in a 50-ml beaker. Don't be embarrassed; everyone makes about 2 liters of saliva over the course of a day! Collect approximately 2 to 3 ml. Discard gum in the disinfectant.
2. Using a 1-μl calibrated loop, streak one line down the center of the sucrose blood agar. Then streak back and forth, perpendicular to that line (Figure 48.3). *Discard the loop in disinfectant.* What volume of saliva did you put on the agar? ________
3. Incubate the plate, inverted, in a CO_2 jar for 72 to 96 hours at 35°C (Figure 48.2).
4. Examine the sucrose blood agar plate for production of large raised mucoid (gumdrop) colonies; these are likely to be *S. salivarius*. Look carefully for glistening colonies surrounded by an indentation of the agar surface that cannot be moved without tearing the agar; these are likely to be *S. sanguinis*. *S. mutans* colonies, if present, may be recognized by the presence of a drop of liquid polysaccharide on top of or surrounding the colony. Count the number of each colony type. Record the number of colony-forming units (cfu) per milliliter of saliva.

$$\text{cfu/ml} = \frac{\text{Number of colonies}}{\text{Amount plated}}$$

Determine whether catalase is produced by suspected streptococci. The catalase test cannot be done directly on agar containing blood. Why not?

____

____

Exercise 48

LABORATORY REPORT

Bacteria of the Mouth

NAME ______________________

DATE ______________________

LAB SECTION ______________________

Purpose

____

____

____

Data

MSB Agar

Species	Number	Description

Sucrose Blood Agar

	Colony Morphology		
	Suspected *S. salivarius*	Suspected *S. sanguinis*	Suspected *S. mutans*
Appearance of colonies			
Number of colonies			
cfu/ml			
Catalase production			

Conclusions

Compare your data with those of your classmates, and draw a conclusion about your oral microbiota. ____________

Questions

1. List three characteristics of streptococci found in the mouth. ____________

2. Studies have shown that both sucrose and bacteria are necessary for tooth decay. Why should this be true?

3. Is MSB agar selective or differential? ____________ Explain. ____________

4. Is sucrose blood agar selective or differential? ____________ Explain. ____________

5. Even if a large number of *S. mutans* is in your saliva, how might you avoid tooth decay? ____________

Critical Thinking

1. *S. salivarius* forms large mucoid colonies on sucrose blood agar. Would you expect the same type of colonies on glucose blood agar? Briefly explain your answer.

2. Some dentists determine a patient's susceptibility to dental caries by measuring the pH of the saliva. What is the rationale for using this technique?

3. One month after having her teeth cleaned by a dental hygienist, an 89-year-old woman with a hip replacement developed bacteremia. Gram-positive, catalase-negative bacteria were isolated from her blood. The bacteria ferment inulin but not mannitol. They do not hydrolyze esculin and are V–P-negative. Use Appendix H to identify the species of this bacterium. What is the role of the artificial hip in this infection?

Exercise 49

Bacteria of the Gastrointestinal Tract

Objectives

After completing this exercise, you should be able to:

1. List the bacteria commonly found in the gastrointestinal tract.
2. Define the following terms: coliform, enteric, and enterococci.
3. Interpret results from triple sugar iron (TSI) slants and phenylethyl alcohol (PEA)–blood agar.

Background

The stomach and small intestine have relatively few microorganisms, as a result of the hydrochloric acid produced by the stomach and the rapid movement of food through the small intestine. But microbial populations in the large intestine are enormous, exceeding 10^{11} bacteria per gram of feces. Most of these intestinal organisms are commensals, and some are in mutualistic relationships with their human hosts. Some intestinal bacteria synthesize useful vitamins, such as folic acid and vitamin K. The normal intestinal microbiota prevent colonization of pathogenic species by producing antimicrobial substances and competition.

The population of the large intestine consists primarily of anaerobes of the genera *Bacteroides*, *Bifidobacterium*, *Enterococcus*, and *Lactobacillus*, and such facultative anaerobes as *Escherichia*, *Enterobacter*, *Citrobacter*, and *Proteus*.

Most diseases of the gastrointestinal system result from the ingestion of food or water that contains pathogenic microorganisms. Good sanitation practices, modern methods of sewage treatment, and the disinfection of drinking water help break the fecal–oral cycle of disease. A number of tests have been developed to identify facultative anaerobes associated with fecal contamination. Gram-negative, facultatively anaerobic rods are a large and diverse group of bacteria that includes the **enteric family** (Enterobacteriaceae).

Media have been developed to differentiate between lactose-fermenting enterics and nonlactose-fermenting enterics. The lactose fermenters are called **coliforms** and are generally not pathogenic. The nonlactose-fermenting group includes such pathogens as *Salmonella* and *Shigella*. One of the most common media is MacConkey agar. **MacConkey agar** is selective in that bile salts are inhibitory to gram-positive organisms; thus, they allow the medium to selectively culture gram-negative organisms. MacConkey agar is differential in that lactose-fermenting organisms (coliforms) give red, opaque colonies and nonlactose fermenters produce colorless, translucent colonies.

After isolation on MacConkey agar, differential screening media such as **triple sugar iron (TSI)** agar can be used to further characterize organisms. TSI contains the following:

0.1% glucose

1.0% lactose

1.0% sucrose

0.02% ferrous sulfate

Phenol red

Nutrient agar

As shown in Figure 49.1, if the organism ferments only glucose, the phenol red indicator will turn yellow in a few hours. The bacteria quickly exhaust the limited supply of glucose and start oxidizing amino acids for energy, giving off ammonia as an end-product. Oxidation of amino acids increases the pH, and the indicator in the slanted portion of the tube will turn back to red. The butt will remain yellow. If the organism in the TSI slant ferments lactose and/or sucrose, the butt and slant will turn yellow and remain yellow for days because of the increased level of acid production. Gas production by an organism can be ascertained by the appearance of bubbles *in* the agar. TSI can also be used to indicate whether hydrogen sulfide (H_2S) has been produced from desulfurization of sulfur-containing amino acids. H_2S reacts with ferrous sulfate in the medium, producing ferrous sulfide, a black precipitate. The key in Appendix H, Figure H.5, shows how enteric bacteria can be identified using MacConkey agar, TSI, and additional tests.

Enterococci are gram-positive, catalase-negative cocci. They are members of the normal intestinal microbiota of humans and other animals. The presence of enterococci can be used to indicate fecal contamination. *Enterococcus faecalis* and *E. faecium* are leading causes of hospital-acquired infections. Enterococci are the second most common cause of surgical wound infections, behind coagulase-negative staphylococci, and second to *E. coli*

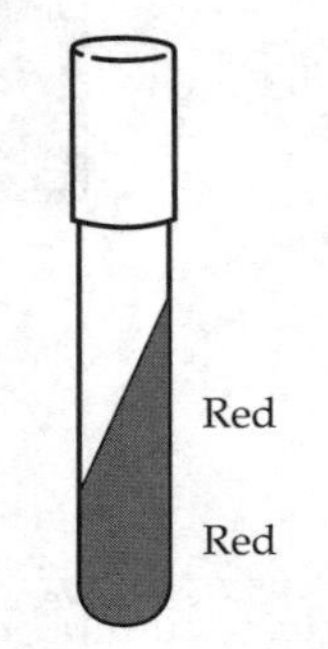

(a) No sugar fermentation

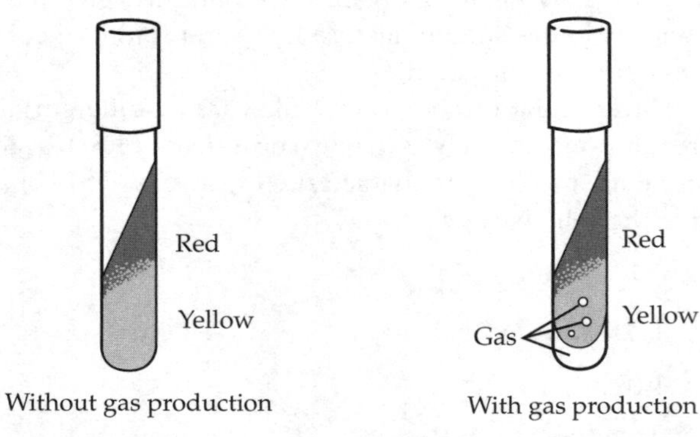

(b) Glucose fermented; lactose and sucrose not fermented

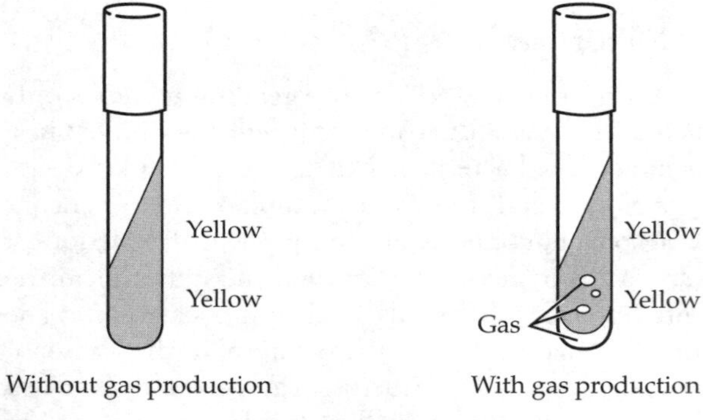

(c) Glucose and lactose and/or sucrose fermented

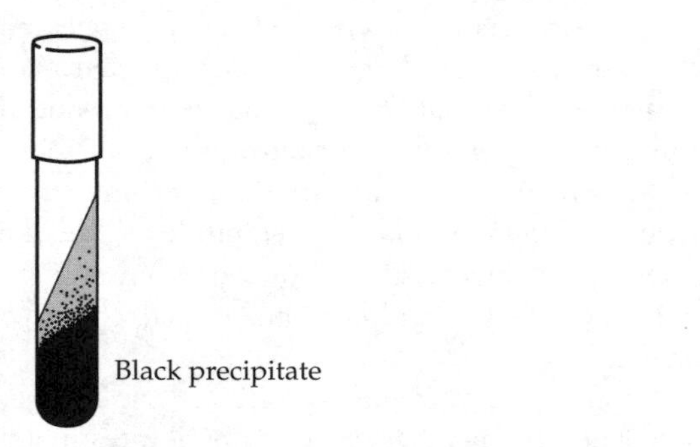

(d) H_2S production can occur in addition to **(a)**, **(b)**, and **(c)**.

Figure 49.1

Reactions in triple sugar iron (TSI) agar after incubating for 24 hours.

as the cause of nosocomial urinary tract infections. Enterococci can be isolated on an enriched medium (containing blood) that inhibits the growth of gram-negative bacteria. Phenylethyl alcohol (PEA) inhibits the growth of gram-negative bacteria. The key in Appendix H, Figure H.6, will help you identify *Enterococcus* species commonly found in the human gastrointestinal tract.

Materials

First Period

Petri plate containing MacConkey agar

Petri plate containing PEA–blood agar

Tube containing sterile cotton swab

10-ml pipette

Sterile water

Fecal sample (animal or human feces; swab feces or the anus after a bowel movement with a sterile swab, and bring the contaminated swab to class in a sterile tube)

Second Period

Petri plate containing PEA–blood agar

Gram-stain reagents

TSI slant

Toothpicks

3% hydrogen peroxide

Fermentation tubes, as needed

Techniques Required

Gram staining, Exercise 5

Plate streaking, Exercise 11

Fermentation tests, Exercise 14

Catalase test, Exercise 17

Blood agar, Exercise 47

Procedure

First Period

1. If the feces are hard, break them apart in a sterile Petri dish and emulsify them in sterile water. Swab a small area on the MacConkey agar with an inoculated swab, and then streak for isolation with a loop (Figure 46.1b).

2. Incubate the plate, inverted, at 35°C for 24 to 48 hours.
3. Using the fecal swab, inoculate PEA–blood agar. Incubate the plate, inverted, at 35°C for 24 to 48 hours.

Discard the test tube, tube of water, and other contaminated materials in the To Be Autoclaved area.

Second Period

1. Observe your MacConkey plate and those of other students. Record the appearance and colors of the colonies. (See Color Plate XII.6.) Prepare a Gram stain from an isolated colony.
2. Select one isolated colony using an inoculating needle and inoculate the TSI slant. *Stab* into the butt of the agar, and then streak the surface of the slant. Incubate the tube at 35°C for 24 hours. Observe and record results from the TSI slant inoculated. (See Color Plate III.15.) Observe other students' TSI slants.
3. Observe the PEA–blood agar for the presence of hemolysis. Perform a Gram stain and a catalase test on isolated colonies. Is *Enterococcus* catalase-positive or catalase-negative? ______________________
   Transfer a colony to a slide with a toothpick, and perform a catalase test. Discard the toothpick in the biohazard container. Why can't the catalase test be done on blood agar?______________________
   ____
   ____
4. Subculture a catalase-negative, gram-positive coccus on another PEA–blood agar plate.
5. Using the key in Appendix H, Figure H.6, identify your isolate. Inoculate the fermentation tubes. Incubate tubes at 35°C for 24 to 48 hours.

Exercise 49

LABORATORY REPORT

Bacteria of the Gastrointestinal Tract

NAME ______________________

DATE ______________________

LAB SECTION ______________________

Purpose ______________________

____

____

Data

MacConkey Agar

Source of inoculum: ______________________

	Colony			
	1	2	3	4
Appearance of colonies				
Lactose fermentation				

TSI

Which colony was used? ______________________

	Appearance of Agar		
	Slant	Butt	Results
Acid from glucose			
Acid from lactose/sucrose			
Gas produced			
H_2S produced			

First PEA–Blood Agar Plate

	Colony		
	1	2	3
Colony description			
Hemolysis			

Which colony was isolated? ______________________

Catalase reaction: ______________________

Gram stain

Reaction: ______________________

Morphology: ______________________

Arrangement: ______________________

Additional Biochemical Tests

____

____

____

Conclusions

1. What genus do you think you have in your TSI slant? ______________________

   What additional information do you need to identify it? ______________________

   ____

2. Did you culture *Enterococcus* on the PEA-blood agar? ______________________

   How do you know? ______________________

3. What enterococci did you identify? ______________________

Questions

1. What tests, if any, would you need to perform to identify bacteria producing a red colony on MacConkey agar?

   ____

   ____

2. Differentiate between a coliform, an enteric that is not a coliform, and enterococci. Give an example of an organism in each group. ______________________

   ____

   ____

   ____

   ____

3. Is MacConkey agar a selective medium? ______________ A differential medium? ______________

   How can you tell? ____

   ____

   ____

4. Is PEA–blood agar a selective medium? ______________ A differential medium? ______________

   How can you tell? ____

Critical Thinking

1. How would you determine whether a colorless colony on MacConkey agar is *Salmonella* or *Shigella*? Why would you want to identify a colorless colony?

2. Tests to determine the presence of *Bacteroides* in clams and oysters are being developed. If these bacteria do not present a health hazard, why are these tests being developed?

3. You are called to investigate an outbreak of diarrheal disease in a child-care center. The symptoms include vomiting, fever, nausea, and cramps. You culture fecal samples from children and find a gram-negative, lactose-negative rod. The bacterium does not produce gas from glucose and makes colorless colonies on MacConkey agar. Use Appendix H to identify the genus of this bacterium.

Exercise 50

Bacteria of the Urogenital Tract

Objectives

After completing this exercise, you should be able to:

1. List bacteria found in urine from a healthy individual.
2. Identify, through biochemical testing, bacteria commonly associated with urinary tract infections.
3. Determine the presence or absence of *Neisseria gonorrhoeae* in a GC smear.

Background

The urinary and genital systems are closely related anatomically, and some diseases that affect one system also affect the other system, especially in the female. The upper urinary tract and urinary bladder are usually sterile. The urethra does contain resident bacteria, including *Streptococcus*, *Bacteroides*, *Mycobacterium*, *Neisseria*, and enterics. Most bacteria in urine are the result of contamination by skin microbiota during passage. The presence of bacteria in urine is not considered an indication of urinary tract infection unless there are ≥1000 bacteria of one species or ≥100 coliforms per milliliter of urine.

Many infections of the urinary tract, such as cystitis (inflammation of the urinary bladder) or pyelonephritis (inflammation of the kidney), are caused by opportunistic pathogens and are related to fecal contamination of the urethra and to medical procedures, such as catheterization.

Standard examination of urine consists of a plate count on blood agar for the total number of organisms, coupled with a streak plate of undiluted urine on MacConkey agar. Why? ______________________________

____

The patient is given a sterile container and instructed to collect a midstream sample, which is obtained by voiding a small volume from the bladder before collection. This washes away skin microbiota.

In the first part of this exercise, you will examine normal urine. In the second part, three gram-negative rods that commonly cause cystitis will be provided to you. *E. coli* and *Proteus* are enterics. *E. coli* is one of the coliforms. What is a coliform? ______________________________

____

Proteus is actively motile and exhibits "swarming" on solid media (see Color Plate XII.7), where the cells at the periphery move away from the main colony. *Pseudomonas* is a gram-negative aerobic rod. *Pseudomonas aeruginosa* is commonly found in the soil and other environments. Under the right conditions, particularly in weakened hosts, this organism can cause urinary tract infections, burn and wound infections, and abscesses. *P. aeruginosa* infections are characterized by blue-green pus. This bacterium produces an extracellular, water-soluble pigment called **pyocyanin** ("blue pus") that diffuses into its growth medium (see Color Plate V.6).

Most diseases of the genital system are transmitted by sexual activity and are therefore called **sexually transmitted diseases (STDs).** Most of the bacterial diseases can be readily cured with antibiotics if treated early and can largely be prevented by the use of condoms. Nevertheless, STDs are a major U.S. public health problem.

The most common reportable communicable disease in the United States is gonorrhea, an STD caused by the gram-negative diplococci *Neisseria gonorrhoeae*, also called gonococci or GC. Gonorrhea is diagnosed by identifying the organism in the pus-filled discharges of patients, as demonstrated with GC smears in the third part of this exercise. Cultures from patients' discharges can be made on Thayer–Martin medium and incubated in a CO_2 jar. An oxidase test is performed on characteristic colonies for confirmation.

Materials

Urine Culture

Sterile widemouthed jar, approximately 50 to 250 ml

Petri plate containing blood agar

Petri plate containing MacConkey agar

Sterile 10-μl calibrated loops (2)

Cystitis

Petri plate containing MacConkey agar

Petri plate containing Pseudomonas agar P

Tubes containing OF-glucose medium (4)

Urea agar slants (2)

Mineral oil

Oxidase reagent (second period)

Gram-stain reagents (second period)

Cultures

Tube containing *Escherichia coli*, *Proteus vulgaris*, and *Pseudomonas aeruginosa*

GC Smears

GC smears

One unknown smear # ________

Techniques Required

Gram staining, Exercise 5

Plate streaking, Exercise 11

Fermentation tests, Exercise 14

Urea hydrolysis, Exercise 15

Oxidase test, Exercise 17

MacConkey agar, Exercise 49

Procedure

Urine Culture

Work only with urine collected from your own body.

1. Collect a "clean-catch" urine specimen, as described by your instructor, using the sterile jars available in the lab. Refrigerate the specimen until you are ready to perform step 2. Why? ________
____
2. Gently shake the urine to suspend the bacteria. Using a 10-μl calibrated loop, streak one line down the center of the blood agar. Then streak back and forth, perpendicular to that line (Figure 48.3). *Discard the loop in disinfectant.* What volume of urine did you put on the agar? ________
3. Using another 10-μl loop, repeat step 2 to inoculate the MacConkey agar. *Discard the loop in disinfectant.*

Discard the jar and pipettes in the appropriate biohazard container.

4. Incubate both plates, inverted, at 35°C for 24 to 48 hours.
5. Count the colonies on the blood agar plate, and determine the number of colony-forming units (cfu) per milliliter of urine:

$$\text{cfu/ml} = \frac{\text{Number of colonies}}{\text{Amount plated}}$$

6. Repeat step 5 to determine the cfu/ml on the MacConkey plate. Examine the MacConkey plate for the presence of possible coliforms. Consult your instructor if you are alarmed by your results.

Cystitis

1. Inoculate a MacConkey plate and Pseudomonas agar P plate with the mixed culture of bacteria.
2. Incubate the plates, inverted, at 35°C for 24 to 48 hours. (See Color Plates V.5, V.6, and XII.6.)
3. Examine the MacConkey plate for lactose-fermenting colonies. Which of the three organisms ferments lactose? ________
Look for swarming. Which organism is actively motile on solid media? ________
Are any small, nonlactose-fermenting colonies present? ________
4. Prepare a Gram stain from each different colony type.
5. Examine the Pseudomonas agar P plate. Can you identify *Pseudomonas*? ________
6. Perform an oxidase test on each different organism.
7. Inoculate four tubes of OF-glucose medium: two with *Proteus* and two with *Pseudomonas*. Plug one tube of each organism with mineral oil. Inoculate a urea agar slant with each organism. Incubate the tubes at 35°C for 24 to 48 hours.
8. Record the results of the OF-glucose and urease production tests. Why isn't it necessary to perform these two biochemical tests on *E. coli*? ________
____

GC Smears

1. Examine the GC smears provided.
2. Determine whether *N. gonorrhoeae* could be present in your unknown GC smear.

Exercise 50

LABORATORY REPORT

Bacteria of the Urogenital Tract

NAME ______________________

DATE ______________________

LAB SECTION ______________________

Purpose ______________________

____

Data

Urine Culture

	Blood Agar	MacConkey Agar
Number of colonies		
Total count (cfu/ml)		
Hemolysis		
Possible coliforms present		
How can you identify coliforms?		

Cystitis

	Organism		
	E. coli	*P. vulgaris*	*P. aeruginosa*
MacConkey agar			
Appearance of colonies	______	______	______
Lactose fermentation	______	______	______
Swarming	______	______	______
Gram stain	______	______	______
Pseudomonas agar P			
Appearance of colonies	______	______	______
Pigmentation	______	______	______
Oxidase reaction	______	______	______
OF-glucose (fermentative or oxidative)	______	______	______
Urease production	______	______	______

GC Smears

Diagram the appearance of a known GC-positive smear.

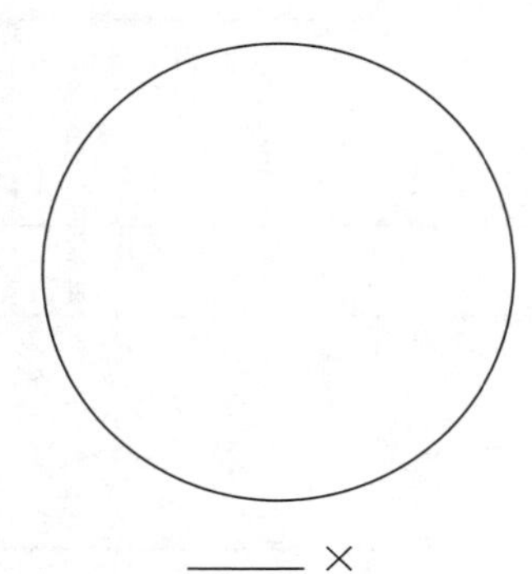

_____ ×

Diagram the appearance of a known GC-negative smear.

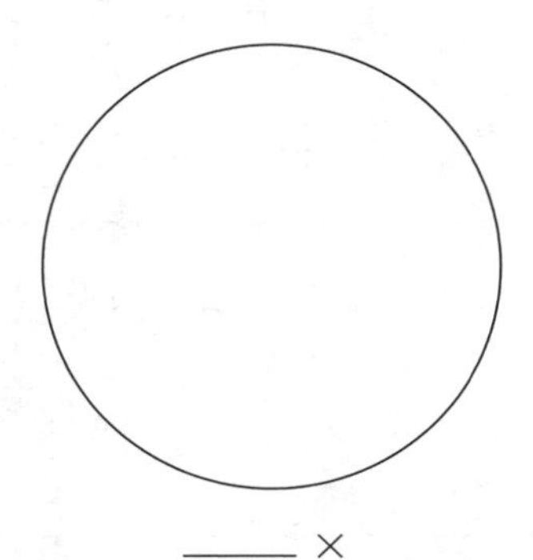

_____ ×

Unknown # _________
Diagram the appearance of your unknown smear.

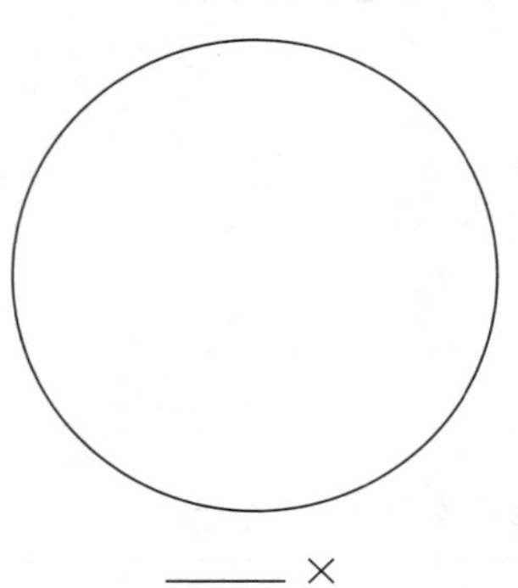

_____ ×

Could *N. gonorrhoeae* be present? _________________

Questions

1. Why is MacConkey agar inoculated with a urine specimen? _________________

____

____

2. With the results obtained on the blood agar and MacConkey agar plates, is a urinary tract disease possible?

____

3. The enterics and pseudomonads look alike microscopically. How can you easily distinguish between these two groups of bacteria? _________________

____

____

4. Why are females more prone to urinary tract infections than males? _________________

____

____

Critical Thinking

1. Differentiate between the pigment of *P. aeruginosa* on Pseudomonas agar P and the "pigment" of *E. coli* on MacConkey agar without referring to the colors.

2. What role does antibiotic treatment have in yeast infections of the urinary tract?

3. An otherwise healthy 22-year-old female was seen at a hospital emergency room because of frequent and painful urination. A urine culture reveals a gram-negative, lactose-positive rod that produces indole but no H_2S. Use Appendix H to identify the genus of this bacterium.

Exercise 51

Identification of an Unknown from a Clinical Sample

So simple did it seem to me
The day t'was first presented,
A Sherlock Holmes of lab technique
No effort be resented.

Now six Gram stains, three cultures later
The first step barely taken
Gram vari rods, both red and blue
My confidence was shaken.

Straight rods 'tis true, yet red and blue
In ones and twos and chain
Somes spores would show the truth of it
But stains were all in vain.

Alas! No spores of vivid green
Would show in fields of red.
Ah ha! Too soon! Spores take some time
Flashed through my seeking head.

With rod in hand, though young it be,
It surely must, said I,
Reveal some trait, another test
Than just a non-cocci.

So while I wait don't waste the time
The sugars show some trait.
So off to these, ferment or not
I can't just sit and wait.

Glucose a plus, sucrose the same
No gas, and lactose zilch.
The spores again with no green spots
It's on to litmus milk.

So mannitol and nitrate too
I'll run, for these should show,
Though spores resist at now day six,
Some little cells should grow.

Day seven's gone and spores still hide
From vision sharply tooled.
But agents stained from tests complete
Have many others ruled.

Divorce now near, by child disowned
On job, the thought "to fire"
I'll take my stand and flunk or not
B. subtilis, *or tests are liars.**

MARY DAVIS

Objectives

After completing this exercise, you should be able to:

1. Separate and identify two bacteria from a sample.
2. Determine the sensitivity of the unknown bacteria to chemotherapeutic agents.

Background

Clinical samples often contain many microorganisms, and a pathogen must be separated from resident normal microbiota. Differential and selective media are used to facilitate isolation of the pathogen, after which it must be identified. Also, antimicrobial sensitivity tests are performed to aid the physician in prescribing treatment. Effective treatment depends on identification of the pathogen; therefore, a clinical laboratory must provide results as quickly and accurately as possible. Microbiology laboratory results are usually available within 48 hours.

In this exercise, you will be provided with a simulated clinical sample containing two bacteria. The procedure will differ from that used in a clinical laboratory in that you will be asked to isolate and identify *both* organisms in the sample. In a clinical laboratory, selective and enrichment techniques are used to identify the pathogen only. After you have separated the two bacteria in pure culture, prepare stock and working cultures. To obtain pure cultures, inoculate a Trypticase soy agar plate *and* a differential medium. Select the differential medium using the type of sample as a clue. For

*M. E. Davis. "Unknown #3." *ASM News* 42(3):164, 1976. Reprinted with the permission of the American Society for Microbiology.

instance, a fecal sample should probably be inoculated onto ______________________ agar. Considering differential media should give the information needed to isolate the organisms, why is a Trypticase soy agar inoculated also? ______________________

____

Materials

Petri plates containing Trypticase soy agar

All stains, reagents, and media previously used

Trypticase soy agar slants (second period)

Mueller–Hinton agar (third period)

Antimicrobial disks (third period)

Culture

Unknown sample # ______________

Techniques Required

Gram staining, Exercise 5

Plate streaking, Exercise 11

Differential and selective media, Exercise 12

Kirby–Bauer technique, Exercise 25

Use of *Bergey's Manual,* Exercise 33

Procedure

First Period

1. Review general information regarding identification of an unknown. Study the identification schemes in Appendix H, Figures H.1 through H.6. Try to identify the bacterial species in Color Plate XIV.
2. You may be able to prepare a Gram stain directly from the unknown if it does not contain a high concentration of organic matter to interfere with the stain results.
3. Streak the unknown onto a Trypticase soy agar plate and an appropriate differential medium. *Do not be wasteful.* Inoculate only the necessary media. Incubate the plates at 35°C for 24 to 48 hours.

Second Period

1. Examine the plates, and select colonies that differ from each other in appearance and that are separated for easy isolation. (See Color Plate XIV.)
2. Streak each organism for isolation onto half of a Trypticase soy agar plate. Incubate the plate at 35°C for 24 to 48 hours. This plate can be your first working culture. Prepare a stock culture of each organism on separate Trypticase soy agar slants (one culture on each slant).

Third Period

1. Prepare a Gram stain of each organism. What should you do if you do not think you have a pure culture? ______________________

   ____

   ____
2. After determining the staining and morphologic characteristics of the two organisms, determine which biochemical tests you will need. Plan your work carefully. The same tests need not be performed on both bacteria.
3. Record which tests were performed and the results.
4. Select six antimicrobials from Table 25.1 to test against your unknowns. Use the Mueller–Hinton agar and antimicrobial disks to determine whether your unknowns are resistant to any antimicrobial agents. (See Color Plate VIII.1.)

Exercise 51

Laboratory Report

Identification of an Unknown from a Clinical Sample

Name ____________________

Date ____________________

Lab Section ____________________

Purpose

____

Data

Unknown # ____________________

Source: ____________________

Preliminary Gram-stain results: ____________________

	Bacterium	
	A	B
Appearance of colonies on Trypticase soy agar		
Gram reaction		
Cell morphology and arrangement		

Make a flowchart of your procedure. Remember, the *same* tests may *not* be required for each bacterium. The following will help you get started:

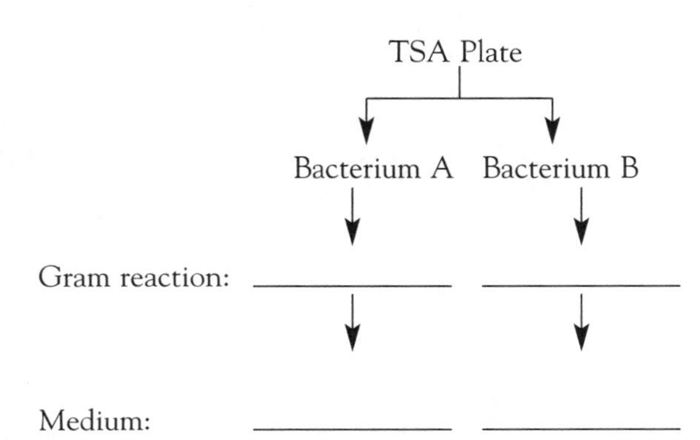

	Zone of Inhibition (mm)	
	Bacterium A	Bacterium B
Antimicrobial sensitivity to		
______________	______________	______________
______________	______________	______________
______________	______________	______________
______________	______________	______________
______________	______________	______________
______________	______________	______________

Conclusions

1. What two bacterial species were in unknown # ______________ ? ______________________________

   ____

2. Are either of your unknowns resistant to any antimicrobials? ______________________________

   If so, which ones? ____

   ____

Questions

1. Why would a clinical microbiology laboratory perform antimicrobial susceptibility tests as well as identify the unknown?

2. Write your rationale for arriving at the identities of your unknowns.

Part Thirteen

Microbiology and the Environment

EXERCISES

Microorganisms are omnipresent in our environment; however, the presence of some microorganisms in certain places, such as food or water, may be undesirable. In Exercises 52 and 53, we will use the presence of nonpathogenic bacteria to indicate water pollution. The presence and number of bacteria in foods can indicate that foods were contaminated during processing; these bacteria may result in food spoilage (Exercise 54).

Harmful microorganisms are only a small fraction of the total microbial population. The activities of most microbes are in fact beneficial. In Exercise 55, we will use selected microorganisms to produce desired flavors in foods and to preserve foods.

The activities of soil bacteria are essential to the maintenance of life on Earth (see the illustration on the next page). Nitrogen fixation (Exercise 56) was first

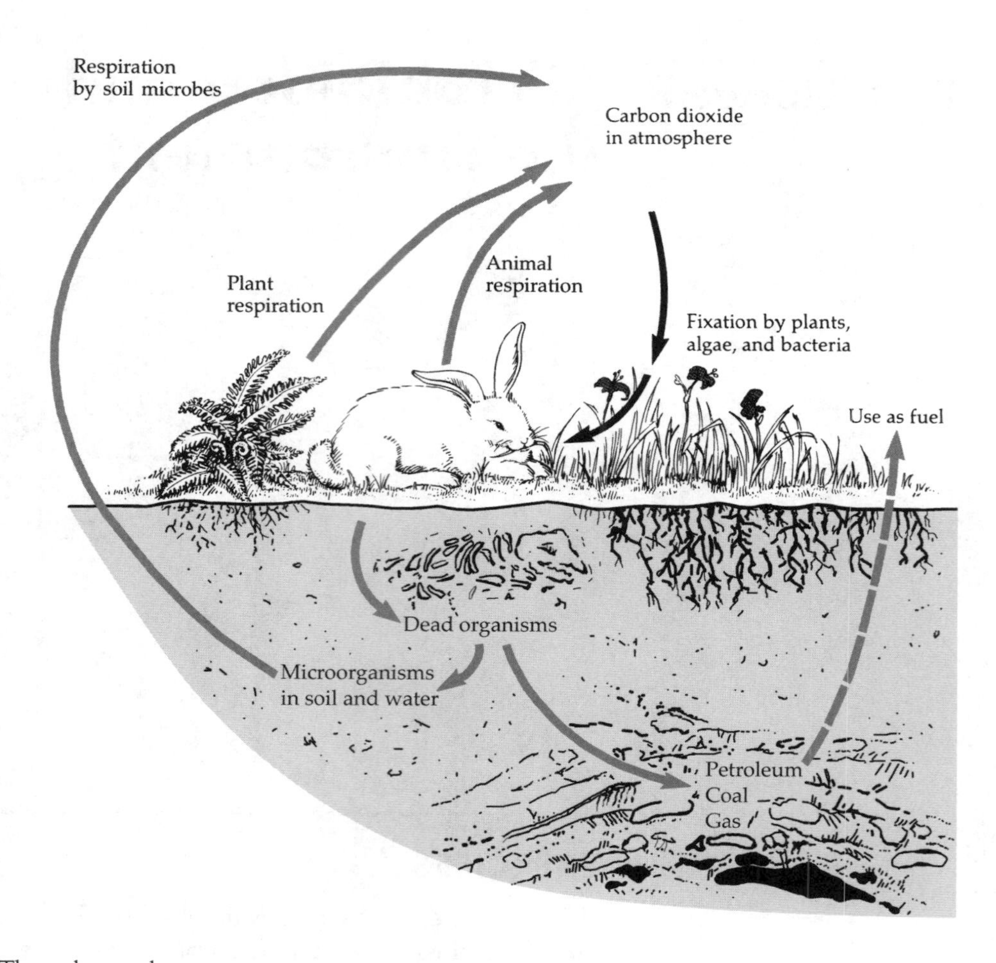

The carbon cycle.

described in 1893 by Sergei Winogradsky, who entered into the study of biogeochemical cycles because he was

> . . . *impressed by the incomparable glitter of Pasteur's discoveries. I started to investigate the great problem of fixation of atmospheric nitrogen. I succeeded without too much difficulty in isolating an anaerobic rod called* Clostridium *that would perform this function.**

Winogradsky named the organism *Clostridium pasteurianum* in honor of Pasteur. In addition to fixing nitrogen and recycling other elements, some microbes are able to degrade dangerous pollutants in the environment (Exercise 57).

---

*Quoted in H. A. Lechevalier and M. Solotorovsky. *Three Centuries of Microbiology*. New York: Dover Publications, 1974, p. 263.

Exercise 52

Microbes in Water: Multiple-Tube Technique

Objectives

After completing this exercise, you should be able to:

1. Define coliform.
2. Provide the rationale for determining the presence of coliforms.
3. List and explain each step in the multiple-tube technique.

Background

Tests that determine the bacteriological quality of water have been developed to prevent transmission of water-borne diseases of fecal origin. However, it is not practical to look for pathogens in water supplies because pathogens occur in such small numbers that they might be missed by sampling. Moreover, when pathogens are detected, it is usually too late to prevent occurrence of the disease. Rather, the presence of **indicator organisms** is used to detect fecal contamination of water. An indicator organism must be present in human feces in large numbers and must be easy to detect. The most frequently used indicator organisms are the coliform bacteria. **Coliforms** are aerobic or facultatively anaerobic, gram-negative, nonendospore-forming, rod-shaped bacteria that ferment lactose with acid and gas formation within 48 hours at 35°C. Coliforms are not usually pathogenic although they can cause opportunistic infections. Coliforms are not restricted to the human gastrointestinal tract but may be found in other animals and in the soil. Tests that determine the presence of fecal coliforms (of human origin) have been developed. The IMViC tests (Exercise 18) historically were used to distinguish coliforms of fecal origin, such as *Escherichia coli,* from other coliforms found in plants and soil.

Established public health standards specify the maximum number of coliforms allowable in each 100 ml of water, depending on the intended use of the water (for example, water for drinking or water-contact sports or treated wastewater for irrigation or for discharge into a bay or river). (See Color Plates XI.1 and XI.2.)

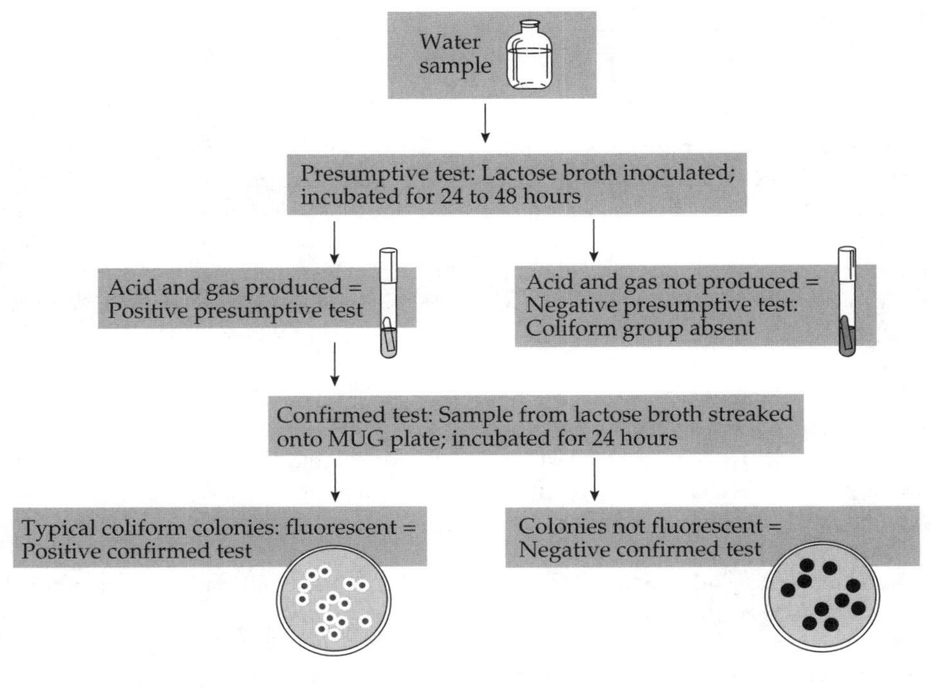

Figure 52.1
Analysis of drinking water for coliforms by the multiple-tube technique.

Table 52.1

Most Probable Number (MPN) Index for Various Combinations of Positive and Negative Results When Three 10-ml Portions, Three 1-ml Portions, and Three 0.1-ml Portions Are Used

Number of Tubes Giving Positive Reaction Out of			MPN Index per 100 ml	Number of Tubes Giving Positive Reaction Out of			MPN Index per 100 ml
3 of 10 ml Each	3 of 1 ml Each	3 of 0.1 ml Each		3 of 10 ml Each	3 of 1 ml Each	3 of 0.1 ml Each	
0	0	0	<3	3	0	0	23
0	0	1	3	3	0	1	39
0	1	0	3	3	0	2	64
1	0	0	4	3	1	0	43
1	0	1	7	3	1	1	75
1	1	0	7	3	1	2	120
1	1	1	11	3	2	0	93
1	2	0	11	3	2	1	150
2	0	0	9	3	2	2	210
2	0	1	14	3	3	0	240
2	1	0	15	3	3	1	460
2	1	1	20	3	3	2	1,100
2	2	0	21	3	3	3	≥2,400
2	2	1	28				

Source: Standard Methods for the Examination of Water and Wastewater, 13th ed. New York: American Public Health Association, 1971.

Coliforms can be detected and enumerated in the **multiple-tube technique** (Figure 52.1). In this method, coliforms are detected in two stages. In the **presumptive test,** dilutions from a water sample are added to lactose fermentation tubes. The lactose broth can be made selective for gram-negative bacteria by the addition of lauryl sulfate or brilliant green and bile. Fermentation of lactose to gas is a positive reaction.

Samples from the positive presumptive tube at the highest dilution are examined for coliforms by inoculating a differential medium in the **confirmed test.** A confirmed test can be done on **MUG agar.** Almost all strains of *E. coli* produce the enzyme β-glucuronidase (GUD). If *E. coli* is added to a nutrient medium containing 4-methylumbelliferone glucuronide (MUG), GUD converts MUG to a fluorescent compound that is visible with an ultraviolet lamp (see Color Plate XII.4).

The number of coliforms is determined by a statistical estimation called the **most probable number (MPN)** method. In the presumptive test, tubes of lactose broth are inoculated with samples of the water being tested. A count of the number of tubes showing acid and gas is then taken, and then is compared to statistical tables such as the one shown in Table 52.1. The MPN number is *the most probable number* of coliforms per 100 ml of water.

Materials

Water sample, 50 ml (Bring your own from a pond or stream.)

9-ml, single-strength lactose fermentation tubes (6)

20-ml, 1.5-strength lactose fermentation tubes (3)

Sterile 10-ml pipette

Sterile 1-ml pipette

Petri plate containing MUG agar (second period)

Techniques Required

Aseptic technique, Exercise 10

Plate streaking, Exercise 11

Fermentation tests, Exercise 14

Pipetting, Appendix A

Procedure

1. Label three single-strength lactose broth tubes "0.1," label another three tubes "1," and label the three 1.5-strength broth tubes "10."
2. Inoculate each 0.1 tube with 0.1 ml of your water sample.
3. Inoculate each 1 tube with 1.0 ml of your water sample.
4. Inoculate each 10 tube with 10 ml of the water sample. Why is 1.5-strength lactose broth used for this step? ____________________

   ____
5. Incubate the tubes for 24 to 48 hours at 35°C.
6. Record the results of your presumptive test (Figure 52.1). Which tube has the highest dilution of the water sample? ____________________

   ____

   Determine the number of coliforms per 100 ml of the original sample using Table 52.1. If a tube has gas, streak the MUG agar with the positive broth. Incubate the plate, inverted, for 24 to 48 hours at 35°C.
7. Examine the plate using an ultraviolet lamp.

Do not look directly at the ultraviolet light, and do not leave your hand exposed to it.

Record the results of your confirmed test (Figure 52.1). (See Color Plate XII.4.) How can you tell whether coliform colonies are present? ____________________

____

Exercise 52 **Laboratory Report**

Microbes in Water: Multiple-Tube Technique

Name ______________________

Date ______________________

Lab Section ______________________

Purpose ______________________

____

____

Data

Water sample source: ______________________

Presumptive Test

Results	Number of Tubes with a Positive Result		
	Inoculum		
	10 ml	1 ml	0.1 ml
Growth			
Acid			
Gas			
Number of tubes positive for all 3			

Possible coliforms present? ______________ MPN: ______________

Confirmed Test

Tube: ______________ Growth: ______________________

Appearance of colonies with white light: ______________________

____

Appearance of colonies with UV light: ______________________

Are coliforms present? ______________

Data from water samples tested by other students:

Sample	Coliforms Present	MPN

Conclusions

What is the MPN of your water sample? ________________ per ________________ ml

Questions

1. Could the water have a high concentration of the pathogenic bacterium *Vibrio cholerae* and give negative results in the multiple-tube technique? Briefly explain. ____

2. Why are coliforms used as indicator organisms if they are not usually pathogens? ______________________

3. Why isn't a pH indicator needed in the lactose broth fermentation tubes? ____________________________

4. If coliforms are found in a water sample, the IMViC tests will help determine whether the coliforms are of fecal origin and not from plants or soil. What IMViC results would indicate the presence of fecal coliforms? ______

Critical Thinking

1. Why didn't we inoculate MUG agar directly and bypass lactose broth?

2. The following table lists the etiologies of illness associated with drinking and recreational waters in 1998.

Agent	Number of Cases
Cryptosporidium hominis	936
Pseudomonas aeruginosa dermatitis	393
Norovirus	106
Escherichia coli O157:H7	9
Giardia intestinalis	2

Which of the water samples would have shown a high coliform count?

Which diseases could have been prevented by chlorinating the water?

3. Use the data shown below to explain how waterborne diseases can best be eliminated.

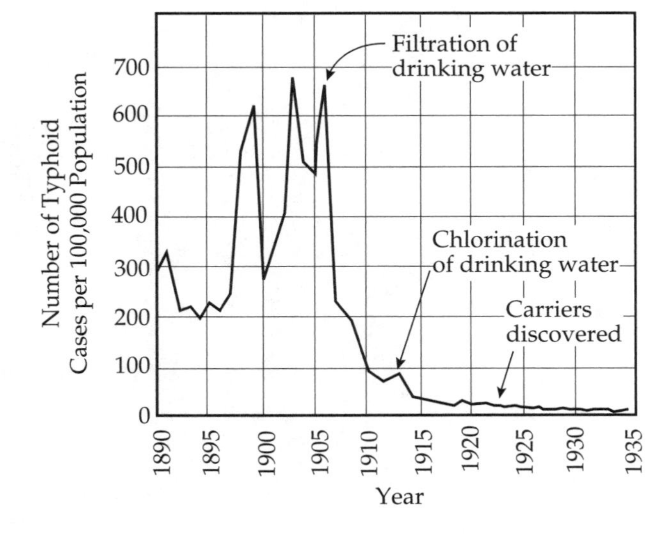

Exercise 53

Microbes in Water: Membrane Filter Technique

Objectives

After completing this exercise, you should be able to:

1. Explain the principle of the membrane filter technique.
2. Perform a coliform count using the membrane filter technique.

Background

Fecal contamination of water can be determined by the number of coliforms present in a water sample, by the multiple-tube technique (Exercise 52). Coliforms can also be detected by the membrane filter technique.

In the **membrane filter technique,** water is drawn through a thin filter (see Appendix F). Filters with a variety of pore sizes are available. Pores of 0.45 μm are used for filtering out most bacteria. Bacteria are retained on the filter, which is then placed on a suitable nutrient medium. In field situations, nutrients are added to a thick absorbent pad on which the filter is placed. Nutrients that diffuse through the filter can be metabolized by bacteria trapped on the filter. Each bacterium that is trapped on the filter will develop into a colony. Bacterial colonies growing on the medium can be counted. When a selective or differential medium is used, desired colonies will have a distinctive appearance.

Eosin methylene blue (EMB) agar is frequently used as a selective and differential medium with the membrane filter technique. The composition of EMB agar is shown in Table 53.1. **EMB agar** is selective because the EMB dyes inhibit the growth of gram-positive organisms, allowing the growth of gram-negative bacteria. EMB is differential in that the colonies of lactose-fermenting bacteria have a pink to blue color, black center, or metallic sheen from the eosin and methylene blue (see Color Plates XII.1, XII.2, XII.3, XII.5, and XII.7).

Materials

Water sample (Bring your own from a pond or stream.)

47-mm Petri plate containing EMB agar

Sterile membrane filter apparatus

Sterile 0.45-μm filter

Blunt-end forceps

Alcohol

Table 53.1

Chemical Composition of EMB Agar

Component	Percentage
Peptone	1.0%
Lactose	0.5%
Sucrose	0.5%
Dipotassium phosphate	0.2%
Agar	1.35%
Eosin Y	0.04%
Methylene blue	0.0065%

Source: Difco & BBL Manual.

Sterile pipette or graduated cylinder, as needed

Sterile rinse water

Techniques Required

Special media, Exercise 12

Pipetting, Appendix A

Membrane filtration, Appendix F

Procedure

1. Set up the filtration equipment (see Appendix F). Remove wrappers as each piece is fitted into place. Why shouldn't all the wrappers be removed at once?

   ____

   a. Attach the filter trap to the vacuum source. What is the purpose of the filter trap? ______

      ____

   b. Place the filter holder base (with stopper) on the filtering flask. Attach the flask to the filter trap.

> ⚠ **Disinfect the forceps by burning off the alcohol. Keep the beaker of alcohol away from the flame.**

   c. Using the sterile forceps, place a filter on the filter holder. Why must the filter be centered exactly on the filter holder? ______________

      ____

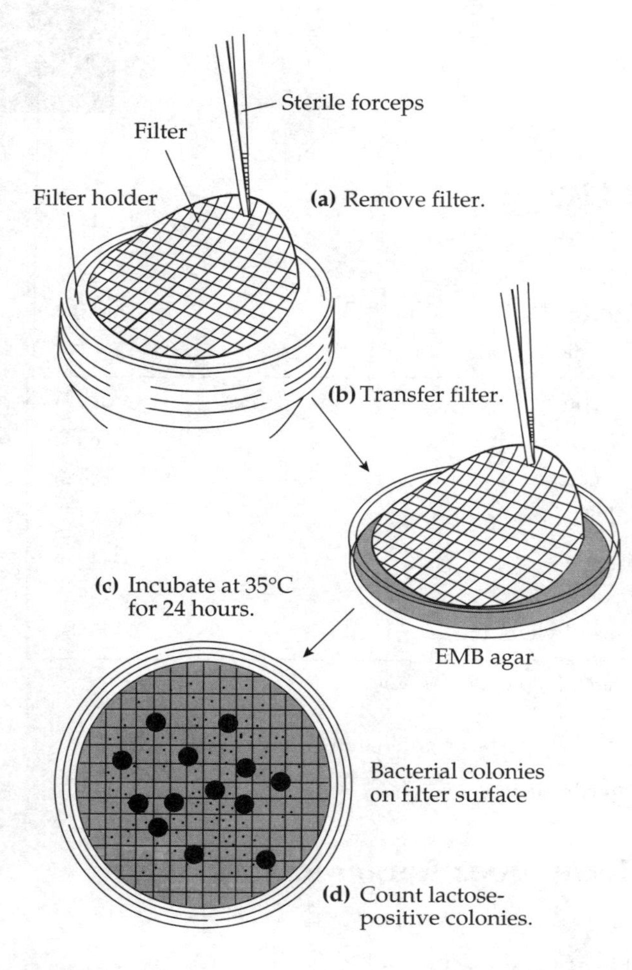

Figure 53.1

Inoculation. Using sterile forceps, remove the filter from the filter holder. Place the filter on the culture medium, gradually laying it down from one edge to the other.

   d. Set the funnel on the filter holder and fasten it in place.

2. Filter the sample.
   a. Shake the water sample well to resuspend all material, and pour or pipette a measured volume into the funnel. Your instructor will help you determine the volume. (*For samples of 10 ml or less, pour 20 ml of sterile water into the funnel first.*)
   b. Turn on the vacuum and allow the sample to pass into the filtering flask. Leave the vacuum on.
   c. Pour sterile rinse water into the funnel. (*Use the same volume as the sample.*) Allow the rinse water to go through the filter. Turn the vacuum off.
3. Inoculate the filter (Figure 53.1).
   a. Carefully remove the filter from the filter holder using sterile forceps. Why does the filter have to be "peeled" off? ______________

   ____
   b. Carefully place the filter on the culture medium. Do not bend the filter; place one edge down first, then carefully set the remainder down. Place the filter on the agar as it was in the filter holder.
4. Invert the plate and incubate it for 24 hours at 35°C.
5. Examine the plates for the presence of coliforms. On EMB, coliforms will form colored colonies, and some may have a green metallic sheen. (See Color Plates XII.1, XII.2, XII.3, XII.5, and XII.7.) Count the number of coliform colonies:

Number of coliforms per 100 ml of water =

$$100 \times \frac{\text{Number of coliform colonies}}{\text{Volume of sample filtered}}$$

Exercise 53

LABORATORY REPORT

Microbes in Water: Membrane Filter Technique

NAME ______

DATE ______

LAB SECTION ______

Purpose

______

Data

Sample tested: ______

Amount of water tested: ______

Describe the general appearance of the colonies on EMB medium: ______

Number of coliform colonies: ______

Number of coliforms/100 ml of water: ______

Use the space below for your calculations.

Data from water samples tested by other students:

Sample	Coliforms/100 ml

Conclusions

1. Which water sample(s) is (are) potable? ____________________

2. Which water sample(s) is (are) contaminated with fecal material? ____________________

Questions

1. What basic assumption is made in this technique if the number of bacteria is determined from the number of colonies?

____

____

2. Why can filtration be used to sterilize culture media? ____________________

   Air? ____________________

3. If you did the multiple-tube technique, list one advantage and one disadvantage of each method of detecting coliforms.

____

____

____

____

4. Why is the membrane filter technique useful for a sanitarian working in the field? ____________________

____

____

Critical Thinking

1. Can you determine the source of contamination (for example, human, domestic animal, or wild animal) from this test?

2. Design an experiment to detect the presence and number of *Acetobacter* in wine. Why would you want to perform such a test?

3. The following table lists the etiologies of illness associated with drinking water in 2002.

Agent	Number of Cases
Norovirus	727
Legionella spp.	80
Copper poisoning	30
Campylobacter jejuni	25
Giardia intestinalis	18
Cryptosporidium hominis	10
Escherichia coli O157:H7	2

Which of the water samples would have shown a high coliform count?

Which etiologies could have been detected by membrane filtration?

Which diseases could have been prevented by membrane filtration?

4. How could you modify the membrane filter test to detect enterococci in water?

Exercise 54

Microbes in Food: Contamination

Objectives

After completing this exercise, you should be able to:

1. Determine the approximate number of bacteria in a food sample using a standard plate count.
2. Provide reasons for monitoring the bacteriologic quality of foods.
3. Explain why the standard plate count is used in food quality control.

Background

Illness and food spoilage can result from microbial growth in foods. The sanitary control of food quality is concerned with testing foods for the presence of pathogens. During processing (grinding, washing, and packaging), food may be contaminated with soil microbes and microbiota from animals, food handlers, and machinery.

Foods are the primary vehicle responsible for the transmission of diseases of the digestive system. For this reason, they are examined for the presence of coliforms because the presence of coliforms usually indicates fecal contamination.

Standard plate counts are routinely performed on food and milk by food-processing companies and public health agencies. The **standard plate count** is used to determine the total number of viable bacteria in a food sample. The presence of large numbers of bacteria is undesirable in most foods because it increases the likelihood that pathogens will be present, and it increases the potential for food spoilage.

In a standard plate count, the number of **colony-forming units (cfu)** is determined. Each colony may arise from a group of cells rather than from one individual cell. The initial sample is diluted through serial dilutions (Appendix B) in order to obtain a small number of colonies on each plate. A known volume of the diluted sample is placed in a sterile Petri dish and melted cooled nutrient agar is poured over the inoculum. After incubation, the number of colonies is counted. Plates with between 25 and 250 colonies are suitable for counting. A plate with fewer than 25 colonies is unsuitable for counting because a single contaminant could influence the results. A plate with more than 250 colonies is extremely difficult to count.

The microbial population in the original food sample can then be calculated using the following equation:

$$\text{Colony-forming units/gram or ml of sample} = \frac{\text{Number of colonies}}{\text{Amount plated} \times \text{dilution*}}$$

A limitation of the standard plate count is that only bacteria capable of growing in the culture medium and environmental conditions provided will be counted. A medium that supports the growth of most heterotrophic bacteria is used.

Materials

Melted standard plate count or nutrient agar, cooled to 45°C

Sterile 1-ml pipettes (8)

Sterile Petri dishes (10)

Sterile weighing dish

Sterile spatula

Sterile 99-ml dilution blanks (3)

Sterile 9-ml dilution blanks (2)

Food samples

Vortex mixer

Techniques Required

Aseptic technique, Exercise 10

Pour-plate technique, Exercise 11

Pipetting, Appendix A

Serial dilution technique, Appendix B

---

*"Dilution" refers to the tube prepared by serial dilutions (Appendix B). For example, if 250 colonies were present on the $1{:}10^6$ plate, the calculation would be as follows:

$$\begin{aligned}\text{Colony-forming units/gram} &= \frac{250 \text{ colonies}}{0.1 \text{ ml} \times 10^{-5}} \\ &= 250 \times 10^5 \times 10^1 \\ &= 250{,}000{,}000 \\ &= 2.5 \times 10^8\end{aligned}$$

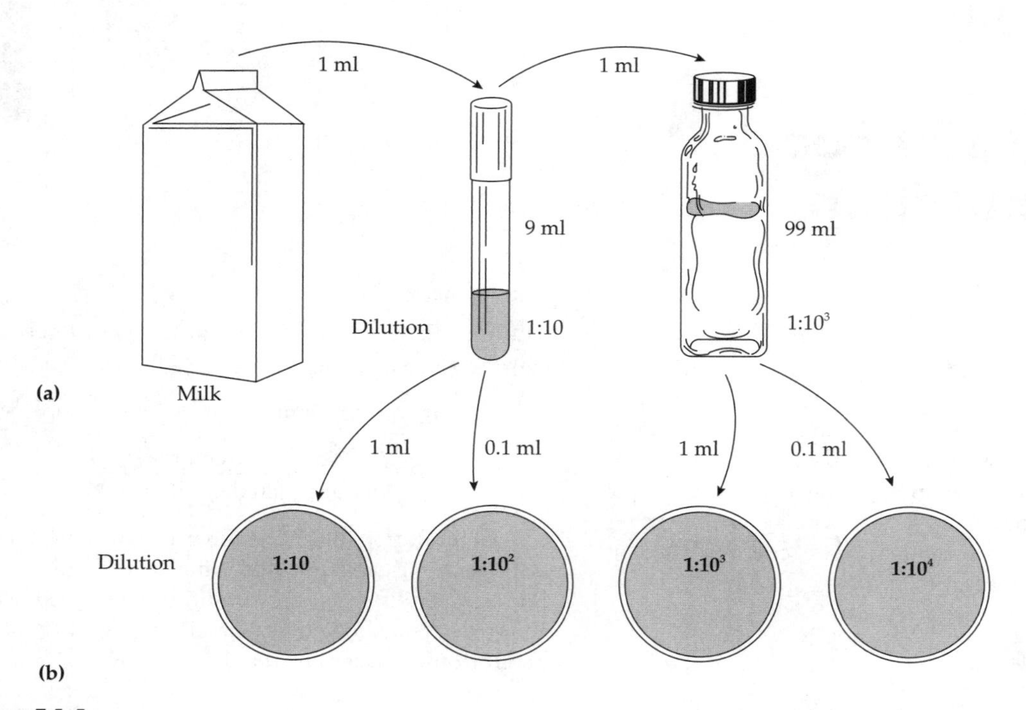

Figure 54.1
Standard plate count of milk. **(a)** Make serial dilutions of a milk sample. **(b)** Mark four Petri dishes with the dilutions.

Procedure

First Period. A: Bacteriologic Examination of Milk

1. Obtain a sample of either raw or pasteurized milk.
2. Using a sterile 1-ml pipette, aseptically transfer 1 ml of the milk into a 9-ml dilution blank; label the tube "1:10" and discard the pipette (Figure 54.1a). Mix the contents of the tube on a vortex mixer.
3. Using a sterile 1-ml pipette, aseptically transfer 1 ml of the 1:10 dilution into a 99-ml dilution blank; label the bottle "$1{:}10^3$" and discard the pipette. Shake the bottle 20 times, with your elbow resting on the table, as shown in Figure 54.2.
4. Label the bottoms of four sterile Petri dishes with the dilutions: "1:10," "$1{:}10^2$," "$1{:}10^3$," and "$1{:}10^4$" (Figure 54.1b).
5. Using a 1-ml pipette, aseptically transfer 0.1 ml of the $1{:}10^3$ dilution into the bottom of the $1{:}10^4$ dish. *Note:* 0.1 ml of the $1{:}10^3$ dilution results in a $1{:}10^4$ dilution of the original sample. Using the same pipette, transfer 1.0 ml of the $1{:}10^3$ dilution into the dish labeled $1{:}10^3$. Pipette 0.1 ml and 1.0 ml from the 1:10 dilution into the $1{:}10^2$ and 1:10 dishes, respectively, with the same pipette (Figure 54.3a). Why is it important to proceed from the highest to the lowest dilution? ________________

____

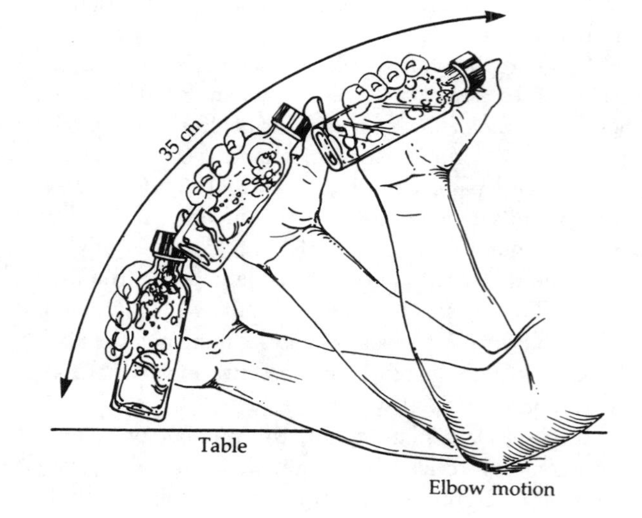

Figure 54.2
Shake the dilution bottle 20 times through a 35-cm arc.

6. Check the temperature of the water bath containing the nutrient agar. Test the temperature of the outside of the agar container with your hand. It should be "baby bottle" warm. Why? ________________

____

7. Pour the melted nutrient agar into one of the dishes (to about one-third full) (Figure 54.3b). Cover the plate and swirl it gently (Figure 54.3c) to distribute

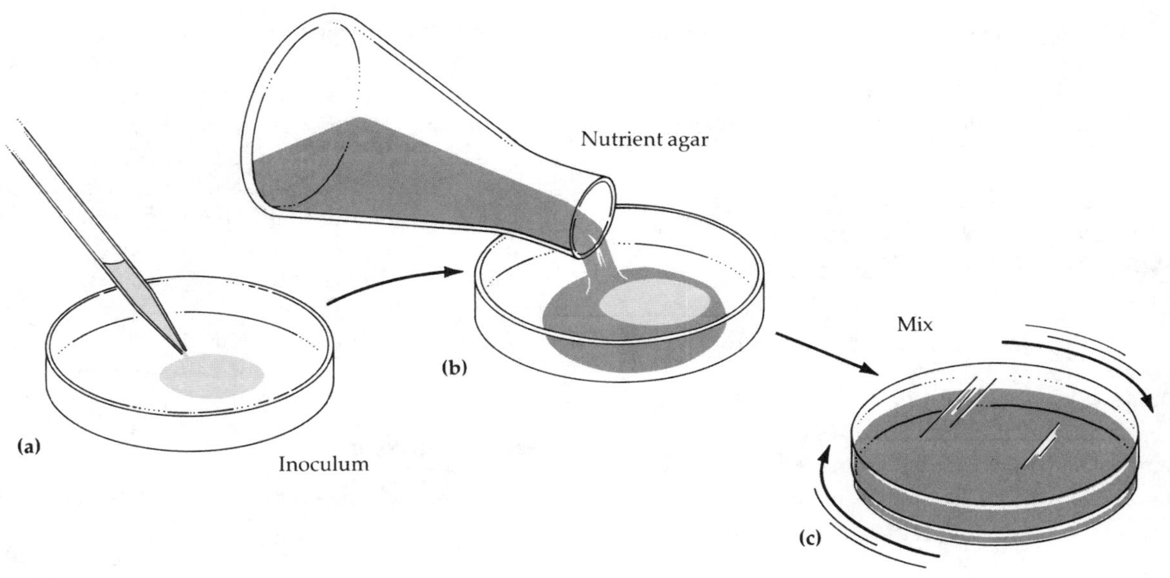

Figure 54.3

Pour plate. **(a)** Pipette inoculum into a Petri dish. **(b)** Add liquefied nutrient agar, and **(c)** mix the agar with the inoculum by gently swirling the plate.

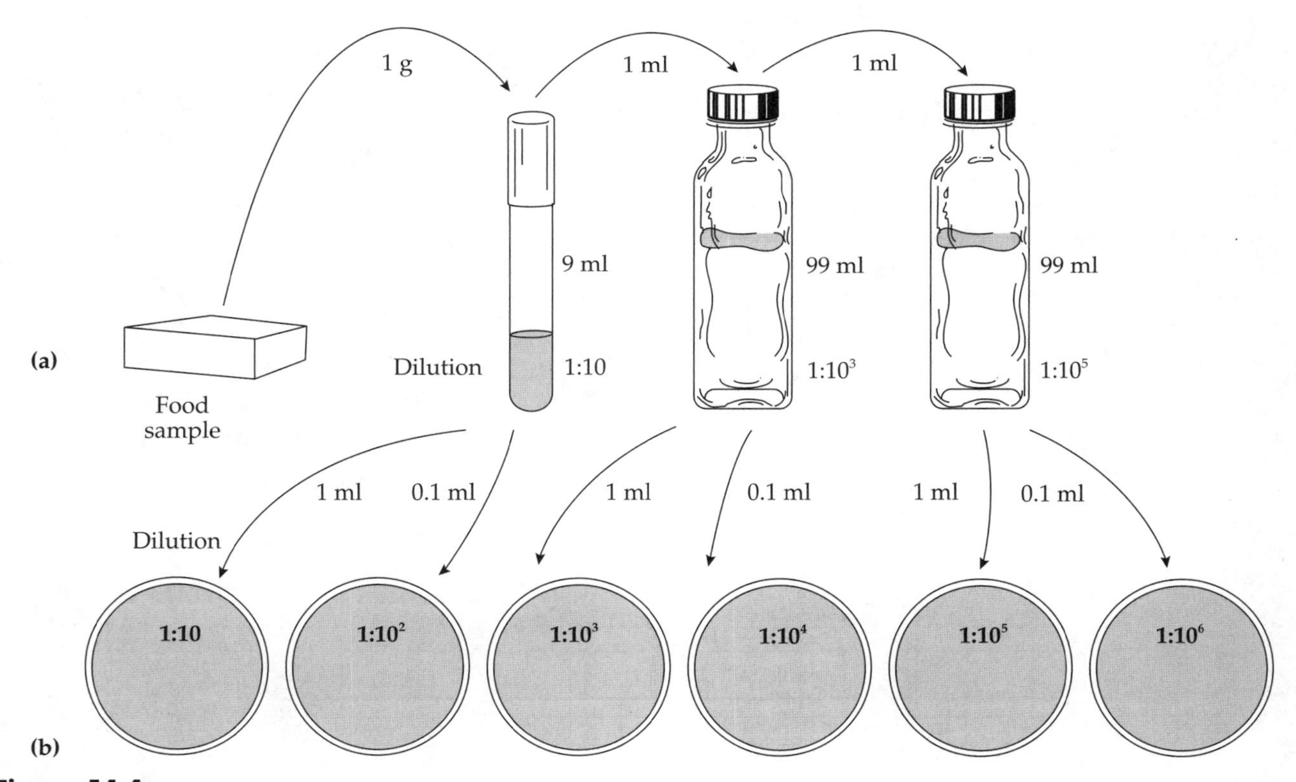

Figure 54.4

Standard plate count on food. **(a)** Prepare serial dilutions of a food sample. **(b)** Label six Petri dishes with the dilutions to be plated.

the milk sample evenly through the agar. Continue until all the plates are poured.
8. When each plate has solidified, invert it, and incubate all plates at 35°C for 24 to 48 hours.

First Period. B: Bacteriologic Examination of Hamburger or Hot Dog and Frozen Vegetables

1. Weigh 1 g of raw hamburger (or hot dog) or frozen, thawed vegetables by using the sterile spatula and weighing dish.
2. Transfer the 1 g of food into a 9-ml dilution blank; label the tube "1:10" (Figure 54.4a). Mix the contents on a vortex mixer. Mixing will shake bacteria off the food and put them in suspension.
3. Using a sterile 1-ml pipette, aseptically transfer 1 ml of the 1:10 dilution into a 99-ml dilution blank; label the bottle "$1{:}10^3$" and discard the pipette. Shake the bottle 20 times, with your elbow resting on the table, as shown in Figure 54.2. Make a $1{:}10^5$ dilution using another 99-ml dilution blank. Shake the bottle as before.
4. Label the bottoms of six sterile Petri dishes with the dilutions: " 1:10," "$1{:}10^2$," "$1{:}10^3$," "$1{:}10^4$," "$1{:}10^5$," and "$1{:}10^6$" (Figure 54.4b).
5. Using a 1-ml pipette, aseptically transfer 0.1 ml of the $1{:}10^5$ dilution into the $1{:}10^6$ dish. *Note:* 0.1 ml of a $1{:}10^5$ dilution results in a $1{:}10^6$ dilution of the original sample. Using the same pipette, repeat this procedure with the $1{:}10^3$ dilution and the 1:10 dilution until all the dishes have been inoculated (Figure 54.3a). Why can the same pipette be used for each transfer? ____________________

____

____

6. Check the temperature of the water bath containing the nutrient agar. Test the temperature of the outside of the agar container with your hand. It should be "baby bottle" warm. Why? ____________________

____

7. Pour the melted nutrient agar into one of the dishes (to about one-third full) (Figure 54.3b). Cover the plate and swirl it gently (Figure 54.3c) to distribute the sample through the agar evenly. Continue until all the plates are poured.
8. When each plate has solidified, invert it, and incubate all plates at 35°C for 24 to 48 hours.

Second Period

1. Arrange each plate in order from lowest to highest dilution.
2. Select the plate with 25 to 250 colonies. Record data for plates with fewer than 25 colonies as *too few to count (TFTC)* and those with more than 250 colonies, *too numerous to count (TNTC)*.
3. Count the number of colonies on the plate selected.
4. Calculate the number of bacteria in the original food. For example, if 129 colonies were counted on a $1{:}10^3$ dilution:

$$\frac{129 \text{ colonies}}{1 \text{ ml} \times 10^{-3}} = 129{,}000$$
$$= 1.29 \times 10^5 \text{ colony-forming units/ml or gram of milk or food}$$

Exercise 54

LABORATORY REPORT

Microbes in Food: Contamination

NAME ______________________

DATE ______________________

LAB SECTION ______________________

Purpose

____

____

____

Data

Milk sample: ______________________

Dilution	Colonies per Plate
1:	
1:	
1:	
1:	

Number of colony-forming units per ml of original milk sample: ______________________

Make your calculations in the space below.

Food sample: ______________________

Dilution	Colonies per Plate
1:	
1:	
1:	
1:	
1:	
1:	

Number of colony-forming units per gram of original food: ______________________

Make your calculations in the space below.

Record data for other foods tested by other students.

Food	cfu/ml (or gram)

Questions

1. What could you do to ensure that the bacteria present in foods do not pose a health hazard? ______________

____

____

2. Why are plates with 25 to 250 colonies used for calculations? ______________

____

____

3. In a quality-control laboratory, each dilution is plated in duplicate or triplicate. Why would this increase the accuracy of a standard plate count? ______________

____

____

4. There are other techniques for counting bacteria, such as a direct microscopic count and turbidity. Why is the standard plate count preferred for food? ______________

____

____

5. Why is ground beef a better bacterial growth medium than a steak or roast? ______________

____

6. Why does repeated freezing and thawing increase bacterial growth in meat? ______________

____

Critical Thinking

1. Assume that a standard plate count indicates that a substantial number of microbes are present in the food sample. What should be done next to determine whether the food sample is a danger to the consumer?

2. Pasteurized milk is allowed 20,000 cfu/ml. How many colonies would be present if 1 ml of a 10^{-3} dilution was plated? ______________ Is 20,000 cfu/ml a health hazard? ______________

3. An outbreak of botulism associated with home-canned chili peppers killed 16 members of one family. Would a standard plate count have detected the etiologic agent in the canned chili peppers? Briefly explain.

Exercise 55

Microbes Used in the Production of Foods

Objectives

After completing this exercise, you should be able to:

1. Explain how the activities of microorganisms are used to preserve food.
2. Define fermentation.
3. Produce an enjoyable product.

Background

Microbial fermentations are used to produce a wide variety of foods. **Fermentation** means different things to different people. In industrial usage, it is any large-scale microbial process occurring with or without air. To a biochemist, it is the group of metabolic processes that release energy from a sugar or other organic molecule, do not require oxygen or an electron transport system, and use an organic molecule as a final electron acceptor. In this exercise, we will examine a lactic acid fermentation used in the production of food.

In dairy fermentations, such as yogurt production, microorganisms use lactose and produce lactic acid without using oxygen. In nondairy fermentations, such as wine production, yeast use sucrose to produce ethyl alcohol and carbon dioxide under anaerobic conditions. If oxygen is available, the yeast will grow aerobically, liberating carbon dioxide and water as metabolic end-products.

Historically, nearly every population used milk that had been fermented by selected microbes. The acids and bacteriocins (antibiotics) produced during fermentation prevented the growth of spoilage bacteria. These "sour" milks have varied from country to country depending on the source of milk, conditions of culture, and microbial "starter" used. Milk from donkeys to zebras has been used, with the Russian *kumiss* (horse milk), containing 2% alcohol, and Swedish *surmjölk* (reindeer milk) being unusual examples. The bacteria yield lactic acid, and the yeast produce ethyl alcohol. Currently, two fermented cow milk products, buttermilk and yogurt, are widely used.

Buttermilk is the fluid left after cream is churned into butter. Today, buttermilk is actually prepared by souring true buttermilk or by adding bacteria to skim milk and then flavoring it with butterflake. *Lactococcus lactis* ferments the milk, producing lactic acid (sour), and neutral fermentation products (diacetyls) are produced by *Leuconostoc*. Yogurt originated in the Balkan countries, goat milk being the primary source. Yogurt is milk in which the milk solids have been concentrated by heating. Then the milk is fermented at elevated temperatures. *Streptococcus* produces lactic acid, and *Lactobacillus* produces the flavors and aroma of yogurt.

Materials

Homogenized milk

Nonfat dry milk

Large sterile beaker

Sterilized stirring rod

Thermometer

Sterile glass test tube containing sterile water

Hot plate or ring stand and asbestos pad

Sterile 5-ml pipette

Styrofoam or paper cups with lids

Second Period

Plastic spoons

Petri plate containing Trypticase soy agar

pH paper

Gram-stain reagents

Optional: jam, jelly, honey, and so on

Culture

Commercial yogurt *or*
Streptococcus thermophilus and *Lactobacillus delbrueckii*

Techniques Required

Gram staining, Exercise 5

Plate streaking, Exercise 11

Pipetting, Appendix A

Procedure

Be sure all glassware is clean.

1. Add 100 ml of milk per person (in your group) to a wet beaker (wash out the beaker first with water to decrease the sticking of the milk). Put a thermometer in a sterile glass test tube containing water before placing it in the beaker.
2. Heat the milk on a hot plate or over a burner on an asbestos pad placed on a ring stand, to about 80°C for 10 to 20 minutes. Stir occasionally. Do not let it boil. Why is the milk heated? ________

   ____
3. Cover and cool the milk to about 65°C. Add 3 g of nonfat dry milk per person. Stir to dissolve. Why is dry milk added? ____________________

   ____
4. Rapidly cool the milk to about 45°C. Pour the milk equally into the cups.
5. Inoculate each cup with 1 to 2 teaspoonfuls of commercial yogurt, or 2.5 ml of *S. thermophilus* and 2.5 ml of *L. delbrueckii*. Cover and label the cups. Make a smear with the inoculum on a slide, heat-fix it, and save it for the second period.
6. Incubate the cups at 45°C for 4 to 18 hours or until they are firm (custardlike).
7. Cool the yogurt to about 5°C. Save a small amount for steps 8 through 10. Then taste it with a clean spoon. Add jam or some other flavor if you desire. Eat and enjoy!

Do not work with other bacteria or perform other exercises while eating the yogurt.

8. Determine the pH of the yogurt.
9. Make a smear next to the smear prepared previously and, after heat-fixing it, prepare a Gram stain. Record your results.
10. Streak for isolation on Trypticase soy agar. Incubate the plate, inverted, at 45°C.
11. After distinct colonies are visible, record your observations. Prepare Gram stains from each different colony.

Exercise 55

LABORATORY REPORT

Microbes Used in the Production of Foods

NAME ____________________

DATE ____________________

LAB SECTION ____________________

Purpose

____

Observations

Yogurt characteristics:

Taste: ____________________

Consistency: ____________________

Odor: ____________________

pH: ____________________

Gram-stain results:

Inoculum: ____________________

Yogurt: ____________________

Streak-plate results: ____________________

Gram stain of isolated colonies: ____________________

Questions

1. How can pathogens enter yogurt, and how can this be prevented? ____________________

2. What could cause an inferior product in a microbial fermentation process? ____________________

3. How are microbial fermentations used to preserve foods? ______________________________

4. When a Gram stain has been performed on the yogurt, what result would indicate that the yogurt is contaminated? ______________________________

Critical Thinking

1. What was the source of the bacteria and yeast originally used in dairy product fermentations and breads?

2. Describe how you would show that *Streptococcus* and *Lactobacillus* are in commercial yogurt.

3. Yogurt has been used as a probiotic for the gastrointestinal tract after extensive antibiotic therapy or surgery. What is the rationale for this use of yogurt?

Exercise 56

Microbes in Soil: The Nitrogen and Sulfur Cycles

Soil is the placenta of life.

PETER FARB

Objectives

After completing this exercise, you should be able to:

1. Diagram the nitrogen cycle, showing the chemical changes that occur at each step.
2. Differentiate between symbiotic and nonsymbiotic nitrogen fixation.
3. Diagram the sulfur cycle as it occurs in a Winogradsky column.
4. Explain the importance of the nitrogen and sulfur cycles.

Background

One aspect of soil microbiology that has been studied extensively is the nitrogen cycle. All organisms need nitrogen for the synthesis of proteins, nucleic acids, and other nitrogen-containing compounds. The recycling of nitrogen by different organisms is called the **nitrogen cycle** (Figure 56.1). Microbes play a fundamental, irreplaceable role in the nitrogen cycle by participating in many different metabolic reactions involving nitrogen-containing compounds. When plants, animals, and microorganisms die, microbes decompose them by proteolysis and ammonification.

Proteolysis is the hydrolysis of proteins to form amino acids. **Ammonification** liberates ammonia by deamination of amino acids or catabolism of urea to ammonia (Figure 15.3). In most soil, ammonia dissolves in water to form ammonium ions:

$$\underset{\text{Ammonia}}{NH_3} + \underset{\text{Water}}{H_2O} \rightarrow \underset{\text{Ammonium hydroxide}}{NH_4OH} \rightarrow \underset{\text{Ammonium ion}}{NH_4^-} + \underset{\text{Hydroxyl ion}}{OH^-}$$

Some of the ammonium ions are used directly by plants and bacteria for the synthesis of amino acids.

The next sequence of the nitrogen cycle is the oxidation of ammonium ions in **nitrification.** Two genera of soil bacteria are capable of oxidizing ammonium ions in the two successive stages shown as follows:

$$\underset{\text{Ammonium ions}}{2NH_4^+} + \underset{\text{Oxygen}}{2O_2} \xrightarrow{\textit{Nitrosomonas}} \underset{\text{Nitrite ions}}{2NO_2^-}$$

$$\underset{\text{Nitrite ions}}{2NO_2^-} + \underset{\text{Oxygen}}{O_2} \xrightarrow{\textit{Nitrobacter}} \underset{\text{Nitrate ions}}{2NO_3^-}$$

These reactions are used to generate energy (ATP) for the cells. Nitrifying bacteria are chemoautotrophs, and many are inhibited by organic matter. Nitrates are an important source of nitrogen for plants.

Denitrifying bacteria reduce nitrates and remove them from the nitrogen cycle. **Denitrification** is the reduction of nitrates to nitrogen gas. This conversion may be represented as follows:

$$\underset{\text{Nitrate ion}}{NO_3^-} \rightarrow \underset{\text{Nitrite ion}}{NO_2^-} \rightarrow \underset{\text{Nitrous oxide}}{N_2O} \rightarrow \underset{\text{Nitrogen gas}}{N_2}$$

Denitrification is **anaerobic respiration.** Many genera of bacteria, including *Pseudomonas* and *Bacillus*, are capable of denitrification under anaerobic conditions.

Atmospheric nitrogen can be returned to the soil by the conversion of nitrogen gas into ammonia, a process called **nitrogen fixation.** Microbes possessing the nitrogenase enzyme can fix nitrogen under anaerobic conditions as follows:

$$\underset{\text{Nitrogen gas}}{N_2} + \underset{\text{Hydrogen ions}}{6H^+} + \underset{\text{Electrons}}{6e^-} \rightarrow \underset{\text{Ammonia}}{2NH_3}$$

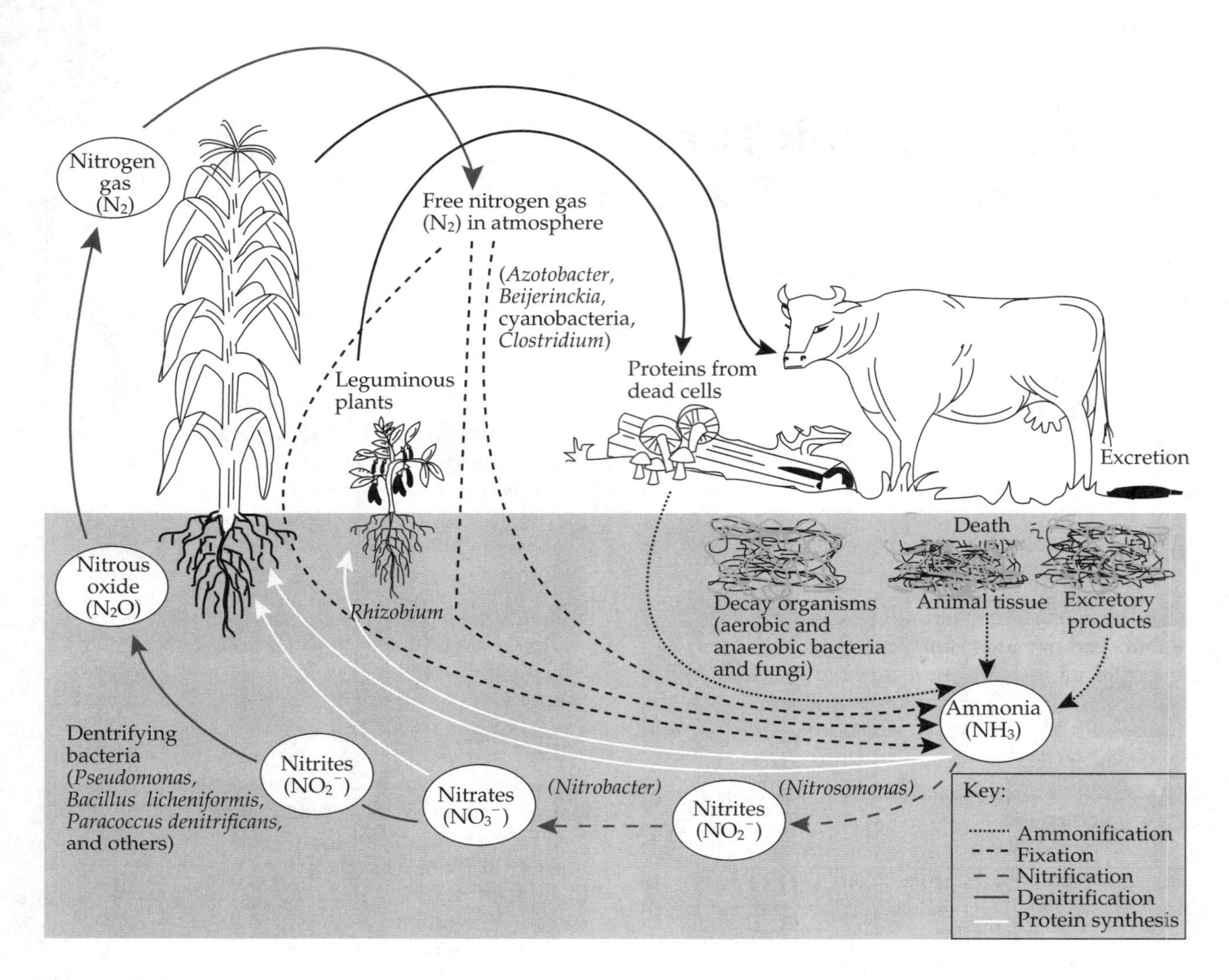

Figure 56.1
The nitrogen cycle.

Some free-living prokaryotic organisms, such as *Azotobacter*, clostridia, and cyanobacteria, can fix nitrogen. However, many of the nitrogen-fixing bacteria live in close association with the roots of grasses in the **rhizosphere,** where root hairs contact the soil.

Symbiotic bacteria serve a more important role in nitrogen fixation. One such symbiotic relationship is a mutualistic relationship between *Rhizobium* and the roots of legumes (such as soybeans, beans, peas, alfalfa, and clover) (Figure 56.2). There are thousands of legumes. Farmers grow soybeans and alfalfa to replace nitrogen in their fields. Many wild legumes are able to grow in the poor soils found in tropical rain forests or arid deserts. *Rhizobium* species are specific for the host legume that they infect. When a root hair and rhizobia make contact in the soil, a root nodule forms on the plant. The nodule provides the anaerobic environment necessary for nitrogen fixation.

Symbiotic nitrogen fixation also occurs in the roots of nonleguminous plants. The actinomycete *Frankia* forms root nodules in alders.

Any break in the nitrogen cycle would be critical to the survival of all life.

Purple and green bacteria are involved in another biogeochemical cycle, the **sulfur cycle** (Figure 56.3).

Green photosynthetic bacteria are colored by bacteriochlorophylls, although they may appear brown because of the presence of red accessory photosynthetic pigments called **carotenoids. Purple photosynthetic bacteria** appear purple or red because of large amounts of carotenoids. Purple bacteria also have bacteriochlorophylls.

Photosynthetic bacteria use **bacteriochlorophylls** to generate electrons for ATP synthesis and use sulfur, sulfur-containing compounds, hydrogen gas, or organic

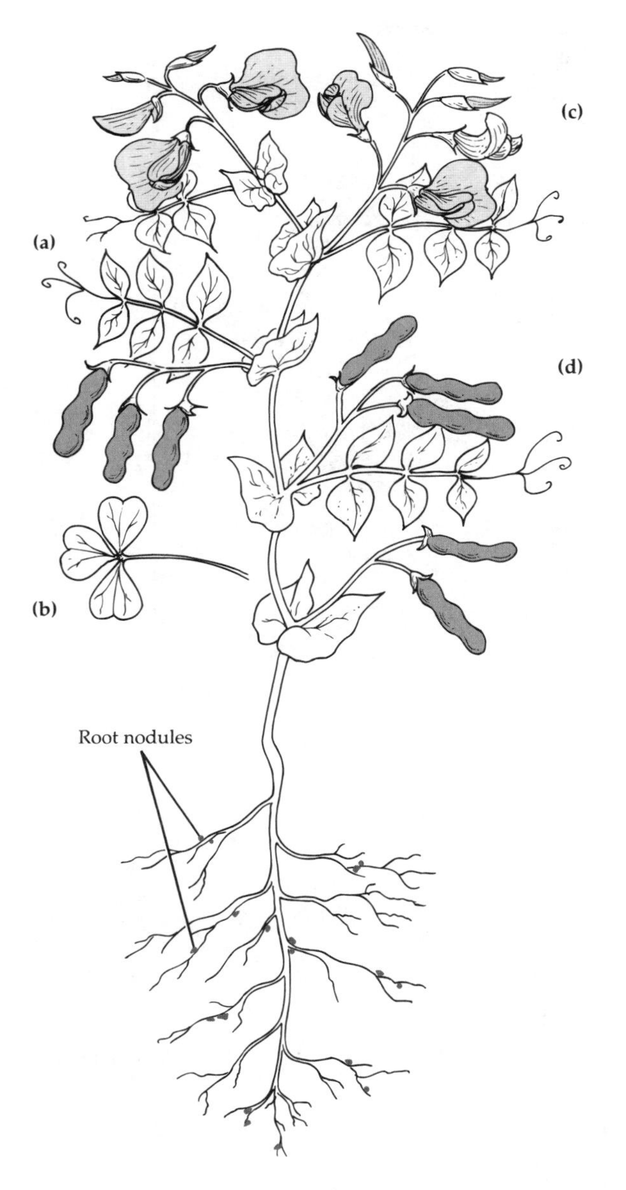

Figure 56.2

Some characteristic features of legumes. The leaves are opposite and may be **(a)** pinnately or **(b)** palmately compound. The flowers **(c)** have five asymmetrical petals, and the fruits **(d)** are pealike.

molecules as electron donors. The generalized equation for anoxygenic bacterial photosynthesis is as follows:

$$\underset{\text{Carbon dioxide}}{6CO_2} + \underset{\text{Hydrogen sulfide}}{12H_2S} \xrightarrow{\text{Bacteriochlorophyll}} \underset{\text{Sugar}}{C_6H_{12}O_6} + \underset{\text{Sulfide ion}}{12S^{2-}} + \underset{\text{Water}}{6H_2O}$$

Some photosynthetic bacteria store sulfur granules in or on their cells as a result of the production of sulfide ions. The stored sulfur can be used as an electron donor in photosynthesis, resulting in the production of sulfates.

In nature, hydrogen sulfide is produced from the reduction of sulfates in anaerobic respiration and the degradation of sulfur-containing amino acids. Sulfates can be reduced to hydrogen sulfide by several genera of **sulfate-reducing** bacteria (the best known of which is *Desulfovibrio*). Carbon dioxide used by photosynthetic bacteria is provided by the fermentation of carbohydrates in an anaerobic environment.

In this exercise, we will use an enrichment culture technique involving a habitat-simulating device called a Winogradsky column to enhance the growth of bacteria involved in the anaerobic sulfur cycle. A variety of organisms will be cultured depending on their exposure to light and the availability of oxygen (Figure 56.4). We will also investigate the steps in the nitrogen cycle.

Materials

Ammonification

Tube containing peptone broth

Soil (Bring your own.)

Second period

Nessler's reagent

Tube containing peptone broth

Ammonium hydroxide

Spot plate

Denitrification

Tubes containing nitrate-salts broth (2)

Soil (Bring your own.)

Second period

Nitrate reagents A and B

Zinc dust

Nitrogen Fixation

Petri plate containing mannitol–yeast extract agar

Methylene blue

Sterile razor blade

Legumes (Bring your own; see Figure 56.2.)

Winogradsky Column

Mud mixture (mud, $CaCO_3$, hay or paper, and $CaSO_4$)

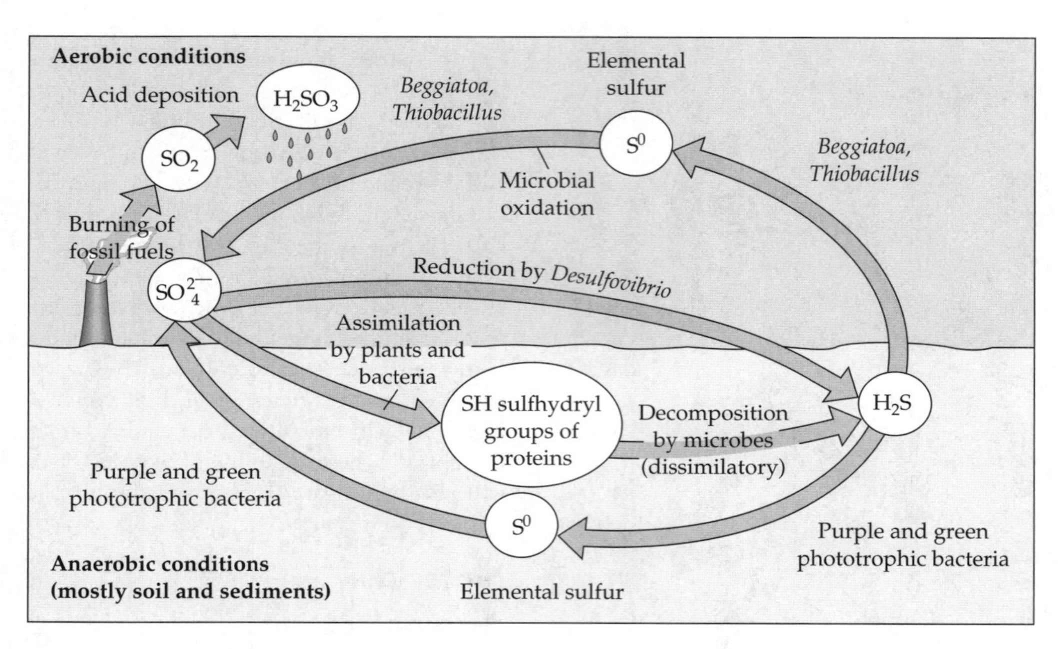

Figure 56.3

The sulfur cycle.

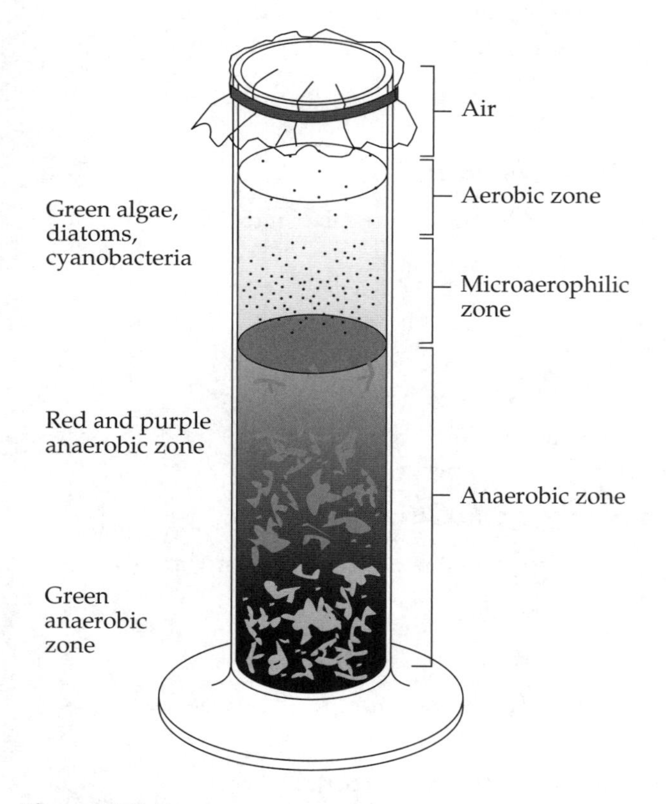

Figure 56.4

Winogradsky column.

Large test tube or graduated cylinder

Glass rod

Plain mud

Winogradsky buffer (NH_4Cl, $Na_2S{\cdot}9\ H_2O$, KH_2PO_4, K_2HPO_4)

Aluminum foil

Opaque tape

Light source

Culture

Pseudomonas aeruginosa

Demonstrations

Rhizobium inoculum used by farmers

Slides of stained root nodules

Techniques Required

Compound light microscopy, Exercise 1

Wet-mount technique, Exercise 2

Simple staining, Exercise 3

Aseptic technique, Exercise 10

Nitrate reduction test, Exercise 17

Procedure

Ammonification

1. Obtain an inoculum of soil on a wetted loop and inoculate the peptone broth. What is peptone? ___
2. Incubate the peptone broth at room temperature and test for ammonia at 2 days and at 7 days.
3. Test for ammonia by placing a drop of Nessler's reagent in a spot plate. Add a loopful of peptone broth, and mix (Figure 56.5). A yellow to brown color indicates the presence of ammonia. Compare your results to a spot to which a drop of ammonium hydroxide and Nessler's reagent have been added. Use an uninoculated tube of peptone broth as a control. Why is a control necessary? ___

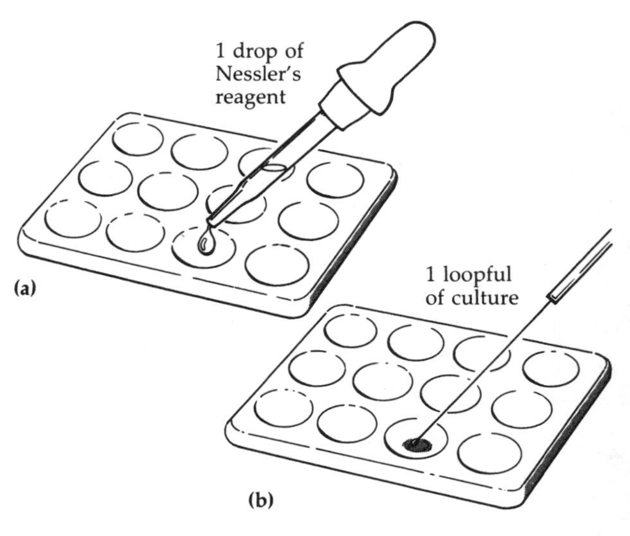

Figure 56.5

(a) Place a drop of Nessler's reagent in a well of a spot plate. **(b)** Add a loopful of the sample to be tested and mix. Observe for a color change.

Denitrification

1. Inoculate one nitrate-salts broth with soil by using a wetted loop to pick up the soil. Inoculate the other tube with *P. aeruginosa*.
2. Incubate both tubes at room temperature for 1 week.
3. Test for nitrate reduction. Remember how? Describe. ___

   Record your results.

Nitrogen Fixation

1. Cut off a nodule from a legume and wash it under tap water. (See Color Plate VI.5.) Observe.
2. Cut the nodule in half with a sterile razor blade. Observe. Crush the nodule between two slides and make a smear by rotating the slides together.
3. Streak a loopful of the crushed nodule on the mannitol–yeast extract agar. Incubate the plate at room temperature for 7 days.
4. Air-dry the slide and heat-fix it. Stain the slide for 1 minute with methylene blue. Rinse the slide and observe it under oil immersion.
5. Observe the demonstrations. What is in the farmer's inoculum? ___
6. Observe the growth on the plate, and make a simple stain of the growth with methylene blue. Compare this stain to the stain prepared from the nodule.

Winogradsky Column

1. Pack a large test tube two-thirds full with the mud mixture. Pack it with the glass stirring rod to eliminate air bubbles. Why? ___
2. Carefully pack a narrow layer of plain mud on top of the first layer.
3. Gently pour the buffer down the side of the tube, taking care not to disturb the mud surface. Fill the tube as full as possible.
4. Cover the top of the tube with foil, and place it in front of a light source. Cover the back of the cylinder with opaque tape, such as duct tape, so one portion of the mud will not be exposed to light. The instructor may assign different light sources (e.g., incandescent, fluorescent, red, or green). Do not place it too close to the bulb. Why not? ___
5. Observe the tubes at weekly intervals for 4 weeks. Buffer may need to be added to the tube. Record the appearance of colored areas. Aerobic mud will be brownish, and anaerobic mud will be black. Why? ___
6. After 4 weeks, remove the tape and prepare wet mounts from the purple or green patches. Record the microscopic appearance of the bacteria and whether sulfur granules are present (see Color Plate VI.4).

Exercise 56

LABORATORY REPORT

Microbes in Soil: The Nitrogen and Sulfur Cycles

NAME ____________________

DATE ____________________

LAB SECTION ____________________

Purpose ____________________

Data

Ammonification

Incubation	Growth	Nessler's Reagent			Ammonia Present
		Color of Inoculated Peptone	Color of NH_4OH	Color of Control	
2 days					
7 days					

Denitrification

How did you test for denitrification? ____________________

Inoculum	Growth	Gas	Nitrate Reduction	Denitrification
Soil				
Pseudomonas				

Nitrogen Fixation

Diagram of root nodule:

Diagram microscopic appearance of bacteria and cells in root nodule.

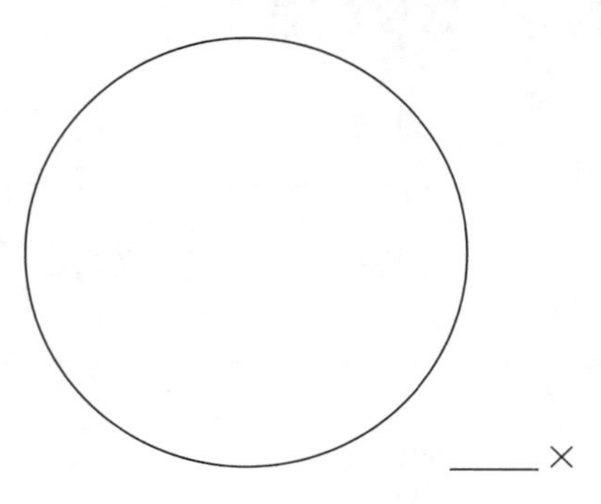

Morphology: ____________________

Diagram microscopic appearance of bacteria from culture.

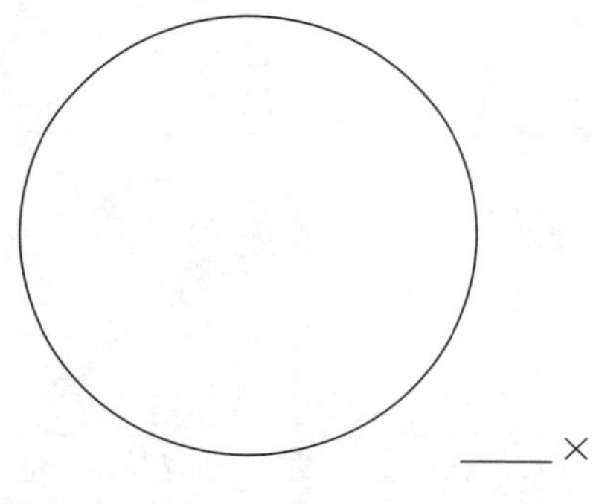

Morphology: ____________________

Demonstration slides:
Label root, nodule, and bacteria.

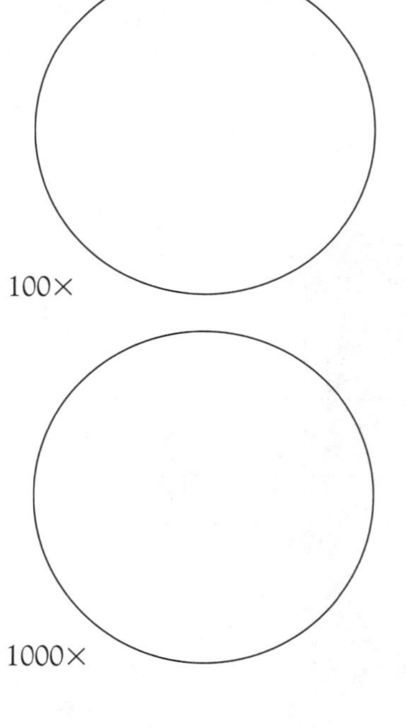

Describe colonies on the mannitol–yeast extract plate. ____

____

____

Winogradsky Column

Diagram the appearance of your enrichment at weekly intervals. Label patches of photosynthetic bacteria.

Light source: ______________________________

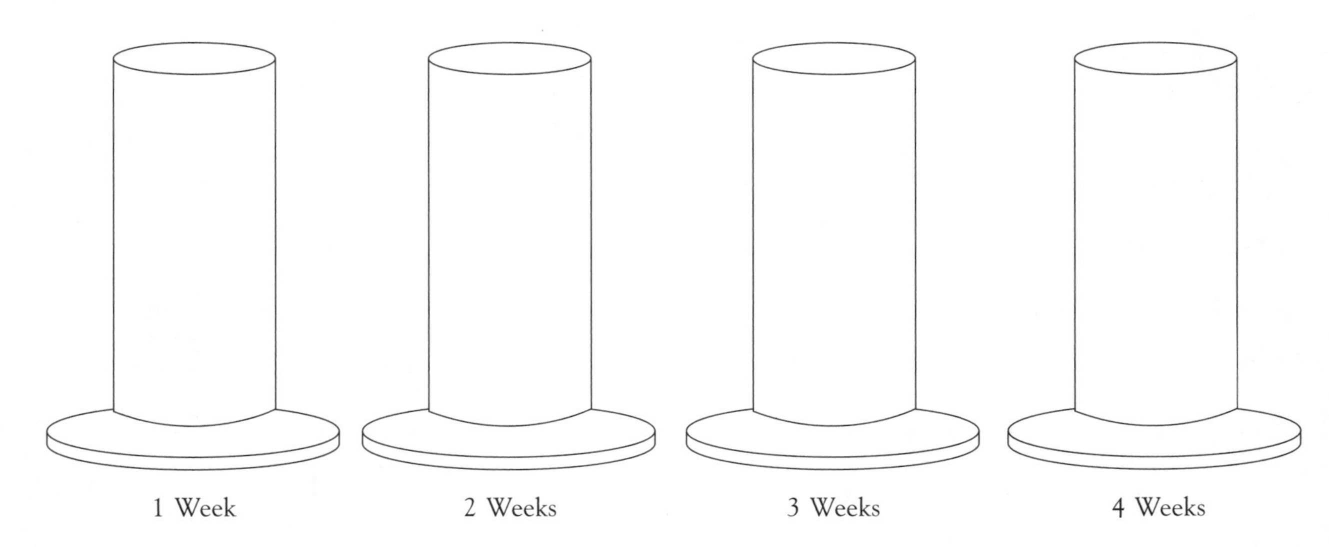

Microscopic observations:

Source	Morphology	Motility	Sulfur Granules

If other light sources were used, compare the appearance of the enrichments after 4 weeks. ______________

____

____

____

Questions

1. Show ammonification and desulfurylation of this amino acid.

$$H_2N-\underset{\underset{\displaystyle SH}{|}}{\underset{\displaystyle CH_2}{\underset{|}{CH}}}-C\begin{matrix}\nearrow O \\ \searrow OH\end{matrix}$$

Of what advantage are these processes to a bacterium? ______

2. Why can't you smell ammonia in the peptone broth tubes? ______

3. Why is denitrification a problem for farmers? ______

4. What gas is in the gas tube of a positive denitrification test? ______

5. Which step(s) in the nitrogen cycle require oxygen? ______

6. Why is the nitrogen cycle important to all forms of life? ______

7. What was the purpose of each of the following chemicals in the Winogradsky enrichment?

$CaCO_3$ ______

$Na_2S{\cdot}9H_2O$ ______

Hay or paper ______

KH_2PO_4 and K_2HPO_4 ______

$CaSO_4$ ______

NH_4Cl ______

8. Indicate the aerobic and anaerobic regions on the diagrams of your enrichment column. How can you tell?

____

9. Is there evidence of any nonphotosynthetic growth in the Winogradsky column? Explain.

____

____

Critical Thinking

1. What genetic engineering project would you propose concerning nitrogen fixation?

2. Why do the bacteria in the root nodule look different from the bacteria cultured from the nodule?

3. Design an experiment that would determine the effect of heat produced by the lights on the growth of bacteria in the Winogradsky column.

Exercise 57

EXERCISE 57

Microbes in Soil: Bioremediation

Objectives

After completing this exercise, you should be able to:

1. Define the following terms: bioremediation, lipid, hydrocarbon.
2. Demonstrate enrichment of oil-degrading bacteria.
3. Describe the degradation or decomposition of lipids.

Background

The use of bacteria to eliminate pollutants is called **bioremediation.** Unlike some forms of environmental cleanup, in which dangerous substances are removed from one place only to be dumped in another, bacterial cleanup eliminates the toxic substance and often returns a harmless or useful substance to the environment. (See Color Plates XI.1, XI.2, and XI.3.)

Many bacteria use exoenzymes such as lipase to hydrolytically decompose lipids. The lipid molecule is first broken down into glycerol and three fatty acids (Figure 57.1). Some bacteria ferment the glycerol, while others oxidize the fatty acids. In the presence of oxygen, the bacteria remove two carbons at a time in a process called *beta-oxidation*. When lipid hydrolysis occurs in decomposition of foods such as butter, it results in a rancid flavor and aroma caused by the fatty acids. Bacterial cultures are being used to degrade grease from restaurants and meat processors before disposal.

On their own, bacteria degrade large, complex molecules in petroleum too slowly to be helpful in cleaning up an oil spill. However, scientists have hit upon a simple way to speed them up. Nitrogen and phosphorous plant fertilizers can be added in a process called *bioenhancement*. This approach was used on an Alaskan beach affected by the 1989 Valdez oil spill; as a result, the beach was free of oil after 1 week.

In this exercise, we will enrich for hydrocarbon-degrading bacteria. Degradation of the hydrocarbon should result in the opaque, emulsified oil becoming small, soluble molecules, such as fatty acids and acetyl groups. (See Color Plate XI.4.)

Materials

Soil slurry (1 g of soil in 9 ml of water)

Enrichment (hydrocarbon–minimal salts broth inoculated with soil slurry and incubated)

Sterile 1-ml pipettes (6)

Propipette or pipette bulb

Sterile 99-ml dilution blanks (4)

Petri plates containing hydrocarbon–minimal salts (4)

Spreading rod

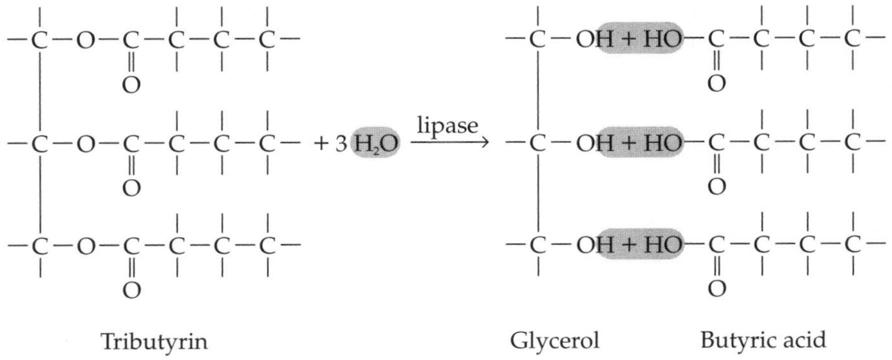

Tributyrin + 3 H_2O —lipase→ Glycerol + Butyric acid

Figure 57.1

Tributyrin digestion by hydrolysis. Lipids are composed of a glycerol and three fatty acid molecules.

Alcohol

Gram-staining reagents

Techniques Required

Compound light microscopy, Exercise 1

Gram staining, Exercise 5

Enrichment techniques, Exercise 12

Spreading rod, Exercise 27

Pipetting, Appendix A

Serial dilution technique, Appendix B

Procedure

First Period

1. Obtain a soil slurry.
2. Using a sterile 1-ml pipette, aseptically transfer 1 ml of the soil slurry to a 99-ml dilution blank; label the bottle "1:10^2" and discard the pipette. Mix the contents of the bottle by shaking it 20 times (Figure 54.2).
3. Make a 1:10^4 dilution by transferring 1 ml of the 1:10^2 dilution to another 99-ml dilution blank. Thoroughly mix the contents of the bottle as before.
4. Label the bottoms of two hydrocarbon agar plates "10^{-4}" and "10^{-5}." Which hydrocarbon is in the plates? ______________________
5. Using a 1-ml pipette, aseptically transfer 0.1 ml of the 1:10^4 dilution onto the agar of the "10^{-5}" plate.

   Using the same pipette, transfer 1 ml of the 1:10^4 dilution onto the agar of the "10^{-4}" plate. Spread the inoculum over the agar using the rod. Disinfect the spreading rod.

Disinfect the spreading rod by dipping it in alcohol. Keep the flame away from the beaker.

   Incubate the plate at room temperature for 24 to 48 hours.
6. Obtain an enrichment of the same hydrocarbon used in the plates in step 4. What is the hydrocarbon? ______________________
7. Repeat steps 2 through 5 to inoculate these plates from the enrichment.

Second Period

1. Observe and describe the growth on the plates. Count the number of hydrocarbon-degrading colonies (Color Plate XI.4). How will you identify them? ______________________

   Calculate the number of hydrocarbon-degrading bacteria using the following formula:

$$\frac{\text{Number of colonies}}{\text{Amount plated} \times \text{dilution}} = \text{cfu/gram of soil}$$

2. Prepare a Gram stain of some of the oil-degrading colonies.

Exercise 57

LABORATORY REPORT

Microbes in Soil: Bioremediation

NAME ____________________

DATE ____________________

LAB SECTION ____________________

Purpose

____

Data

Before enrichment (soil slurry):

Color Description	Oil Degrader	Morphology and Gram Reaction	Number of This Type

Number of hydrocarbon-degrading cfu/gram of soil: ____________________

After enrichment:

Color Description	Oil Degrader	Morphology and Gram Reaction	Number of This Type

Number of hydrocarbon-degrading cfu/gram of soil: ____________________

Conclusions

1. Did you isolate any oil-degrading bacteria? How can you tell? ______________________________

____

____

2. Compare the results before and after enrichment. ______________________________

____

____

Questions

1. What is an enrichment? ______________________________

2. Besides clearing, how else could you determine whether lipids had been hydrolyzed? ______________________________

____

____

3. The enrichment broth consisted of a hydrocarbon and minimal salts. What was the purpose of the hydrocarbon?

____

Of the minimal salts? ______________________________

____

Critical Thinking

1. Show beta-oxidation of the following hydrocarbon:

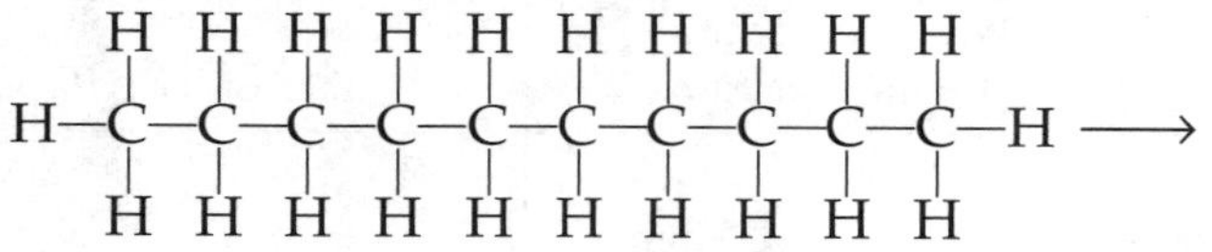

```
   H  H  H  H  H  H  H  H  H  H
   |  |  |  |  |  |  |  |  |  |
H—C—C—C—C—C—C—C—C—C—C—H ⟶
   |  |  |  |  |  |  |  |  |  |
   H  H  H  H  H  H  H  H  H  H
```

2. Here are the partial formulas of two detergents that have been manufactured. Which of these would be readily degraded by bacteria? Why?

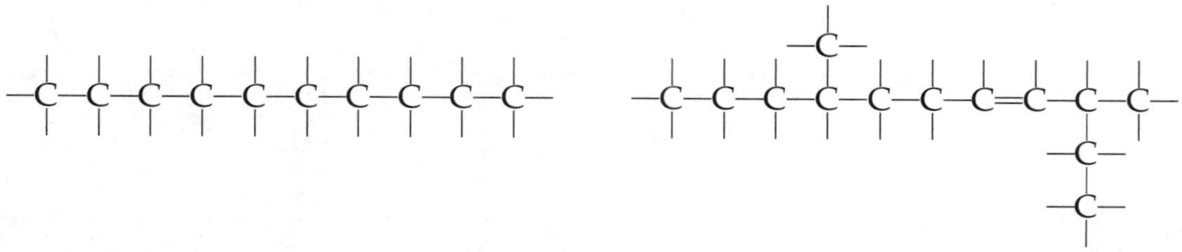

```
                                            |
                                           —C—
 |  |  |  |  |  |  |  |  |  |       |  |  |  |  |  |        |  |
—C—C—C—C—C—C—C—C—C—C—     —C—C—C—C—C—C—C=C—C—C—
 |  |  |  |  |  |  |  |  |  |       |  |  |  |  |  |        |  |
                                                            —C—
                                                             |
                                                            —C—
                                                             |
```

3. How could the oil-degrading bacteria you grew be used commercially in a detergent or drain cleaner?

4. Design an experiment to isolate mercury-utilizing bacteria from soil.

Appendices

APPENDICES

Appendix A

Pipetting

Pipettes are used for measuring small volumes of fluid. They are usually coded according to the total volume and graduation units (Figure A.1). To fill a pipette, use a bulb or other mechanical device, as shown in Figure A.2.

Never use your mouth to fill a pipette.

Draw the desired amount of fluid into the pipette. Read the volume at the bottom of the meniscus (Figure A.3). Fill the pipette to above the zero mark, and then allow it to drain just to the zero. The desired amount can then be dispensed.

Microbiologists use two types of pipettes. The **serological pipette** is meant to be emptied to deliver the total volume (Figure A.4a). Note that the graduations stop above the tip. The **measuring pipette** delivers the volume read on the graduations (Figure A.4b). This pipette is not emptied, but the flow must be stopped when the meniscus reaches the desired level.

Aseptic use of a pipette is often required in microbiology. Bring the entire closed pipette container to your work area. Lay down the canister, as shown in Figure A.5. If the pipettes are wrapped in paper, open the wrapper at the end opposite the delivery end; in a canister, the delivery end will be at the bottom of the canister. Do not touch the delivery end of a sterile pipette. Remove a pipette, attach the pipette aspirator, and with your other hand, pick up the sample to be pipetted. Remove the cap from the sample with the little finger of the hand holding the pipette. Fill the pipette, and replace the cap on the sample.

After pipetting, place the contaminated pipette in the appropriate container of disinfectant. If it is a disposable pipette, discard it in a biohazard bag or any container designated for biohazards.

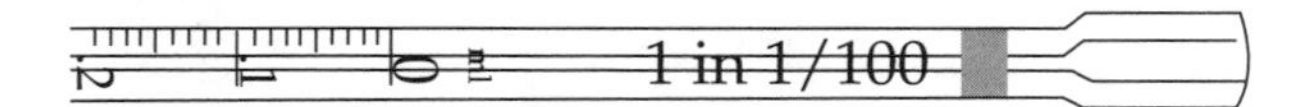

Figure A.1

This pipette holds a total volume of 1 ml when filled to the zero mark. It is graduated in 0.01-ml units.

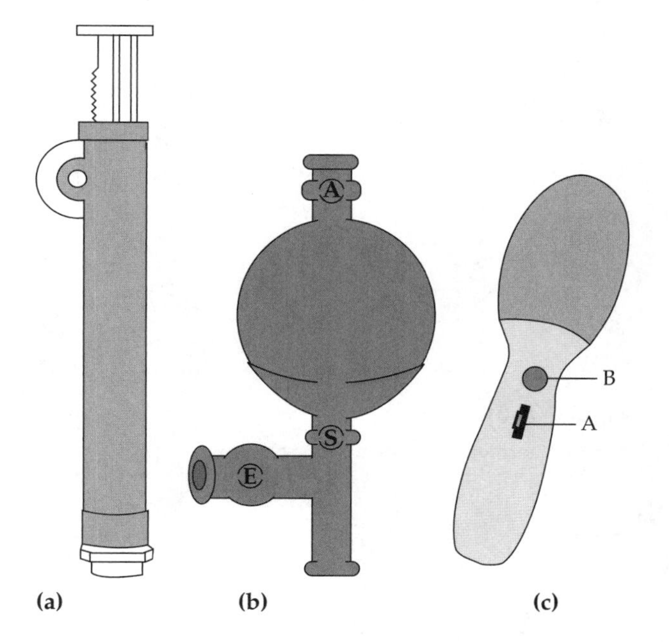

Figure A.2

Three types of pipette aspirators. **(a)** Attach this plastic pump to the pipette, and turn the wheel to draw fluid into the pipette. Turning the wheel in the other direction will release the fluid. **(b)** Insert a pipette into this bulb. While pressing the A valve, squeeze the bulb, and it will remain collapsed. To draw fluid into the pipette, press the S valve; to release fluid, press the E valve. **(c)** Insert a pipette into the stem. Raise the lever **(A)** to draw fluid up; lower the lever to release the fluid. Pushing the button **(B)** will release the last drop.

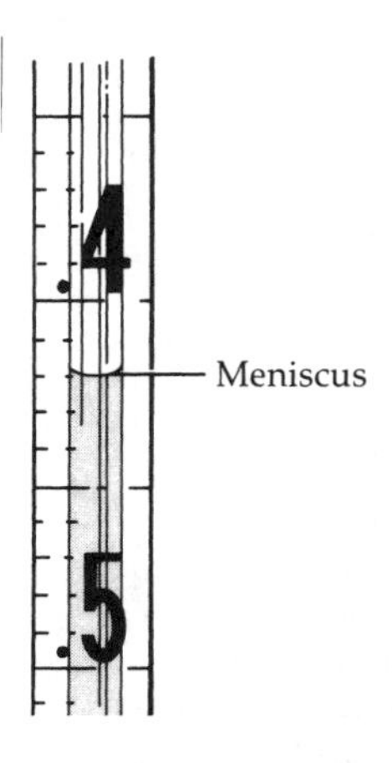

Figure A.3

Read the fluid volume at the lowest level of the meniscus.

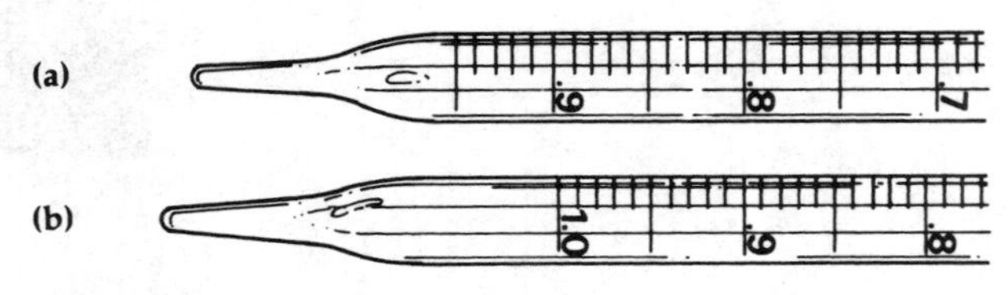

Figure A.4
(**a**) A serological pipette. (**b**) A measuring pipette.

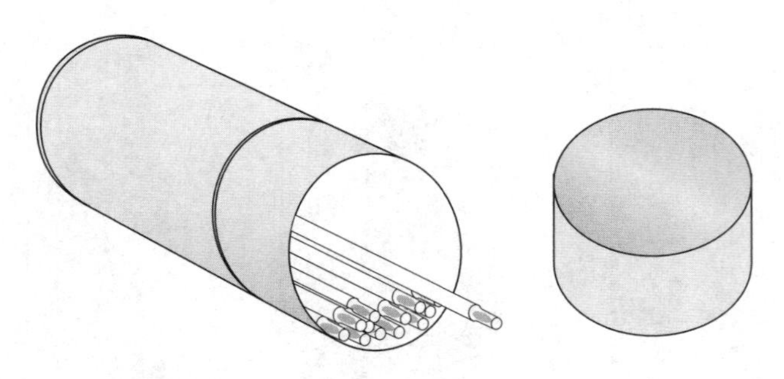

Figure A.5
Proper placement of a pipette canister.

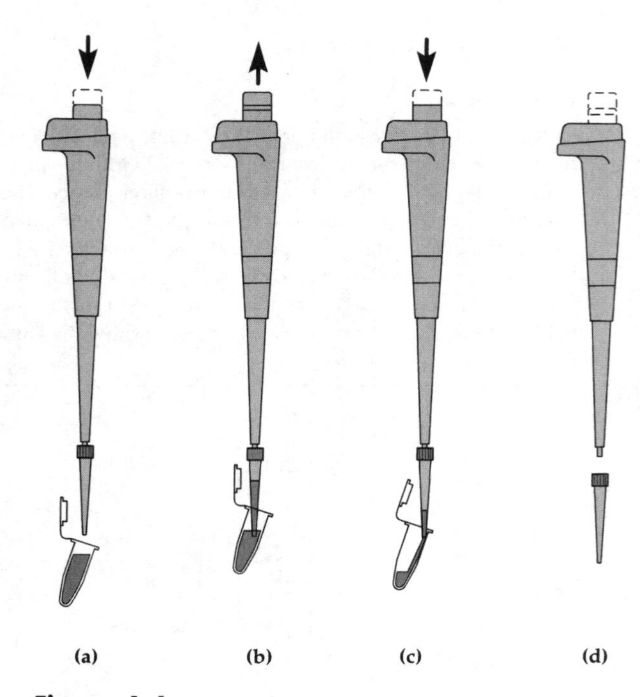

Figure A.6
Generalized use of a micropipette. (**a**) Depress the control button, and insert the tip into the liquid. (**b**) Smoothly release the button to allow the liquid to enter the tip. (**c**) Place the tip against the inside of the receiving tube, and depress the button. (**d**) Eject the used tip by pressing the eject button or by pressing the control button to the final stop.

Micropipettes are used to measure volumes of less than 1 ml. The design and operation of micropipettes vary according to the manufacturer. Most use a disposable tip to hold the fluid, as shown in Figure A.6. To use a micropipette, select one that holds the volume you need. The micropipette is labeled with its range—for example, 0.5–10 μl, 1–10 μl, 10–100 μl, 100–1000 μl. Set the desired volume on the digital display (Figure A.7a) by turning the control knob (Figure A.7b).

Questions

1. Which micropipette would you use to measure 15 μl?

____

2. If this is the display for the micropipette you chose for question 1, write in the 1 and 5 to show 15 μl.

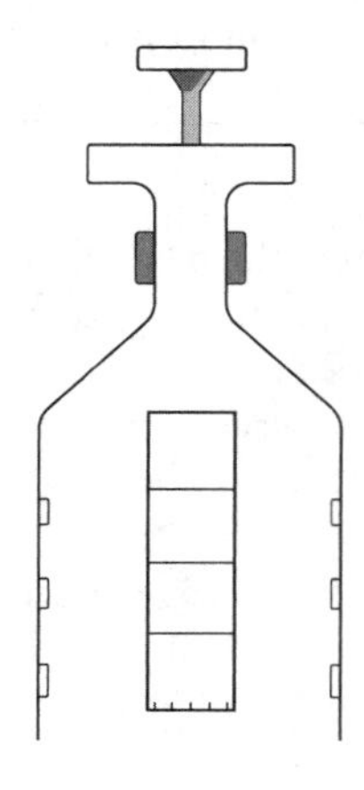

3. How do you eject the tip on your micropipette?

____

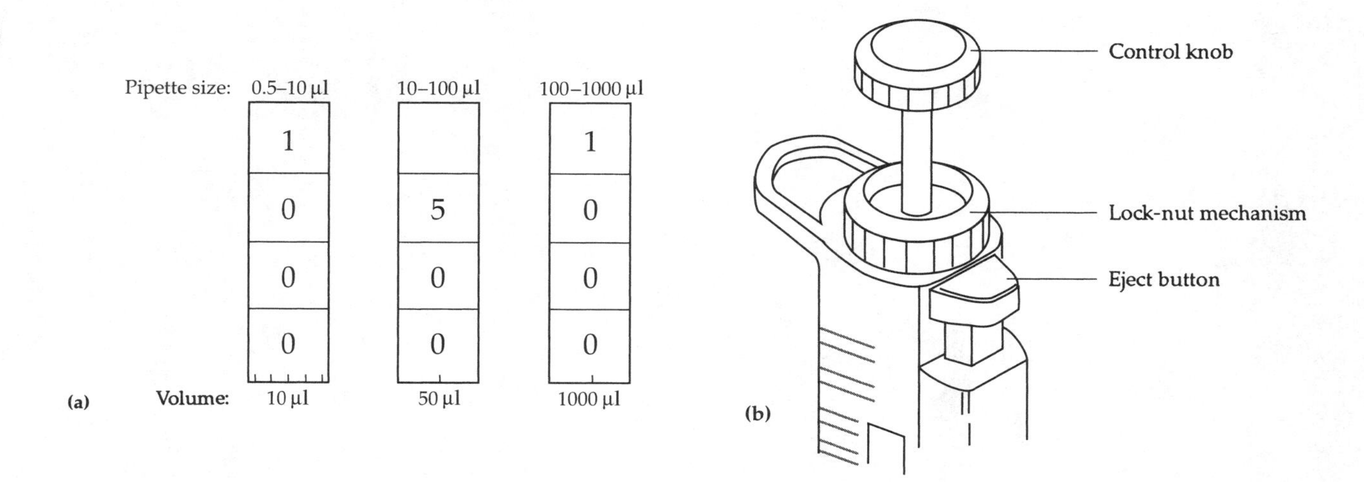

Figure A.7

Adjusting a micropipette. **(a)** Reading the display. **(b)** Turning the control knob on this micropipette changes the volume picked up when the control knob is depressed.

Appendix B

APPENDIX B

Dilution Techniques and Calculations*

Bacteria, under good growing conditions, will multiply into such large populations that it is often necessary to dilute them to isolate single colonies or to obtain estimates of their numbers. This requires mixing a small, accurately measured sample with a large volume of sterile water or saline, which is called the **diluent** or **dilution blank.** Accurate dilutions of a sample are obtained through the use of pipettes. For convenience, dilutions are usually made in multiples of 10.

A single dilution is calculated as follows:

$$\text{Dilution} = \frac{\text{Volume of the sample}}{\text{Total (volume of the sample + the diluent)}}$$

For example, the dilution of 1 ml into 9 ml equals

$$\frac{1}{1+9}, \text{ which is } \frac{1}{10} \text{ and is written 1:10}$$

The same formula applies for all dilutions, regardless of the volumes. A dilution of 0.1 ml into 0.9 ml equals

$$\frac{0.1}{0.1+0.9}, \text{ which is } \frac{0.1}{1} = \frac{1}{10} = 1{:}10$$

*Adapted from C. W. Brady. "Dilutions and Dilution Calculations." Unpublished paper. Whitewater, WI: University of Wisconsin, n.d.

A dilution of 0.5 ml into 4.5 ml equals

$$\frac{0.5}{0.5+4.5}, \text{ which is } \frac{0.5}{5.0} = \frac{1}{10} = 1{:}10$$

Experience has shown that better accuracy is obtained with very large dilutions if the total dilution is made out of a series of smaller dilutions rather than one large dilution. This series is called a **serial dilution,** and the total dilution is the product of each dilution in the series. For example, if 1 ml is diluted with 9 ml, and then 1 ml of that dilution is put into a second 9-ml diluent, the final dilution will be

$$\frac{1}{10} \times \frac{1}{10} = \frac{1}{100} \text{ or } 1{:}100$$

To facilitate calculations, the dilution is written in exponential notation. In the example above, the final dilution 1:100 would be written 10^{-2}. Remember,

$$1{:}100 = \frac{1}{100} = 0.01 = 10^{-2}$$

(See the section Exponents, Exponential Notation, and Logarithms, on page 418.) A serial dilution is illustrated in Figure B.1.

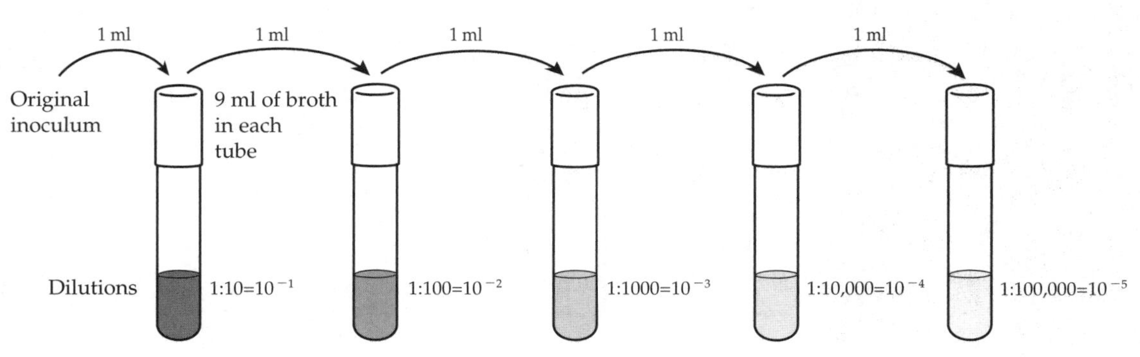

Figure B.1
A 1-ml sample from the first tube will contain 1/10 the number of cells present in 1 ml of the original sample. A 1-ml sample from the last tube will contain 1/100,000 the number of cells present in 1 ml of the original sample.

Twofold dilutions are commonly used to dilute patient's serum to measure antibodies. The same formula applies: a dilution of 100 μl of sample into 100 μl of saline equals

$$\frac{100}{100 + 100}, \text{ which is } \frac{100}{200} = \frac{1}{2} = 1{:}2$$

If 100 μl of this 1:2 dilution is put in 100 μl of saline, the final dilution is

$$\frac{1}{2} \times \frac{1}{2} = \frac{1}{4}$$

Procedure

1. Aseptically pipette 1 ml of sample into a dilution blank.
   a. If the dilution is into a tube, mix the contents on a vortex mixer or by rolling the tube back and forth between your hands.
   b. If the dilution is into a 99-ml blank, hold the cap in place with your index finger and shake the bottle up and down through a 35-cm arc (see Figure 54.2).
2. It is necessary to use a fresh pipette for each dilution in a series, but it is permissible to use the same pipette to remove several samples from the same bottle, as when plating out samples from a series of dilutions.

Problems

Practice calculating serial dilutions using the following problems.

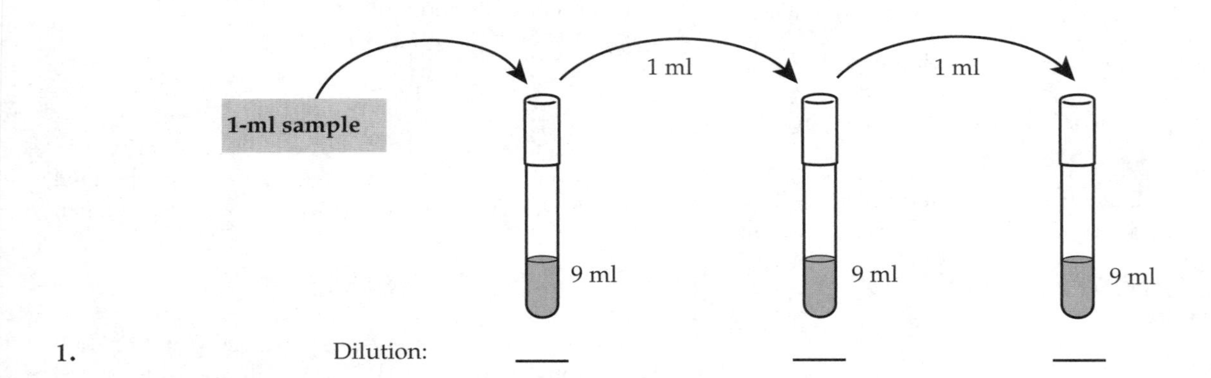

1. Dilution: ____ ____ ____

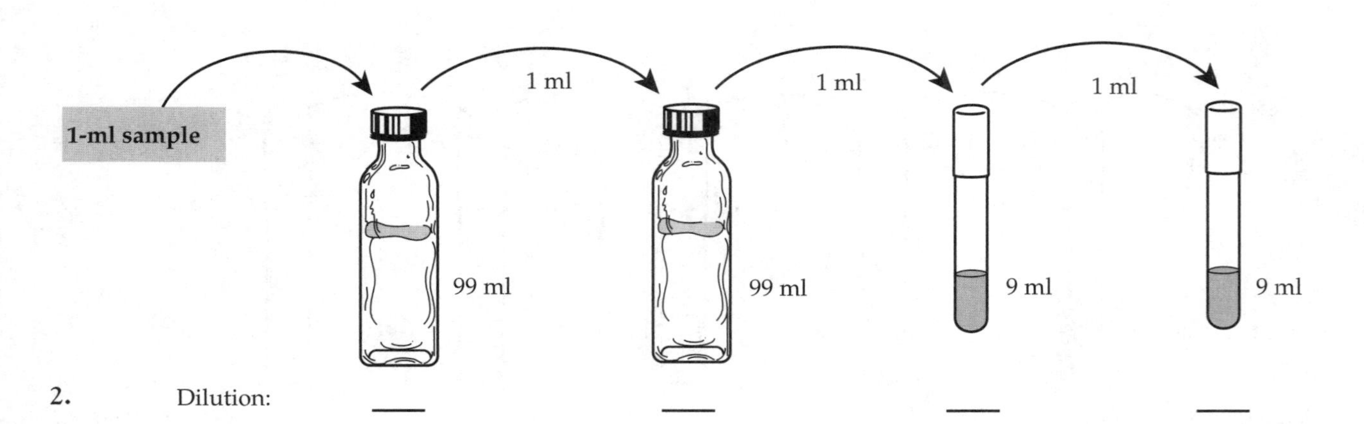

2. Dilution: ____ ____ ____ ____

3. Design a serial dilution to achieve a final dilution of 10^{-8}.

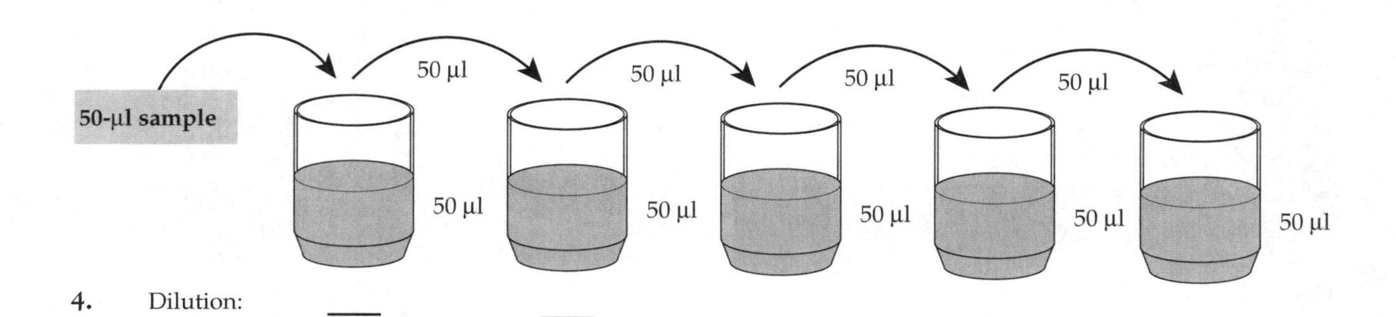

4. Dilution: ___ ___ ___ ___ ___

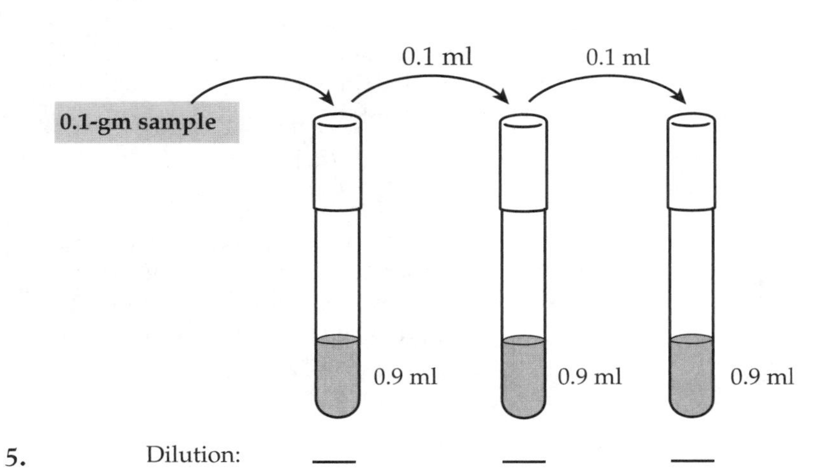

5. Dilution: ___ ___ ___

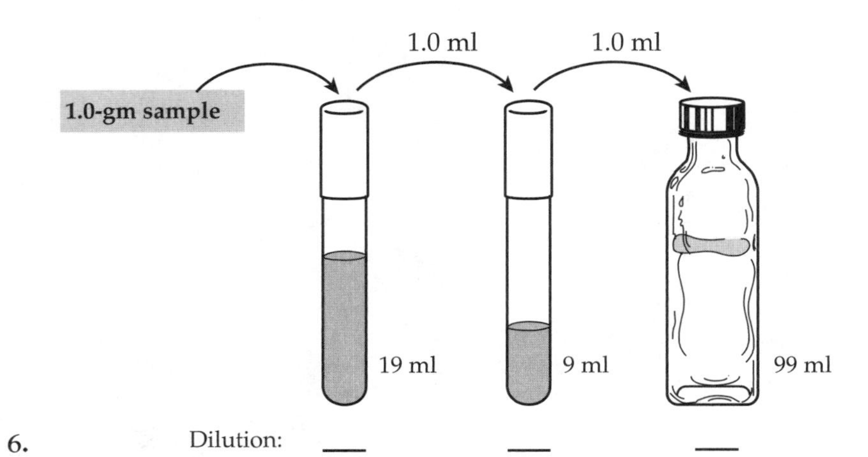

6. Dilution: ___ ___ ___

Exponents, Exponential Notation, and Logarithms*

Very large and very small numbers—such as 4,650,000,000 and 0.00000032—are cumbersome to work with. It is more convenient to express such numbers in exponential notation—that is, as a power of 10. For example, 4.65×10^9 is written in **standard exponential notation,** or **scientific notation;** 4.65 is the **coefficient,** and 9 is the power, or **exponent.** In standard exponential notation, the coefficient is a number between 1 and 10, and the exponent is a positive or negative number.

To change a number into exponential notation, follow two steps. First, determine the coefficient by moving the decimal point so you leave only one nonzero digit to the left of it. For example:

$$0.00000032$$

The coefficient is 3.2. Second, determine the exponent by counting the number of places you moved the decimal point. If you moved it to the left, the exponent is a positive number. If you moved it to the right, the exponent is negative. In the example, you moved the decimal point 7 places to the right, so the exponent is −7. Thus

$$0.00000032 = 3.2 \times 10^{-7}$$

Now suppose we are working with a very large number instead of a very small number. The same rules apply, but our exponential value will be positive rather than negative. For example:

$$4{,}650{,}000{,}000 = 4.65 \times 10^{+9}$$

$$= 4.65 \times 10^{9}$$

To multiply numbers written in exponential notation, multiply the coefficients and *add* the exponents. For example:

$$(3 \times 10^4) \times (2 \times 10^3) = (3 \times 2) \times 10^{4+3} = 6 \times 10^7$$

To divide numbers written in exponential notation, divide the coefficients and *subtract* the exponents. For example:

$$\frac{3 \times 10^4}{2 \times 10^3} = \frac{3}{2} \times 10^{4-3} = 1.5 \times 10^1$$

Microbiologists use exponential notation in many kinds of situations. For instance, exponential notation is used to describe the number of microorganisms in a population. Such numbers are often very large. Another application of exponential notation is to express concentrations of chemicals in a solution—chemicals such as media components, disinfectants, or antibiotics. Such numbers are often very small. Converting from one unit of measurement to another in the metric system requires multiplying or dividing by a power of 10, which is easiest to carry out in exponential notation.

A **logarithm** is the power to which a base number is raised to produce a given number. Usually we work with logarithms to the base 10, abbreviated $\mathbf{log_{10}}$. The first step in finding the $\log_{10}$ of a number is to write the number in standard exponential notation. If the coefficient is exactly 1, the $\log_{10}$ is simply equal to the exponent. For example:

$$\log_{10} 0.00001 = \log_{10}(1 \times 10^{-5}) = -5$$

If the coefficient is not 1, as is often the case, a calculator must be used to determine the logarithm.

Microbiologists use logs for pH calculations and for graphing the growth of microbial populations in culture.

**Source:* G. J. Tortora, B. R. Funke, and C. L. Case. *Microbiology: An Introduction*, 9th ed. San Francisco, CA: Benjamin Cummings, 2007, Appendix D.

Appendix C

Use of the Spectrophotometer

In a **spectrophotometer,** a beam of light is transmitted through a bacterial suspension to a photoelectric cell (Figure C.1). As bacterial numbers increase, the broth becomes more turbid, causing the light to scatter and allowing less light to reach the photoelectric cell. The change in light is registered on the instrument as **percentage of transmission,** or **%T** (the amount of light getting through the suspension) and **absorbance (Abs.)** (a value derived from the percentage of transmission). Absorbance is a logarithmic value and can be used to plot bacterial growth on a graph.

Always wipe the surface of the spectrophotometer tube with a low-lint, nonabrasive paper such as a Kimwipe before placing the tube in the spectrophotometer.

Operation of the Spectronic 20 and Meter-Model Spectronic 21 Spectrophotometers (Figure C.2)

1. Turn on the power and allow the instrument to warm up for 15 minutes.
2. Set the wavelength for maximum absorption of the bacteria and minimal absorption of the culture medium.
3. For the Spectronic 20, *zero* the instrument by turning the zero control until the needle measures 0% transmission.
4. To read a metered scale, look directly at the meter so the needle is superimposed on its reflection in the mirror.
5. Place an uninoculated tube of culture medium (*control*) in the sample holder. To *standardize* the instru-

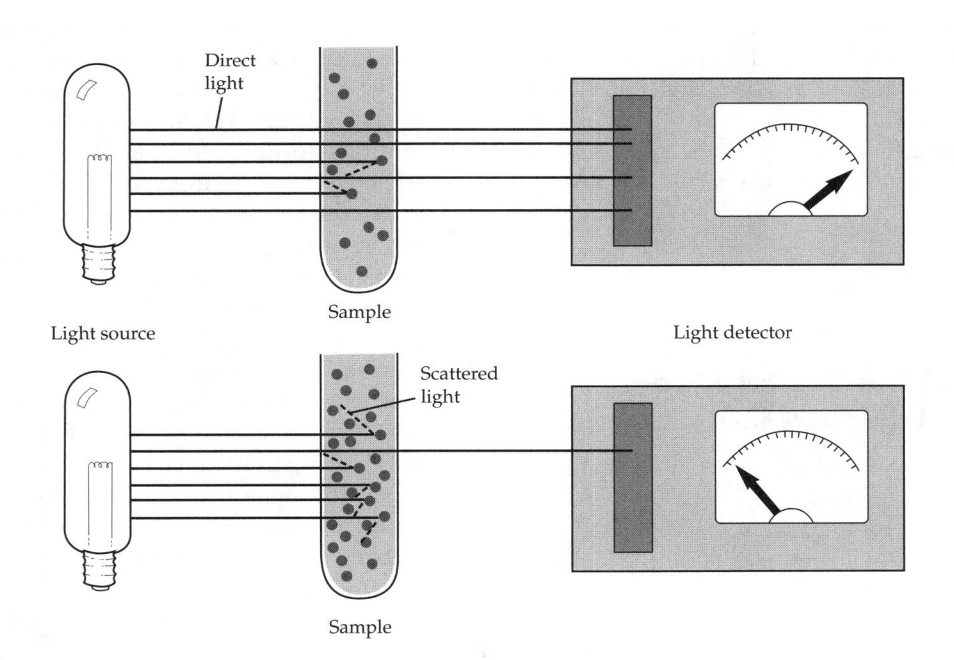

Figure C.1

Estimation of bacterial numbers by turbidity. The amount of light picked up by the light detector is proportional to the number of bacteria. The less light transmitted, the more bacteria in the sample.

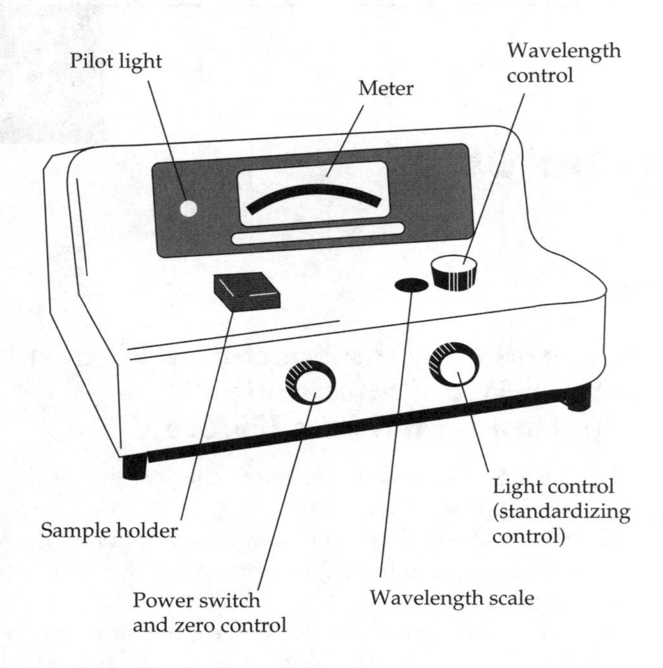

Figure C.2
Bausch and Lomb Spectronic 20.

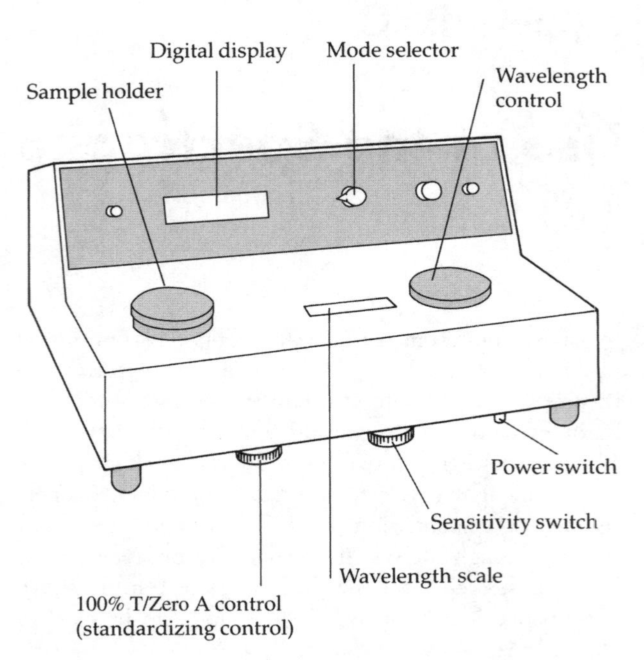

Figure C.3
Bausch and Lomb Spectronic 21, digital model.

ment, turn the light control until the needle registers 100% transmission.

6. To take a sample measurement, place an inoculated tube of culture medium in the sample holder, wait about 45 sec, and record a reading. There will be some fluctuation in turbidity depending on the alignment of particles. Waiting longer than 45 sec is not recommended because the tube will heat up, creating convection currrents that cause rod-shaped cells to line up.
7. %T is usually read on the Spectronic 20 because it is the more accurate scale. Calculate the absorbance as follows:

$$\text{Absorbance} = -\log\left(\frac{\%T}{100}\right)$$

Operation of the Digital-Model Spectronic 21 (Figure C.3)

1. Turn on the power and allow the instrument to warm up for 15 minutes.
2. Set the wavelength for maximum absorption of the bacteria and minimal absorption of the culture medium.
3. Select the operating mode: A (absorbance) or %T (transmission). Absorbance data are more linear and usually preferable.
4. Set the sensitivity switch to LO.
5. Place an uninoculated tube of culture medium (*control*) in the sample holder. To *standardize* the instrument, turn the 100% T/Zero A control until the display registers 000 A. If 000 A can't be reached, turn the sensitivity to M or HI. A U in the upper-left corner of the display indicates that you are under the range of the light detector, and an O in the display indicates that you are over the range. In either case, use the sensitivity switch and the standardizing control to make the necessary adjustments.
6. To take a sample measurement, place an inoculated tube of culture medium in the sample holder, wait about 45 sec, and record a reading. There will be some fluctuation in turbidity depending on the alignment of particles. Waiting longer than 45 sec is not recommended because the tube will heat up, creating convection currrents that cause rod-shaped cells to line up.

Appendix D

Graphing

A **graph** is a visual representation of the relationship between two variables. Whenever one variable changes in a definite way in relation to another variable, this relationship may be graphed.

Microbiologists work with large populations of bacteria and frequently use graphs to illustrate the activity of these populations. The horizontal, or **X-axis,** is a linear scale. The X-axis is used for the **independent variable**—that is, the variable not being tested. The dependent variable is marked off along the vertical, or **Y-axis.** The **dependent variable** changes in relation to the independent variable. The intersection point of the X-axis and Y-axis is called the **origin,** and all units are marked off from this point.

Use a computer-graphing application to graph the numbers of bacteria at each temperature, as shown in Table D.1. First enter the data into a spreadsheet. Next, convert the numbers of bacteria to their log values. At times when no readings were taken, leave the cells blank. The independent variable (X-axis) is time, and the dependent variable (Y-axis) is the number of bacteria. Select the columns with time and log number of cells, and choose an XY (scatter) graph. This format will distribute the points on the X-axis at their correct intervals. (A line graph will simply distribute the X-axis data evenly so the distance between 0 and 4 hours will be the same as the distance between 4 and 5 hours.)

A graph of the numbers of bacteria in the culture incubated at 20°C is shown in Figure D.1. A best-fit line is drawn through the points. Use the log trend line in your graphing application or draw the line by hand. The line should be straight and does not have to connect all the points. When the points do not fall in a straight

Table D.1

Numbers of *E. coli* in Three Cultures Incubated at Different Temperatures

Time (hr)	Number of Bacteria at:		
	15°C	20°C	35°C
0	1.50×10^5	5.00×10^5	5.40×10^5
4	1.50×10^5	*N*	*N*
5	*N*	4.60×10^5	*N*
6	*N*	*N*	5.20×10^5
10	5.62×10^5	5.78×10^5	6.00×10^5
11	*N*	8.80×10^5	1.23×10^6
12	*N*	1.05×10^6	2.58×10^6
13	*N*	1.16×10^6	4.93×10^6
14	*N*	2.32×10^6	9.00×10^6
15	1.62×10^6	3.80×10^6	*N*
16	*N*	7.00×10^6	*N*
17	*N*	8.10×10^6	*N*
18	*N*	9.60×10^6	*N*
19	*N*	1.95×10^7	*N*
20	2.70×10^6	*N*	*N*

N = No reading at that time.

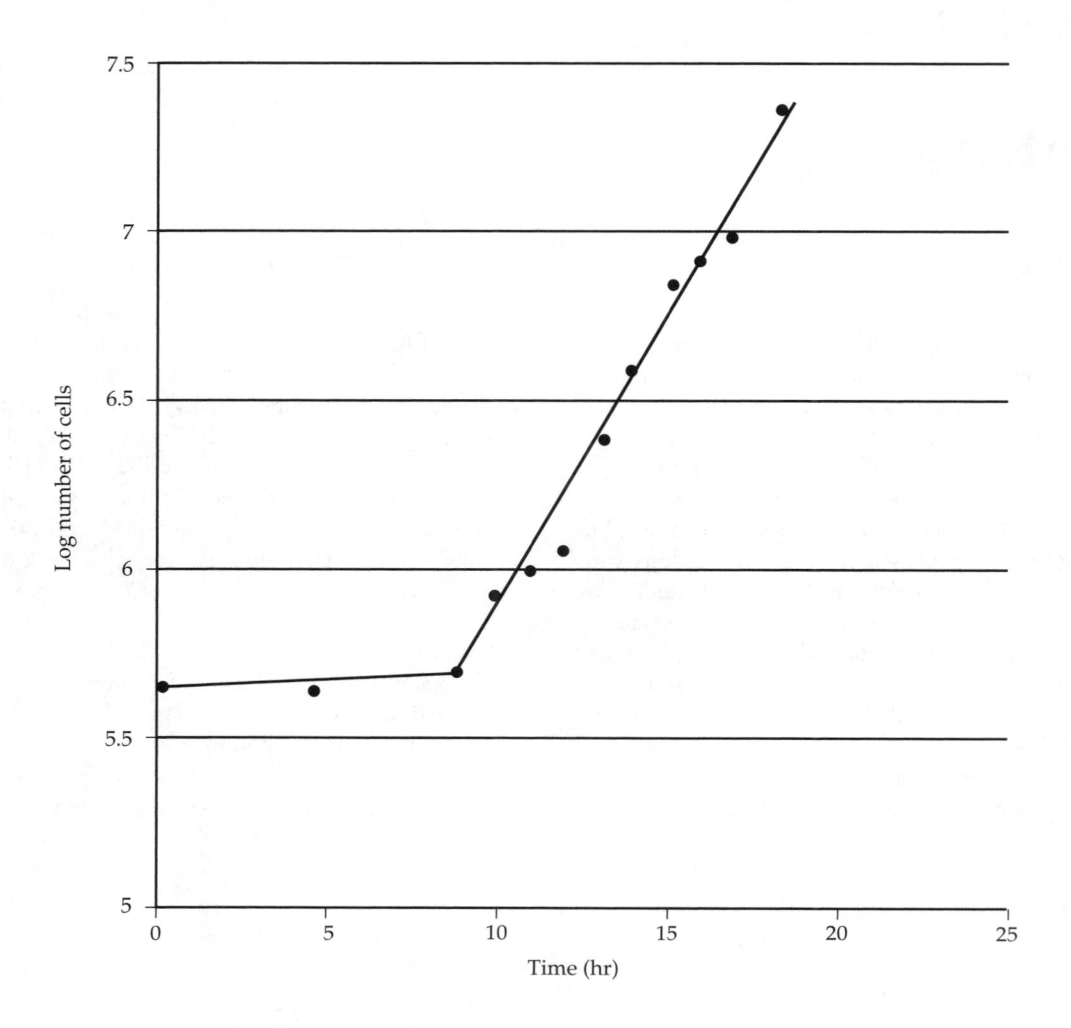

Figure D.1
Growth of *E. coli* at 20°C.

line, draw a line between the points, leaving an equal number of points above and below the line.

When graphs are drawn that compare the same variables under different conditions (for example, bacterial growth rates at different temperatures), it is best to draw all the graphs on the same paper so comparisons are obvious and easily made. Graphs should be titled to explain the data presented.

The data for cultures incubated at 15°C and 35°C (Table D.1) can be plotted on the same graph. It is not necessary to have bacterial numbers for the same "time points" in each graph. The missing points will be filled in by the line. This is possible because the line between each point is an interpolation of what was happening between measurements.

Activity

Using the data provided in Table D.1, graph the growth curves on graph paper and compare the effect of temperature on the rate of growth. Because *rate* is a change over time, the rate can be interpreted by the slope of the line. What will you title this graph? ____________
____

Appendix E

Use of the Dissecting Microscope

The **dissecting microscope** (Figure E.1) is useful for observing specimens that are too large for the compound microscope and too small to be seen with the naked eye. The dissecting microscope is also called a **stereoscopic** microscope because the image is three-dimensional.

Dissecting microscopes are usually equipped with 10× ocular lenses and a 1× objective lens. Some dissecting microscopes have a rotating nosepiece that houses 1× and 2× objective lenses or a zoom objective lens (1×–7×).

The dissecting microscope may have a built-in light source or a substage mirror and auxiliary lamp. Illumination can be adjusted on microscopes with two built-in lamps and/or by using a mirror. For transparent specimens, the light should be directed through the specimen from under the stage. For opaque specimens, light should be directed onto the specimen from the top. Both types of lighting can be used for observing a translucent specimen.

The specimen is placed on the stage. Some microscopes have an alternate black stage plate for use with opaque specimens. The stage clips can be moved to the side when large specimens, such as Petri plates, are used. While looking through the microscope, adjust the width of the eyepieces so you see one field of vision. If two circles or fields are visible, the ocular lenses are too far apart; if two fields overlap, the ocular lenses are too close together. To focus, turn the adjustment knob.

Questions

1. What is the magnification of your dissecting microscope? ____________________
2. Using a ruler, measure the size of the field of vision at low power: ____ mm. At high power: ____ mm.
3. How can you adjust the direction of illumination on your microscope? ____________________

____

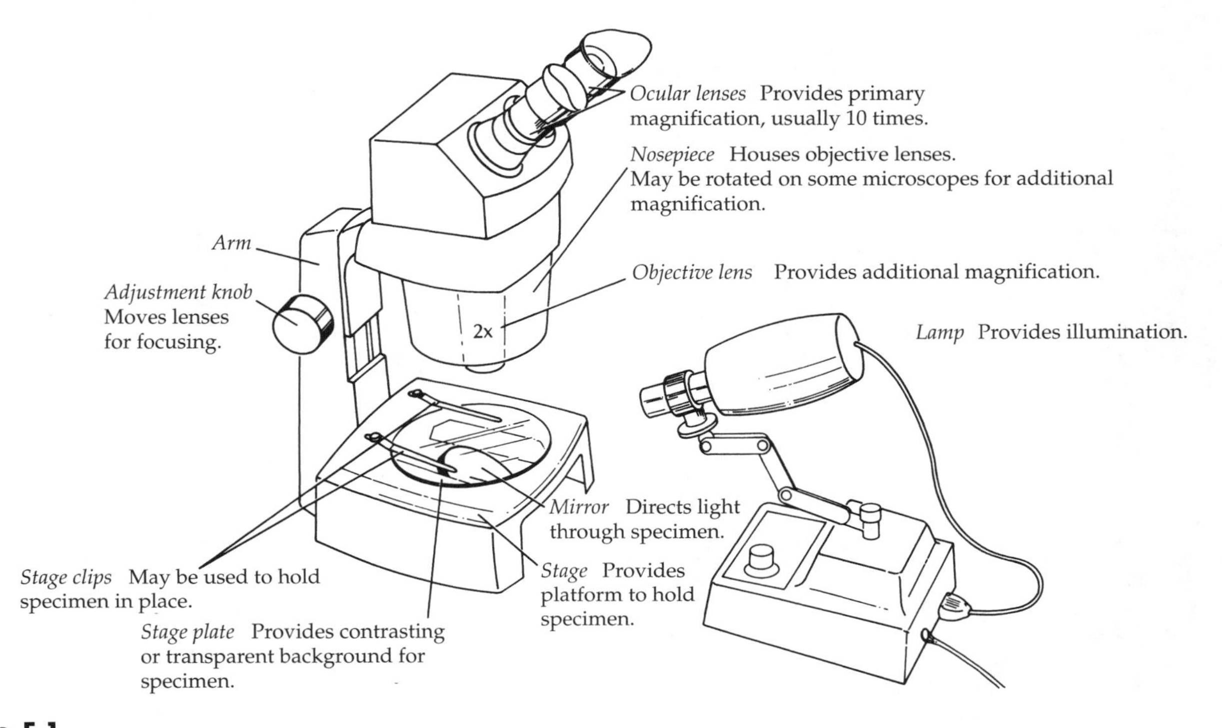

Figure E.1

The dissecting microscope: principal parts and their functions.

Appendix F

Use of the Membrane Filter

Membrane filtration can be used to separate bacteria, algae, yeasts, and molds from solutions. Nitrocellulose or polyvinyl membrane filters with 0.45-μm pores are commonly used to trap microorganisms and generate sterile solutions. Membrane filtration is used to separate viruses from their host cells because the viruses will pass through the filter. Membrane filters are used to trap bacteria found in water in order to count the bacteria.

Membrane filters have many uses in industrial microbiology. They are used to filter wine, soft drinks, air, and water to detect microorganisms and particulate matter. In addition, membrane filters can be used to separate large molecules, such as DNA, from solutions.

The receiving flask, filter base, and filter support can be wrapped in paper and sterilized by autoclaving (Figure F.1). Presterilized, disposable polystyrene membrane filtration units are available. Membrane filters can also be sterilized by autoclaving. For use, the filter base must be unwrapped aseptically. With sterile, flat forceps, a sterile filter is placed on the filter support, and the filter funnel is clamped or screwed into place.

Gravity alone will not pull a sample through the small filter pores, so the filtering flask is attached to a vacuum source (Figure F.1). Because air pressure is then greater outside the flask, the sample is pushed into the vacuum inside the flask. A vacuum pump or vacuum line is usually used to provide the necessary vacuum. An aspirator trap is placed between the vacuum source and the receiving flask to prevent the flow of solution into the pump or vacuum line. With small filters and small volumes of fluid, a syringe can be used to supply the needed vacuum or to push fluid through the filter.

Following filtration, the filtrate in the flask is free of all microorganisms larger than viruses. For observation of cells trapped on the filter, the dried filter can be dipped into immersion oil to make it translucent and then mounted on a microscope slide. Microorganisms can be cultured from the filter by placing the filter on a solid nutrient medium.

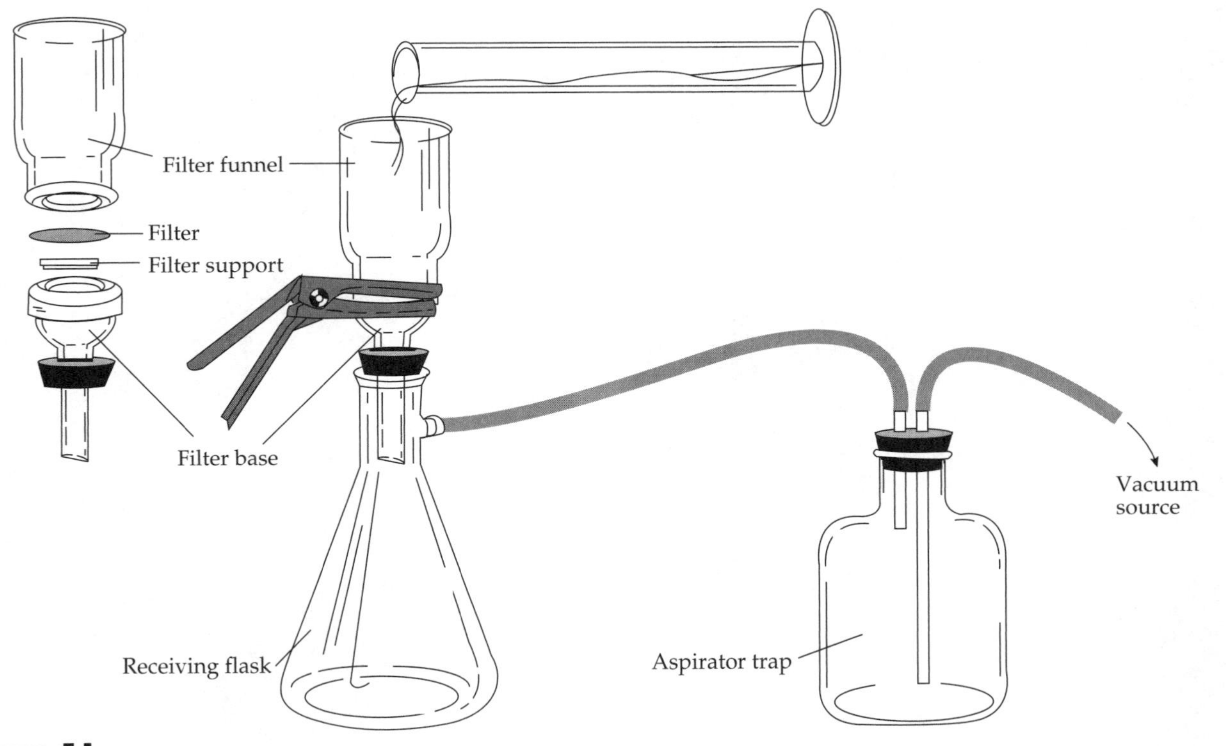

Figure F.1

Membrane filtration setup. (Disposable polystyrene flasks and funnels can be used.)

Appendix G

Electrophoresis

Charged macromolecules can be separated according to size and charge by a process called **electrophoresis.** The material to be tested is placed at one end of a gel matrix. Polyacrylamide gel or agarose is used to separate proteins and small pieces of DNA and RNA. Agarose is the usual matrix for separation of nucleic acids. A buffer (salt solution) is used to conduct electric current through the matrix. When an electric current is applied, the molecules start to move. Molecules with different charges and sizes will migrate at different rates.

Procedure

1. Place the well-forming comb in position on a casting tray (Figure G.1). Seal the open ends of the casting tray with masking tape. Or, if the tray has movable gates, raise the gate at each end. Pour melted agarose onto the casting tray until the agar surrounds the teeth. The agar should be about 5 mm deep and have a flat surface. Let the gel harden for about 20 minutes. When the gel is solid, carefully remove the comb and then remove the tape or lower the end gates.

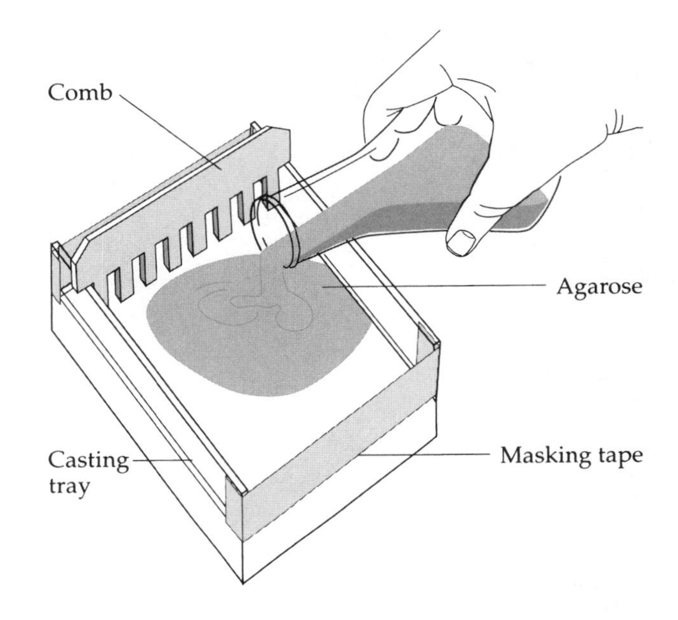

Figure G.1
Preparation of an agarose minigel.

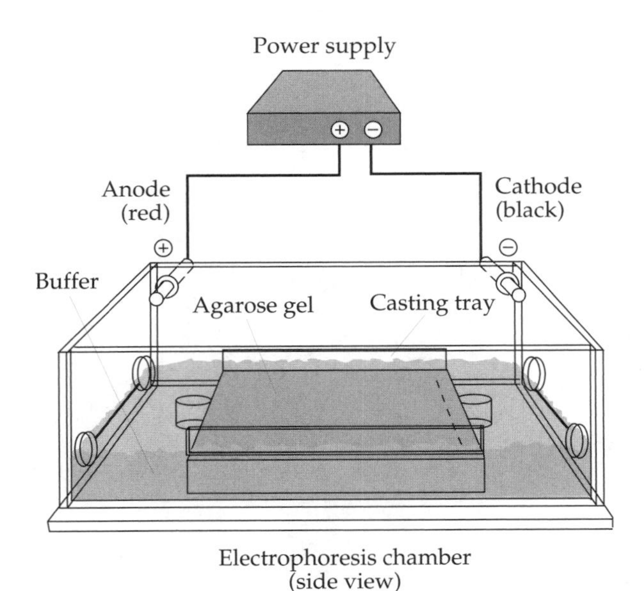

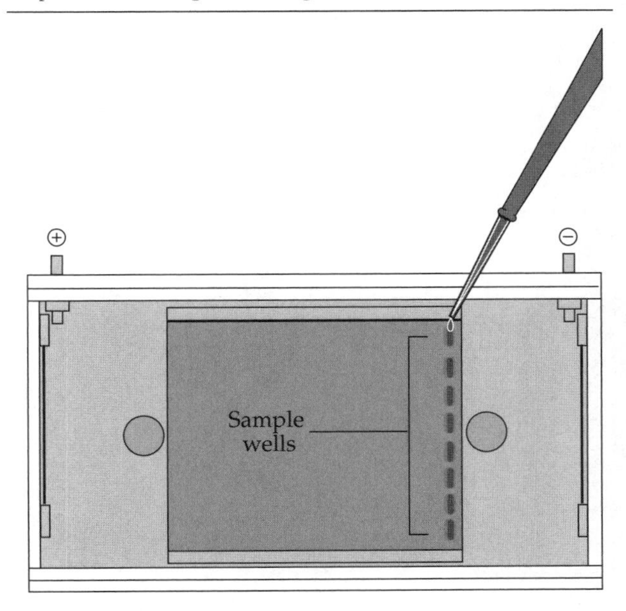

Figure G.2
Setup of electrophoresis equipment.

2. Fill the buffer chamber with the appropriate buffer. Add just enough buffer to cover the gel. Place the agar in the buffer chamber with or without the casting tray.
3. Fill the wells with samples to be tested. Close the lid and apply current (Figure G.2).
4. During electrophoresis, a visible marker dye will migrate in front of the sample. If the dye isn't moving in the right direction, you have accidentally reversed the polarity of the electric field. If the marker doesn't move and if there is little current, you have an open circuit or have not added the correct buffer. In this case, you must start over.
5. The two components of the tracking dye migrate at different rates. Why? ________________________
   Turn off the power when the two dyes are 3 or 4 cm apart. Remove the gel. The gel can now be stained to locate the macromolecules. Store stained or unstained gels in sealed plastic bags in the refrigerator.

Appendix H

Keys to Bacteria

Appendix H deals with the identification of bacteria. The information is arranged in **dichotomous keys.** In a dichotomous key, identification is based on successive questions, each of which has two possible answers (*di-* means two). After answering each question, the investigator is directed to another question until an organism is identified. There is no single "correct" way to write a key; the goal is to conclude with one unambiguous identification.

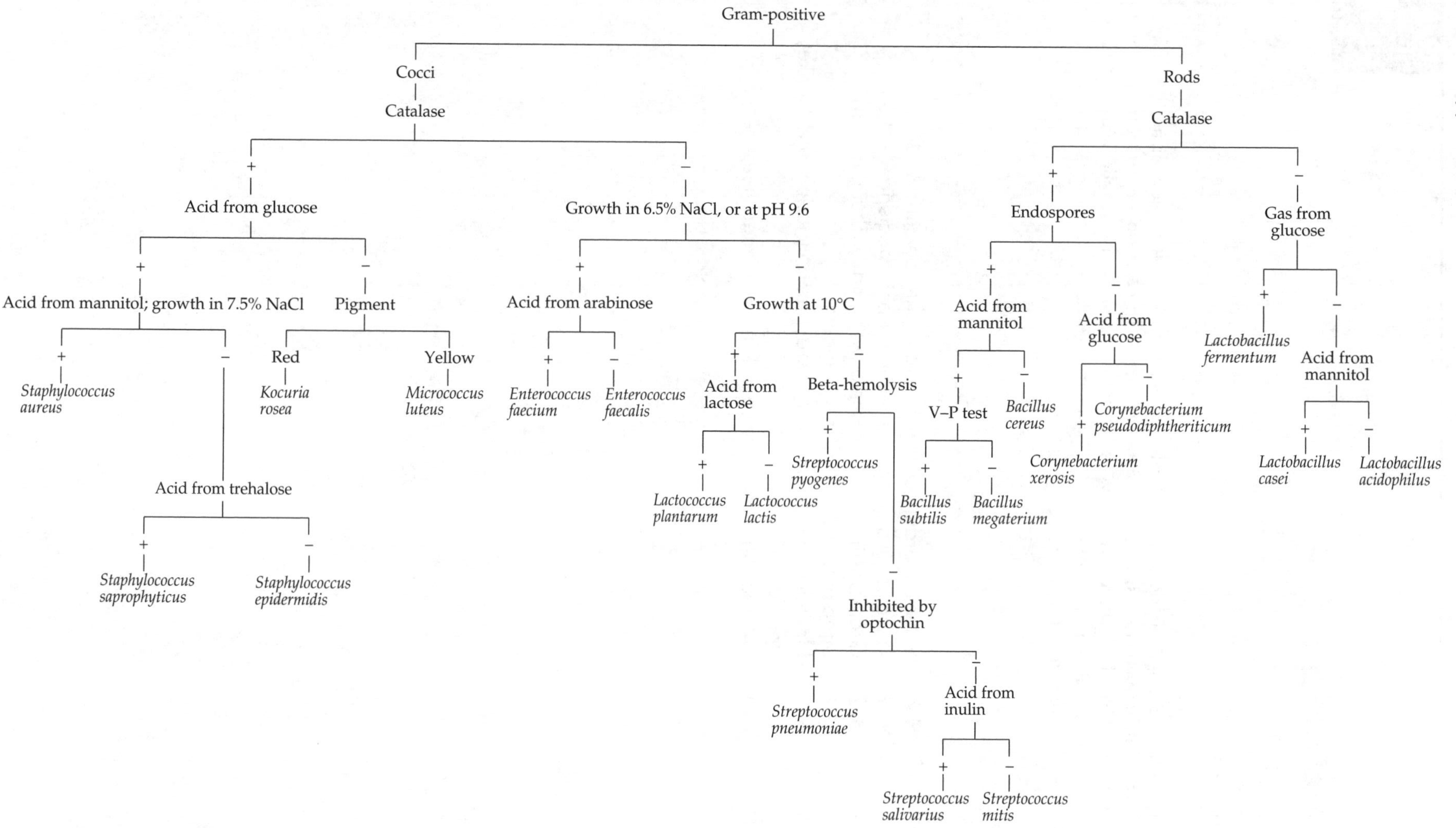

Figure H.1
Key to selected gram-positive heterotrophic bacteria.

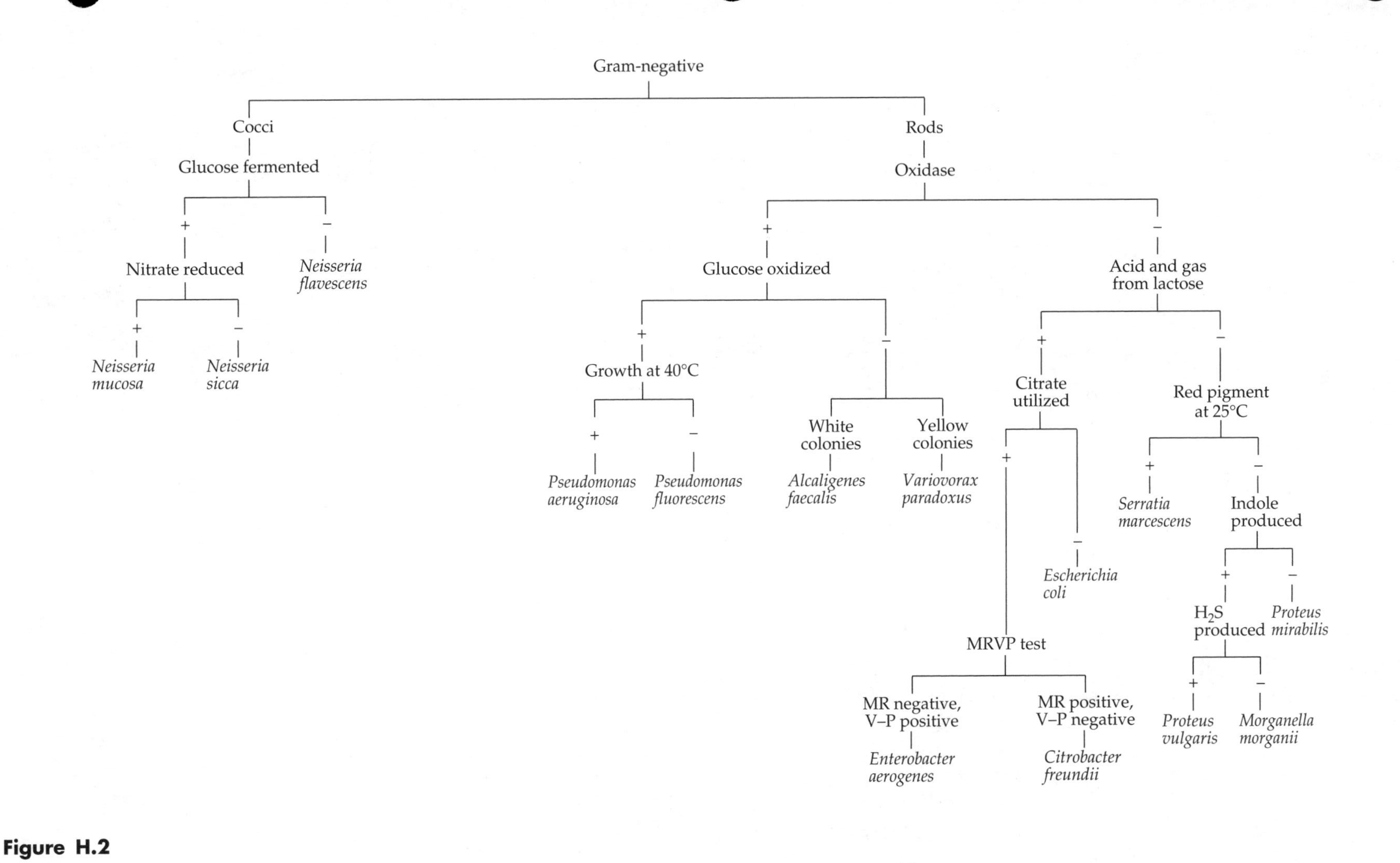

Figure H.2
Key to selected gram-negative heterotrophic bacteria.

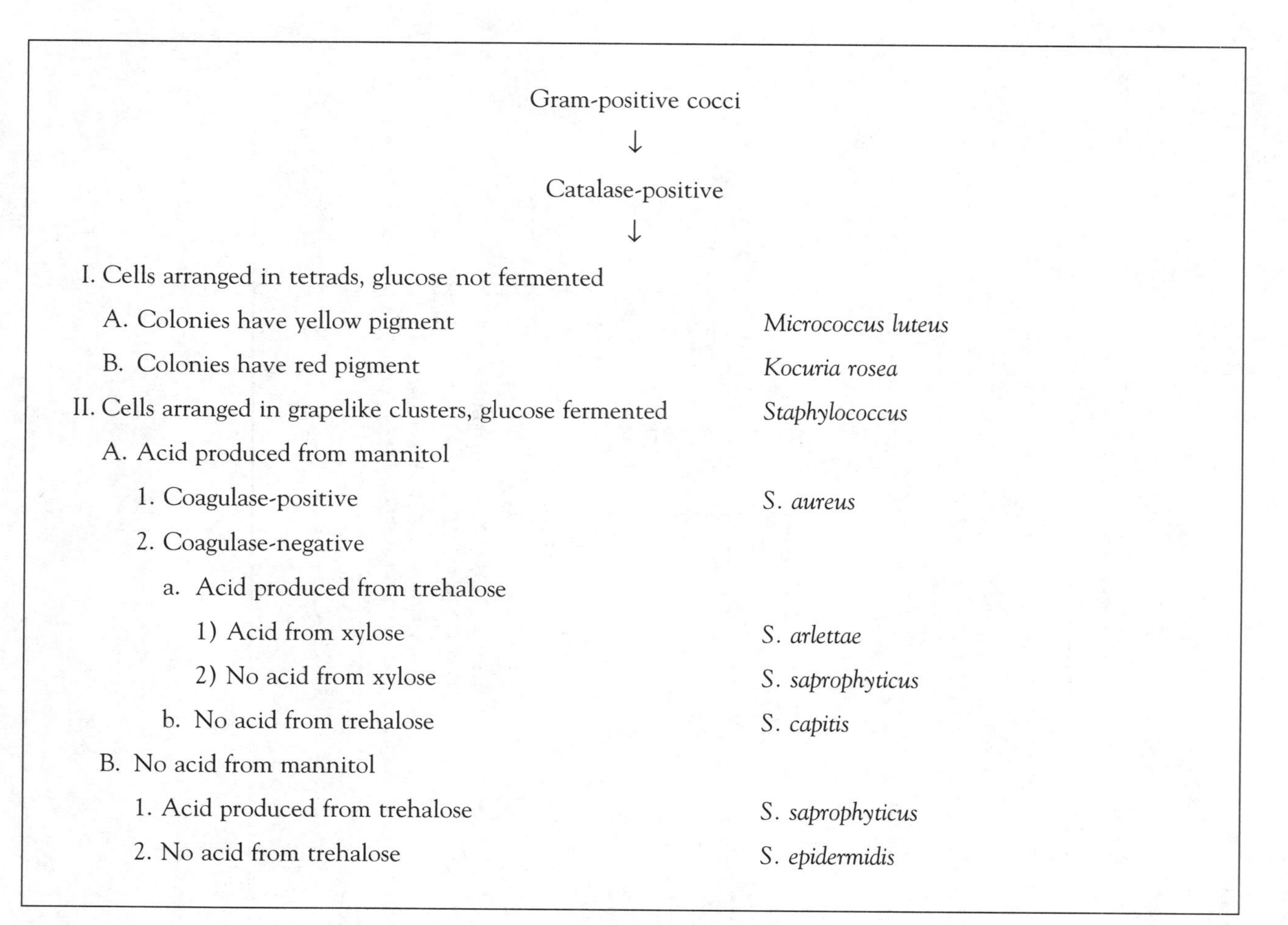

Gram-positive cocci

↓

Catalase-positive

↓

I. Cells arranged in tetrads, glucose not fermented	
A. Colonies have yellow pigment	*Micrococcus luteus*
B. Colonies have red pigment	*Kocuria rosea*
II. Cells arranged in grapelike clusters, glucose fermented	*Staphylococcus*
A. Acid produced from mannitol	
1. Coagulase-positive	*S. aureus*
2. Coagulase-negative	
a. Acid produced from trehalose	
1) Acid from xylose	*S. arlettae*
2) No acid from xylose	*S. saprophyticus*
b. No acid from trehalose	*S. capitis*
B. No acid from mannitol	
1. Acid produced from trehalose	*S. saprophyticus*
2. No acid from trehalose	*S. epidermidis*

Figure H.3

Key to gram-positive cocci commonly found on human skin.

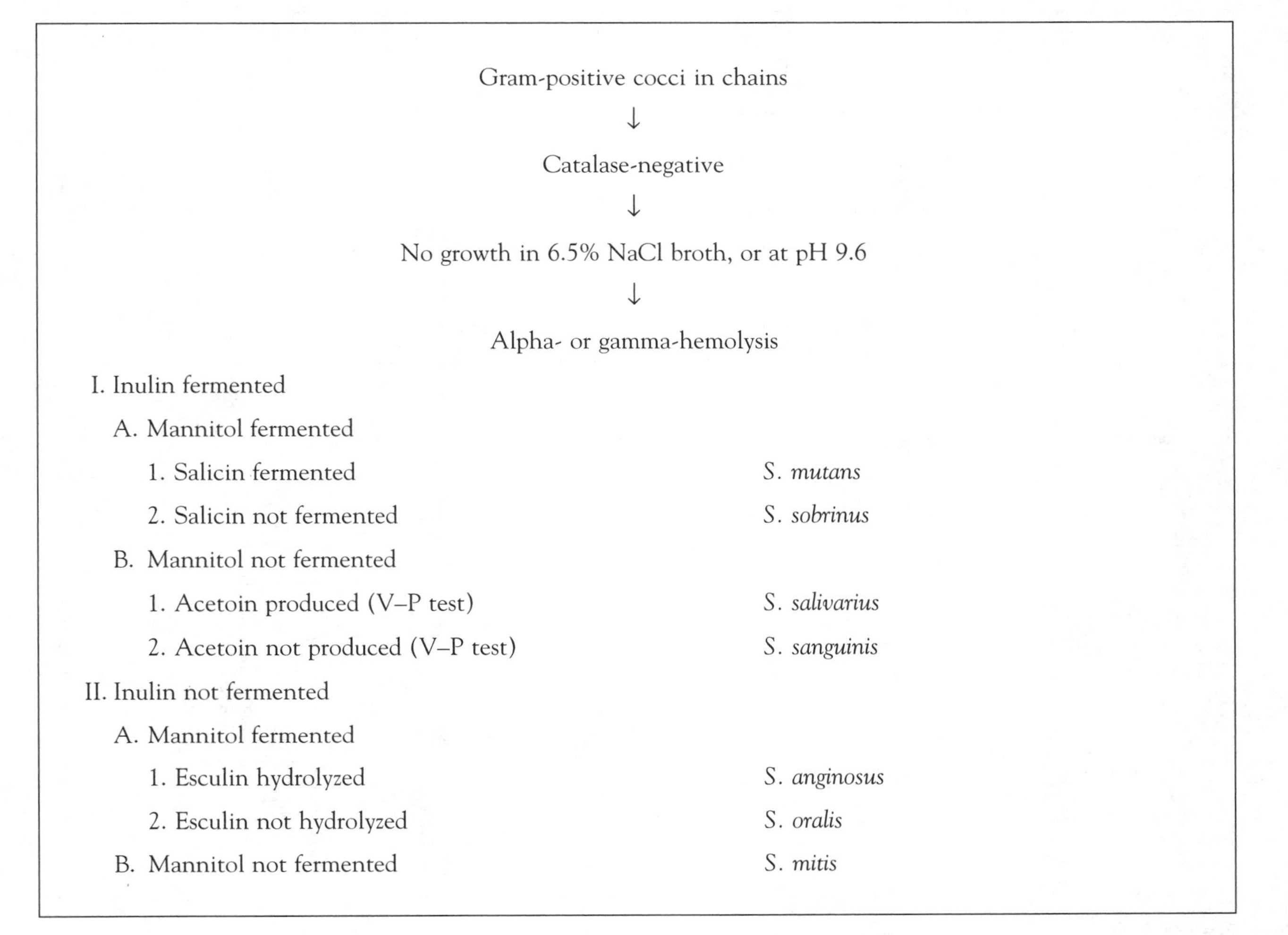

Gram-positive cocci in chains
↓
Catalase-negative
↓
No growth in 6.5% NaCl broth, or at pH 9.6
↓
Alpha- or gamma-hemolysis

I. Inulin fermented
- A. Mannitol fermented
  - 1. Salicin fermented — *S. mutans*
  - 2. Salicin not fermented — *S. sobrinus*
- B. Mannitol not fermented
  - 1. Acetoin produced (V–P test) — *S. salivarius*
  - 2. Acetoin not produced (V–P test) — *S. sanguinis*

II. Inulin not fermented
- A. Mannitol fermented
  - 1. Esculin hydrolyzed — *S. anginosus*
  - 2. Esculin not hydrolyzed — *S. oralis*
- B. Mannitol not fermented — *S. mitis*

Figure H.4
Key to streptococci found in the mouth.

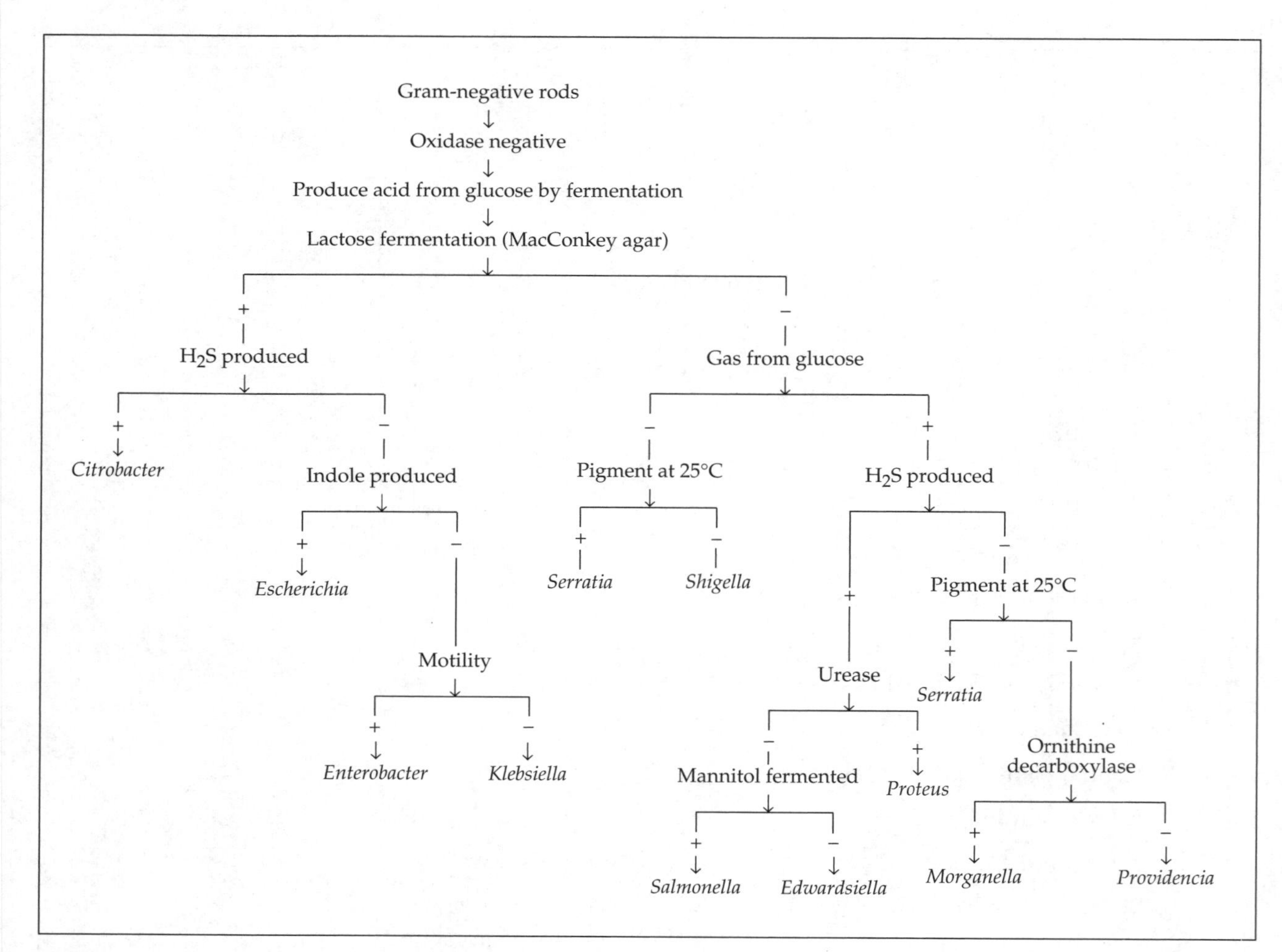

Figure H.5

Identification scheme for enteric genera primarily using MacConkey agar and TSI.

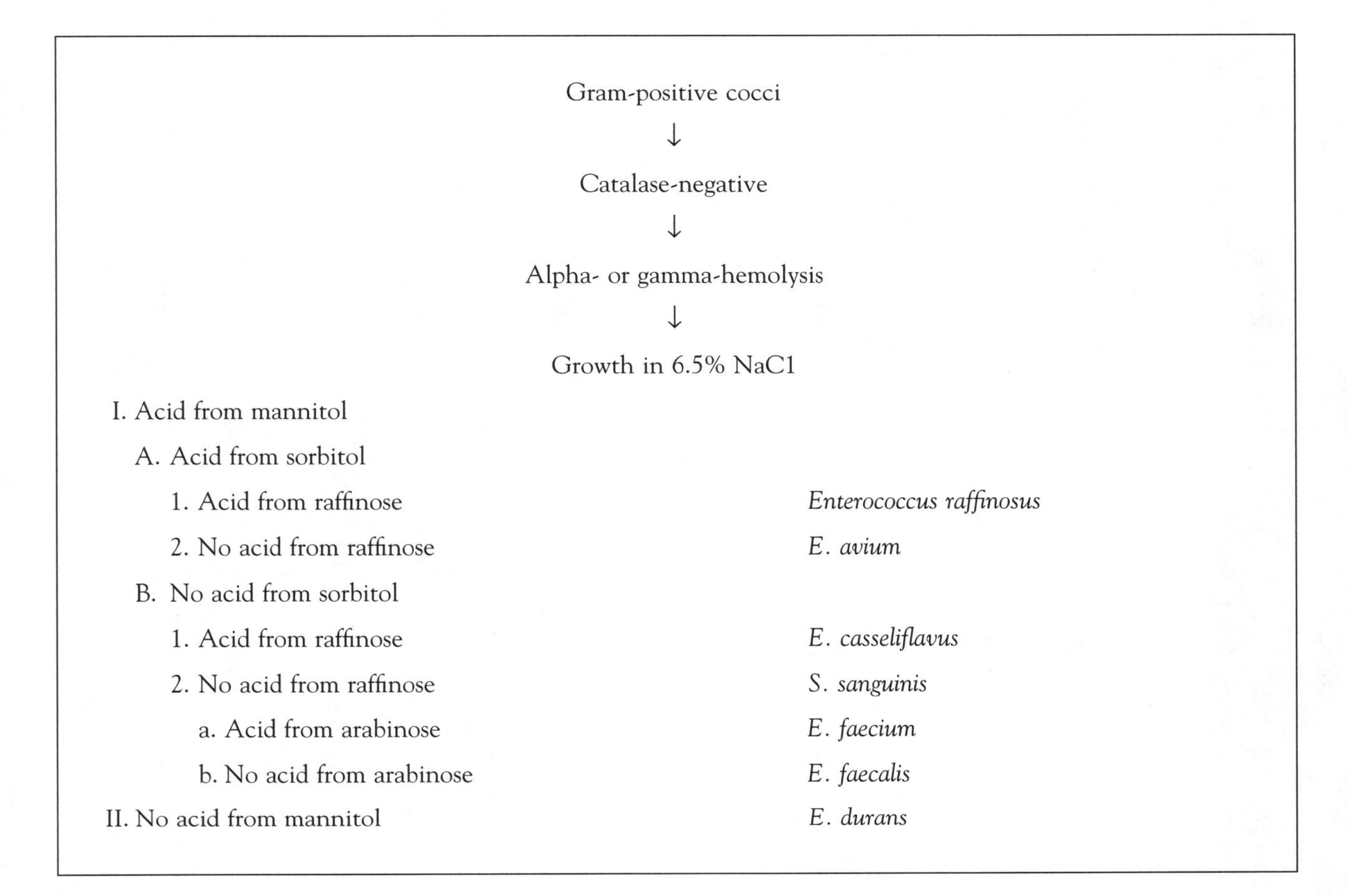

Gram-positive cocci

↓

Catalase-negative

↓

Alpha- or gamma-hemolysis

↓

Growth in 6.5% NaC1

I. Acid from mannitol	
A. Acid from sorbitol	
1. Acid from raffinose	*Enterococcus raffinosus*
2. No acid from raffinose	*E. avium*
B. No acid from sorbitol	
1. Acid from raffinose	*E. casseliflavus*
2. No acid from raffinose	*S. sanguinis*
a. Acid from arabinose	*E. faecium*
b. No acid from arabinose	*E. faecalis*
II. No acid from mannitol	*E. durans*

Figure H.6
Key to enterococci commonly found in the human intestine.

Art and Photography Credits

Art Credits

Figure 14.1: Adapted from *Microbiology*, 9e, by Tortora, Funke, and Case, F5.18, © 2007 Pearson Education, Inc.

Figure 20.2: The Left Coast Group, Inc.

Figure 21.1: Adapted from *Microbiology*, 8e, by Tortora, Funke, and Case, F27.11, © 2004 Pearson Education, Inc.

Figure 21.2: The Left Coast Group, Inc.

Figure 21.3: The Left Coast Group, Inc.

Figure Part 12: The Left Coast Group, Inc.

Figure A.2c: The Left Coast Group, Inc.

Figure A.5: The Left Coast Group, Inc.

Figure D.1: The Left Coast Group, Inc.

Figure H.6: The Left Coast Group, Inc.

Photography Credits

All photographs by Christine Case unless otherwise noted.

Figure 1.1a: Leica Microsystems

Figure Part 11: J. G. Hadley, Battelle-Pacific Northwest Labs/BPS

Color Plate I.3: Centers for Disease Control and Prevention

Color Plate I.4: Centers for Disease Control and Prevention/Dr. George P. Kubica

Color Plate I.6: Steven R. Spilatro, Marietta College

Color Plate XI.3: Courtesy of Randall von Wedel

Color Plate XIV.2: Centers for Disease Control and Prevention

Index

Page references followed by *fig* indicate illustrated figures; followed by *t* indicate a table; followed by *ph* indicate a photograph; color plate photographs are indicated in the index as numbered colored plates.